INDUSTRIAL
BOILER
MANAGEMENT

An Operator's Guide

I N

Don Gresswell Ltd., London, N.21   Cat. No. 1207   DG 02242/71

Industrial Press Inc.

Library of Congress Cataloging-in-Publication Data

Oliver, Kenneth G.
    Industrial boiler management : an operator's guide /
Kenneth G. Oliver.—1st ed.
      p. cm.
    Bibliography: p.
    Includes index.
    ISBN 0-8311-3018-0
    1. Steam-boilers—Handbooks, manuals, etc. I. Title.
TJ289.045 1989   88-35259
621.1'94—dc19   CIP

Industrial Press Inc.
200 Madison Avenue
New York, New York 10016-4078

Composition, printing and binding by
Edwards Brothers, Incorporated, Ann Arbor, MI

2 4 6 8 7 5 3

DEDICATION

With the expectation that we shall increase their operating effectiveness, we dedicate this book to the thousands of industrial boiler plant operators in this country, most of whom perform their duties with a minimum of training.

PREFACE

This book was originally produced as a textbook for a course entitled "Small Boiler Management," started by Parker Schools for Industry in the Southern California area in 1982. Owing to the huge success of both the course and the textbook, the decision was made to produce this book for nationwide distribution.

There is obviously a great thirst for knowledge among operators of our industrial plants that use steam in their processes and heating systems. This thirst is especially acute in these times of rising energy costs, as the first level of attack on plant energy costs lies in the efficiency with which the raw material—including fuel—is utilized. This is shown in the many surveys that have been made over the past decade throughout the country's industrial plants and basic institutional establishments. One of the astounding discoveries was that the average plant that converts fuel and electricity to the plant's use makes proper use of less than 50% of the energy that enters its premises. Many plants had an efficiency as low as 25%.

Fortunately, efficiency is rising steadily, but there is still much energy going to waste in our country.

Industrial Boiler Management has as its main goal the further education of the operators of those steam plants that are relatively small users of energy, but that make up the majority of the steam plants in the industrial complexes of this nation. Consequently, the reader of this book, and the student attending the course for which it was produced, will find the emphasis has been placed mainly on the package type of boiler rated at about a maximum of 750 boiler horsepower.

However, the basic principles are the same, regardless of the size of the industrial boiler, whether 10 hp or the larger central station suppliers of steam to process plants or commercial establishments. This applies also to the energy conservation measures advocated in this book. The equipment complexities vary, but the basic concepts do not.

Regarding the fuels burned, the same approach applies here, also. The discussion herein is based on gas or oil fuels, but most of the comments, calculations, etc., may easily be applied to coal or wood.

The material presented here may also be considered as a basic course for those just entering the industry through the boiler plant door, as well as serving as a refresher for those more advanced.

The author of this textbook does not in any way intend to replace the mass of instructional material supplied by the boiler plant equipment manufacturers. The student and reader is advised to read such free material over very carefully, and follow it religiously. The material we have presented here is to serve as a general guide only and to provide further instructions that may have been omitted or glossed over by the writers of the equipment manufacturers.

We have also attempted to give the "why" as well as the "how" in the use of your equipment.

The student and reader will find the material presented here to be fairly complete. There are very few sources where one may go to find all of this material gathered into one convenient source. We recommend that it be studied very carefully and that the book be retained in your place of employment, handy for looking up data which you may need from time to time.

This book was written based on codes, products, and practices existing and in common use at the time of writing, all of which are subject to change from time to time. However, the subjects covered in this book probably represent about 95% of the existing codes, products, and practices in use at the time of writing. A product may no longer be available from the supplier, but the item in question may be installed and giving good service in thousands of plants and, therefore, is a legitimate subject for discussion in this book.

It is the responsibility of the steam plant operator continually to keep abreast of current changes in those governing codes pertinent to his or her plant and also to trends in operating practices.

Also, the fact that your plant was designed based on codes, products, or practices no longer in existence does not necessarily mean that you are working in unsafe conditions, but it is to your advantage to check into the latest changes and determine for yourself whether or not your equipment should be updated. Your boiler inspector will be able to help you in this respect.

As there are literally thousands of combinations of boilers, burners, valves, controls, and steam-using equipment in today's industrial plant, we can only stress generalities. As it is impossible to cover all the safety factors necessary to the operation of that equipment, we can assume no responsibility for the safety of any industrial plant being operated by any person who has either attended one of our schools or who has read any of the textbooks, supplements, or miscellaneous hand-out materials associated with the school.

Consequently, we stress the point that the ultimate responsibility for the plant safety is in the hands of others who are involved in the operation of such plant. It is the responsibility of the plant managers, the insuring agencies, and, in the final analysis, the operators of the plant equipment to safeguard the plant and personnel under their jurisdiction.

With that, we leave you to your search for knowledge, happy to know that we have contributed in a small way to your advancement up the ladder to a successful and rewarding occupation—operating and managing the basic energy plant in one of this country's great industrial complexes.

ACKNOWLEDGMENTS

Any textbook of this nature, covering the use, maintenance, and philosophy of manufactured equipment, must, of necessity, refer considerably to those products of specific suppliers in our society. Those suppliers have been most helpful in the assembly of the material for this book, and we wish to give them all our heartiest thanks. We trust that those who read this book or take the course for which it was originally produced, will remember the following firms when it comes time for installing equipment in their plant:

American Boiler Manufacturer's Association

American Society of Mechancial Engineers

Automatic Steam Products Corp.

Bacharach Instrument Co.

Bailey Controls Co.

Boiler Efficiency Institute

BTU Consultants

Clark-Reliance Corp.

Clayton Manufacturing Co.

Cleaver-Brooks Co.

Combustion Engineering

Copes-Vulcan Inc.

Crane Co.

Energy Conversion Systems

Fireye Division, Allen-Bradley Co.

Foster Wheeler Ltd.

Hartford Steam Boiler Inspection and Insurance Co.

Industrial Risk Insurers

ITT Fluid Handling Division

C. M. Kemp Manufacturing Co.

Los Angeles City Department of Building and Safety

Magnetrol International, Inc.

Mercoid Corp.

Nalco Chemical Co.

National Board of Boiler and Pressure Vessel Inspectors

National Bureau of Standards

North American Manufacturing Co.

Olin Corp., Chemicals Division

Parker Supply Co.

Penberthy Houdaille, Inc.

Permutit Co.

Platecoil Division, Tranter, Inc.

Spirax/Sarco Co., Inc.

South Coast Air Quality Management District

Sur-Lite Corp.

U.S. Department of Energy

E. J. Walsh & Co., Ltd.

Zurn Industries, Inc., Energy Division

All of the preceding firms contributed in some way, either in offering the use of their material directly or in the education and documentation for this work. If we have omitted mention of any firm who should be on this list, we apologize.

In addition, the author was helped immeasurably in the preparation and editing by several friends and acquaintances:

Tom Atkins

John Delmazzo

Morrie Marwede

Ken Minor

Joseph Parker

Dale Rycraft

To all of the above, the author extends his thanks for their help and encouragement.

ABBREVIATIONS

AC	Alternating Current
ASME	American Society of Mechanical Engineers
ANSI	American National Standards Institute
atmos.	Atmosphere, atmospheric
BD	Blowdown
BF	Boiler Feedwater, Boiler Feed
bhp	Boiler horsepower = boiler output of 33,475 Btu/hr per BHP
BTU	British Thermal Units: The amount of heat required to raise 1 lb of water 1°F.
cfm	Cubic Feet Per Minute
cu ft, ft^3	Cubic Feet
CV	Control Valve
C.W.	Cooling Water
DA, D.A.	Deaerator: a device for removing gas from a liquid
DC	Direct Current
Dia.	Diameter
F.D.	Forced Draft
F.O.	Fuel Oil
ft	Feet
FT	Flow Transmitter
FW	Feedwater
gal/hr	Gallons per hour
gpm	Gallons per minute
H	Heat, heat content, usually in BTUs
Hd	Head, usually in feet of water: pressure applied by a column of water or other liquid

Hz Cycles per second (hertz)

Hg Mercury

HHV Higher Heating Value = gross heating value of a substance, usually in BTU/pound

hp Horsepower, or high pressure

hr Hour

HSBII Hartford Steam Boiler Inspection and Insurance Co.

I.D. Induced Draft

IR Infrared rays

kW Kilowatts

lb/hr Pounds per hour

LC Level Controller

LP, lp Low Pressure

LPG Liquified Petroleum Gases

LT Level Transmitter

max Maximum

M.H. Manhole

min Minimum

NPT National Pipe Thread

pH Measure of acid or base quality of a liquid; 1 is a strong acid, 7 is a neutral solution, and 14 is a strong base solution, with intermediate numbers graded accordingly

ppb Parts per billion

ppm Parts per million

Press. Pressure

PRV Pressure Regulating Valve

psi Pounds per square inch

psia Pounds per square inch absolute = psig + 14.7 = pressure above absolute zero pressure

psig Pounds per square inch gauge = pressure above atmospheric pressure

PT Pressure Transmitter

sat Saturated

sp. gr. Specific Gravity = weight of a substance relative to water at 62°F

sp. vol. Specific Volume = cubic feet occupied by 1 lb of gas or vapor at a common established condition of pressure and temperature

SSV   Safety Shutoff Valve
TCV   Temperature Control Valve
temp   Temperature
UV   Ultraviolet Rays
vac   Vacuum = any pressure less than atmospheric pressure
wc   Inches of Water Column
wt   Weight

CONTENTS

Appendices

C H A P T E R 1
THE STEAM PLANT

A. Purpose of the Steam Plant

Let us trace the flow of the steam generated in the boiler to the heat-using devices and back to the boiler. By applying heat, produced by the burning of the fuel in the furnace, to the heating surface of the boiler, the water in the boiler is changed to steam at the temperature of the boiling point of the water corresponding to the pressure existing in the boiler. This temperature is known as the saturated temperature of the steam. We choose a working pressure for the boiler that gives a steam temperature slightly in excess of the temperature required by our heat exchangers, or that will give us positive and economical transmission.

The steam is conducted from the boiler to the heat exchanger through a system of insulated piping. The steam pipes are always sloped in the same direction the steam flows so that any entrained moisture will be carried along with the steam to suitable low points or drip legs, where steam traps are provided to remove the condensate as formed.

The branches or steam take-off lines are always taken from the top of the steam line to ensure dry steam, since any condensate in the steam line will be found in the bottom of the pipe.

A suitable arrangement of valves and strainers will control the flow of steam to the heat exchanger. Suitable strainers, valves, and steam traps are connected to the lowest point of the heat exchanger to drain air, noncondensable gases, and condensate out of the steam space as fast as it forms. Air vents and vacuum breakers are required on some installations.

The condensate is piped back to a receiver, usually located in the boiler room, in a sufficiently sized return pipe line.

Many installations use heat exchangers to condense the flash steam in the return line and use the available heat to heat water for process or boiler-feed. These exchangers should be near trap outlets. Little or no back pressure is desirable in the condensate return pipe system.

From the condensate receiver, the hot condensate is pumped back into the boiler through a suitable boiler-feed pump and controls.

As you no doubt surmised from the preceding description, the main purpose for the entire system is simply to transport heat in a usable form from one central storage point to the various points throughout the plant where it is required. See Fig. 1-1.

It is possible, of course, to do this without the use of steam, but there are several advantages to using steam, which are summarized below:

1. Steam is relatively safe to pipe and use, as opposed to the piping of such highly flammable materials as gas or oil.

2. Steam is fairly safe to generate in the

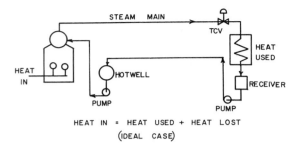

HEAT IN = HEAT USED + HEAT LOST
(IDEAL CASE)

Fig. 1-1. Boiler plant heat balance, ideal case.

boiler, providing reasonable care is exercised.

3. It is easy to control the process temperatures with the multitude of equipment, valves, and accessories on the market for use with steam.

4. Steam is well known, it has gained respect among the plant operators throughout the world, and it is highly predictable and stable.

5. Water is one of the cheapest and most abundant materials on the face of the earth, and at the pressures encountered in the average industrial boiler, treatment and preparation costs for use in the boiler are moderate.

6. The generation, use, and behavior of steam are basic processes upon which the Industrial Revolution was founded, and, therefore, it is almost an institution and part of our way of life.

B. Steam System Heat Balance

The boiler plant as a whole operates on the ideal heat balance equation:

Heat input = Heat used + Heat lost

The heat input to the system is simply the heating value released from the fuel being burned in the furnace of the boiler. In the case of oil fuel, there is approximately 18,000 BTU/

lb of heat energy stored in the oil. In the case of natural gas, there is approximately 1100 BTU in each cubic foot of gas, when measured at atmospheric conditions. There may be isolated cases of other fuels being burned in the small to medium-sized industrial boilers, but this will be rare. See Fig. 1-1.

In some parts of the country, coal is a favorite fuel among larger industrial boiler plants as well as in many central station power plants. The heating value of coal is not as uniform or as easily released as with oil or gas.

The heat lost in the system is often difficult to track down, with the exception of one huge loss, which is the stack loss. The average industrial boiler loses a generous portion of the fuel's heat up the stack in the form of hot products of combustion. This loss may vary from about 15% to 30%, depending on the condition of the equipment and the vigilance of the operator.

Another large heat loss is in the form of radiation and conduction through the walls of the boiler and the piping, as well as through the walls of the heat-using equipment. Keep in mind that a bare steam pipe may lose three to four times as much heat to the surrounding air as when the pipe is properly insulated.

Other sources of heat loss are steam leakage from the system, steam being deliberately blown to waste, boiler blowdown losses, and condensate being discharged to waste.

This discussion should clarify the advice often given to plant operators and maintenance personnel to always be on the watch for steam and condensate leaks in their walks around the plant. The preceding basic system heat balance equation shows that all heat lost from the system must be made up by the burning of additional fuel in the furnace.

The efficiency of the individual industrial boiler plant system taken overall is rather difficult to predict or determine. A good, new system, in tight and well-maintained condition, may attain an overall efficiency probably between 60% and 80%, which means that from one-fifth to two-fifths of the heat in the fuel is lost.

C. System Operating Principles

Let us examine how the boiler plant as a whole operates under the terms of the heat balance equation. First, a few basic concepts are in order.

If the system is at any one moment in complete balance according to the terms of the heat balance, then the system is in equilibrium, or it is considered to be stable. Under this condition, the fuel is being burned at just the right rate to make up the heat lost and the heat used in the system, and every portion of the system is working steadily and continuously.

If any part of the system is not in a steady condition, but is in a changing condition, then the system is not in balance, and the system is hunting, or seeking its balanced condition.

Very seldom will the average boiler plant be in complete equilibrium at any one moment, since the controls and the equipment are usually sized such that a hunting condition is required for the plant to operate, mostly for economic reasons. However, over a period of time, the heat balance equation will be satisfied in all boiler plants. In such case, the heat balance becomes

Average heat input
  = Average heat lost + Average heat used

For illustration, let us now assume that the system is in complete balance. Fuel is being burned at a steady rate, losses are steady, and the steam is being used in the process at a steady rate. The system is in equilibrium, or in a stable condition, similar to Fig. 1-1 conditions.

Let us now increase the steam requirement in some manner in the heat-using equipment. The control valve opens up to permit the steam flow to increase, the flow of steam in the pipe line increases in velocity, the pressure drops in the steam supply line, and there is a drop in steam pressure in the boiler over the saturated water surface. Since the boiler water is at a saturation temperature above the saturation temperature of the reduced steam pressure, more water boils off into steam. If the steam pressure drops low enough, the pressure control operates to increase the flow of fuel to the burner. The increased boiling lowers the water level in the boiler, so that the level control operates to restore the proper level.

At the other end of the system, the increased condensate flowing into the return system causes the velocity in the return lines to increase, and the return pumps (if used) to operate more often. The increased flow into the deaerator or hotwell tends to increase the level in that vessel. However, as the boiler level control is calling for more water, the boiler feed pumps should be pumping at a higher rate, so that the increased flow into the vessel should be handled almost immediately.

This description shows what happens to the ordinary system when an upset takes place somewhere along the steam end of it.

As nearly all level, pressure, and temperature controls in such a system do not work or operate instantaneously, but have a definite floating action or dead area in which the control does not operate to produce a change, there has to be a number of accumulators to take care of the lags produced in operation. The boiler is normally one of the chief storage points, as well as the steam main, the condensate lines, the condensate pump receivers, and the deaerator or hotwell. It is easy to see what difficulty can result if these elements are not properly sized to coordinate their individual capacities with the differential of the controls.

A classical example of this last statement may be found in a steam boiler having 100 gal of water capacity between the high and low contacts on the boiler level control, when the feed pump has a capacity of 200 gpm, and the boiler steaming rate is 70 gpm. Unfortunately, this is the trend with some boiler manufacturers today: high steaming rates per square foot of surface, small water volumes in the boiler, and low differentials in the boiler level controls. More about this later, however.

A safe, conservative, properly designed system will have ample surge or storage capacities in the boiler, the condensate receivers, and the deaerator or hotwell, to permit the system to operate and seek its balance without rapid fluctuations, carryover from the boiler, and pump or control chattering.

Some high-speed steam generators having little or no water volume are supplied with sensitive modulating controls, instead of the usual slow, two-positioned or differential style. The more sensitive a control is, the closer it has to be watched and the more maintenance it will require. Also, the noisier the system will probably be, and the shorter life you can expect out of the equipment.

Throughout most of the plant the average small industrial steam plant uses the on–off style of controls with the possible exception of the pressure-reducing valves and the temperature-control valves, which may be of the modulating style.

Figure 1-2 illustrates this type of steam plant control, whereas Fig. 1-1 illustrates the ideal situation, which we referred to in the last section.

Now let us take another look at that heat balance equation. There are several places in the steam plant where it can come in very handy.

We have used it to illustrate the entire boiler plant principle, but we will use the boiler alone as an example. Assume that the boiler is an isolated part of the system, and that we wish to estimate the fuel requirements for a certain size boiler. The heat balance equation shows us that the heat input to the boiler must equal the heat output in the steam leaving the boiler, plus the stack losses, plus the miscellaneous losses; thus

$$H_{in} = H_{steam} + H_{stack} + H_{misc}$$

For our purposes, we may take the total heat losses as equal to 25% of the fuel input, since this would be a fair average to expect from a new plant. This is what we were discussing a few paragraphs back, under the subject of boiler plant efficiency.

If previous calculations have shown what we need in the plant for the process steam, the final equation becomes

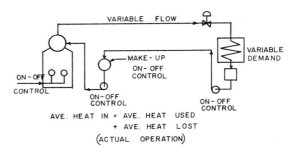

Fig. 1-2. Boiler plant heat balance, actual operation.

$$\text{Heat input} = \cfrac{\text{Total heat required for the plant}}{\cfrac{\text{Steam plant efficiency}}{100}}$$

And suppose, further, that we have already determined the useful output of the boiler is required to be 3,780,000 BTU/hr. Then the heat input to the boiler must be:

$$\text{Heat input} = \cfrac{3{,}780{,}000 \text{ BTU/hr}}{\cfrac{75}{100}}$$

$$= \frac{3{,}780{,}000 \text{ BTU/hr}}{0.75}$$

$$= 5{,}040{,}000 \text{ BTU/hr}$$

If we are going to use oil as fuel, then the pounds per hour of oil required is

$$\text{lb/hr} = \frac{5{,}040{,}000 \text{ BTU/hr}}{18{,}000 \text{ BTU/lb}} = 280 \text{ lb/hr}$$

To carry the example a little further, if we wish to convert the lb/hr to gal/hr, we would simply do the following:

$$\text{gal/hr} = \frac{\text{lb/hr}}{8.3 \times \text{sp.gr.}} = \frac{280 \text{ lb/hr}}{8.3 \times 0.94}$$

$$= 35.9 \text{ gal/hr}$$

This is assuming the use of #5 fuel oil, which has a specific gravity of 0.94, as shown in the Appendix.

The denominator in the above formula, 8.3 × sp.gr., converts the weight of a gallon of water, which is 8.3 lb, to the weight of a gallon of No. 5 fuel oil. If the actual weight of a gallon of the fuel oil is known, then it may be used in place of the 8.3 × sp.gr., since they should be identical in value.

D. Cycling

We referred earlier to a "hunting" operation under the description of the steam plant operating principle. By "hunting" we mean that method of operation in which the process or operation is not steady, but varies considerably from moment to moment, in an attempt to match the output to the demand. This variation may be performed smoothly in a modulating manner, or it may be an on and off operation, or at times a combination of both.

You no doubt have read in the popular literature about the ideal of any manufacturing process being a continuous smooth flow of goods, from the raw materials in at one end of a plant and proceeding on a continuous assembly line until, at the other end of the plant, a finished product emerges all packaged and ready to be shipped.

Such a system is, as we stated, purely the ideal, and would require no storage at any point in its travels.

It is a wonderful goal to aim for, but for all practical purposes, it is a goal that is impossible to reach. Even the most sophisticated and complex of assembly line operations, such as automobile manufacturing, do not achieve it.

So, getting down to the practical aspects of the average plant, we find that nearly all plant operations consist of the following functions, or variations of them:

1. An operation is performed.

2. The product is transported to the next operation.

3. It is stored, or accumulated, with others, awaiting its turn to go through the next operation.

Thus, we see that each item being sent through the line undergoes the same "cycle" of operations, which is a repetition of work–travel–wait–work–travel–wait, etc.

Why is this so? Simply because no two operations being performed along the assembly line can be performed at the same rates and still achieve top efficiency for each operation. Also, any attempt to do so would require very complex and expensive control and layout arrangements.

Perhaps you noted that each "travel" portion of the cycle was followed, or preceded, by a "wait" or "accumulate" element. This is the important part of such a process, and we shall discuss it further. We shall start by stating an axiom of materials handling in the plant: whenever there is a change in speed of operation along a continuous process line, there must be assembly or accumulation points at the junctures of the changes to keep the line performing properly. Referring now to Fig. 1-3, note that there are "surge" storages at each change in flow rates.

Now the important things about this whole setup, and the two elements required to keep it moving properly, are:

1. Each storage or accumulation must be controlled, preferably, automatically controlled.

2. Each storage or accumulation must be sized properly to absorb the "surges" or the "slack" as required to take care of the changes in flow rates under control.

Referring to Fig. 1-4, you will see another version of the same principle, which we will explain to illustrate the subject, and bring it closer to the subject of this book.

Let us assume that we have a tank of water, or any other liquid, into which there is a steady flow, as shown. Now let us assume that there is a process requiring this water at intermittent intervals, and at intervals that are not only repetitive, but identical. Furthermore, let us also assume that the demand is controlled by an on and off valve on the outlet of the tank, and the "on" time is equal to the "off" time.

Fig. 1-3. Handling variable flow rates.

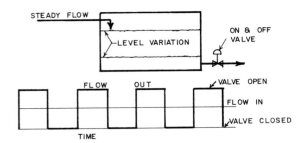

Fig. 1-4. Purpose of surge storage.

The top part of the figure shows the diagrammatic arrangement, while the lower portion of the figure shows the control function. If you will study this control function for awhile, you will note that although the valve is full on for a certain period of time, it is also full off for the same period of time. And the average flow out of the tank must equal the steady flow into the tank. Also, there must be sufficient storage volume between the tank outlet line and the top of the tank to supply each "on" cycle.

This example has been simplified to permit us to make our point clearer. Seldom will you find an actual case as simple as this. For example, it is customary to ignore the amount of water flowing into the tank during the period the valve is open, when calculating the amount of storage. This allows a safety factor, which does no harm, usually.

You are probably wondering just what this discussion has to do with your steam plant. Perhaps, if you look back earlier in the chapter, you will readily see that we were talking about just this subject when we described the hunting as opposed to continuous operation of the steam plant.

Thus, you see that any steam plant operating on the usual control system found in the average industrial steam plant is continually hunting, or cycling, and that the controls and the "surge" or "slack" capacities of the elements of the steam plant are very important.

As we discuss various portions of the boiler plant later in this book, you should be able to recognize the principles that we have expounded so thoroughly here. Some of the more prominent examples where this principle is necessary are:

1. Sizing fuel oil tanks.

2. Sizing fuel valves and lines.

3. Sizing burners.

4. Sizing the boiler.

5. Sizing the steam lines and valves.

6. Sizing the steam traps.

7. Sizing condensate receivers.

8. Sizing condensate pumps.

9. Sizing boiler feed and treatment systems.

In short, just about every part of the steam plant makes use of the principles explained in the discussion on cycling operation.

There is just one more facet of this subject that we should mention before leaving it. We gave you the general method for sizing the "surge" or "slack" volumes, based on the known calculated data available to you, and as a result the final calculated volumes will be no more accurate than the figures used to make the calculations.

There are many things that can cause upsets in the final design, regardless of how carefully you size your equipment. In a system such as the average steam plant, momentary conditions that are contrary to your intentions have a way of creeping in and causing trouble. We term these conditions "transient" conditions, since they are unpredictable and uncalculable.

To provide for such an emergency, we usually allow an additional safety factor in the final design, something on the order of 10% or 15% added to the capacities of the valves, lines, tank volumes, or pump capacities.

As an example, if you own a super-duper new automobile that performs beautifully on average roads, but lacks that one last bit of power required to get you out of an emergency when climbing mountain roads, you will readily understand what we mean by "transient" or emergency conditions.

CHAPTER 2
HEAT AND STEAM

A. Heat

We know how temperature is measured, and we now understand that it is simply a measure of the intensity of molecular activity in a substance, which is evidenced by the reaction of a column of mercury, or other appropriate substance, to that temperature.

Heat may also be measured by quantity. If we heat one pound of water, and raise its temperature one degree Fahrenheit, we have added one British Thermal Unit of heat to the water, providing this is done with the water at atmospheric pressure, and the temperature is about 40°F. This quantity of heat, commonly referred to simply as "BTU," is the basis of all heat calculations in our system of engineering measurements.

When working with large quantities of heat, it is usually more convenient to express the quantities in "therms," which is simply the term for 100,000 BTU, i.e., one therm is equal to 100,000 BTU. You will find this term used mostly in fuel calculations.

Every substance—gas, vapor, liquid, or solid—has the capability of absorbing or giving up heat, similar to the example of water given here. However, the amount of heat, in BTUs, which one pound of a substance requires to raise its temperature one degree Fahrenheit, varies considerably from substance to substance. This amount of heat in BTUs required to raise the temperature of one pound of a substance one degree Fahrenheit is called the "specific heat" of that substance. Page 401 in the Appendix contains a column marked "sp ht" next to the column on specific gravity.

Since a substance absorbs heat, it will also give up heat, and the result will be just the reverse of adding heat. One pound of water will give up to 1 BTU to its surroundings for each degree Fahrenheit drop in temperature. If Bunker C fuel oil loses heat to its surroundings, its temperature would drop 2.5°F for each BTU it loses, since its specific heat is 0.40.

There is one more basic principle concerning heat which we should point out here, and that is in order to move heat from one body to another, or from one substance to another substance, it is necessary that there be a difference in temperature. Heat will only flow from a high temperature to a lower temperature. This is true regardless of what the temperatures are, whether it be in the boiler or in the refrigeration system. If we wish to obtain a temperature of, for example, 0°F in a room, we must produce a temperature inside the cooling coils several degrees below that temperature, in order to induce the room heat to flow from the room into the refrigerant inside the cooling coils.

It is possible for heat to be carried from one place to another by moving currents of air, gas, vapor, or liquid, but in order for heat to flow from one substance to another, it must be at a higher temperature than the substance to which it is to flow.

Heat is energy, and is the source of power in all of our conventional engines and power producers. Steam turbines, diesel engines, gasoline engines, and gas turbines, all operate on heat energy. They all extract as much of the usable heat energy from the fuel or heat carrier as practical, and convert this energy into a more convenient form for our use.

Work can be converted directly into heat with nearly 100% efficiency, but the reverse is not true. There is usually considerable loss when we attempt to make use of the energy in heat. Theoretically, 1 BTU is the equivalent of 778 foot-pounds of work, and 1 horsepower is the equivalent of the expenditure of 2,547 BTU per hour. To give you an idea of the expected efficiencies of conversion from heat energy to useful work, we list here some average values for some of our common heat engines and heat-to-power conversion systems:

Steam turbines, condensing, power station use	35–40%
Diesel-electric locomotives	25–35%
Gasoline-engine-driven automobiles, overall	15–25%
Gas-turbine-driven generators	20–25%
Power station boilers	85–90%
Industrial boilers, small to medium size	70–80%
Steam locomotive	8%
Steamship, turbine driven, oil fired	5–10%

The remainder of the heat that is not utilized appears in the form of frictional heat, heat radiated to the surroundings, and heat discharged in the exhaust and into the cooling air or water.

We shall study this subject of heat in more detail later in this book.

B. Water and Heat

A cubic foot of water weighs 62.4 lb at maximum density of 39.2°F. A cubic foot of ice at 32°F weighs 57.5 lbs. In other words, the water expands as it changes from water at 32°F into ice at 32°F.

There are three states of matter—solid, liquid, and vapor. Heat addition produces a rise in temperature as it enters a substance and is called sensible heat, as previously noted. In the boiler, the heat that passes from the furnace into the water, first raises the water temperature. This is always known as sensible heat; its full name is "sensible heat of the liquid." All steam engineering calculations are based on 32°F as being the minimum. Any temperature on the thermometer above 32°F represents sensible heat in the water up to 212°F at atmospheric pressure. Further addition of heat will not raise the temperature of the water at atmospheric pressure. The additional heat is known as the latent heat of evaporation; this extra heat is converting the water into steam.

Now, for a minute or two, we will imagine that our boiler has a spout on it like the kitchen kettle and that the spout is an outlet for the steam, which is now being made in the boiler. Although more and more heat is passing from the furnace into the water, the temperature of the water will not rise above 212°F. This additional heat is known as latent heat and is the heat required to change the state of the liquid at 212°F into steam at 212°F at atmospheric pressure. This is also known as the saturated temperature of the steam and water.

C. Steam

This book is purposely geared to a rudimentary understanding of steam fundamentals. Our purpose here is to convey specific, concrete information about steam, its properties and terms, its generation, etc. We do not wish to get too technical and risk the possibility of your getting lost in technical terms. We shall try to reduce every detail to understandable terms and stress only the "basics."

We now know the boiling point of water is 212°F at atmospheric pressure at sea level. The normal atmospheric pressure at sea level is 14.7 psia or 29.92 in. Hg on the barometer.

The steam boiler or generator usually found in our plants is a strongly constructed, closed vessel made of steel that confines the steam as it is formed and causes the pressure in the boiler to rise. As the pressure in the boiler rises, the boiling point of the water in the boiler also rises. For each pressure there is a corresponding boiling temperature. This is known as the saturated temperature of the steam, and is shown graphically in Fig. 2-1.

Steam that contains entrained moisture is called wet saturated steam. Steam that does not have entrained moisture is known as dry saturated steam, if it is at saturation temperature.

We might explain the term saturation temperature a litle more at this time. As the feed water is put into the boiler at a temperature below the boiling point, sensible heat is added to the water until it reaches the boiling point. The water is now said to be saturated with sensible heat. Further addition of heat causes the water to change from water at the boiling point into steam at the same temperature.

As we stated before, there is a corresponding temperature for each pressure. The required pressure at which our boiler must operate is determined by the temperature required in our process work and the pressure required to transmit the steam.

Such steam equipment as air-heating coils and storage water heaters in office buildings can operate with low-pressure steam—2 to 5 psig. Steam at 2 psig has a temperature of 219°F. Laundry machinery requires a much higher temperature. Boilers in laundries may operate at 125 psig, in which case the corresponding temperature would be 353°F.

Most machines and processes in our plants are heat-using devices. We should think of steam as a medium to carry the heat to the heat-using machine.

Any heat lost from saturated steam causes some of the steam to condense back to water or condensate. The condensate should be removed from the steam line by a system of drip pockets and traps to ensure that the driest steam possible is getting to our heat-using machines.

To reduce this condensation to a minimum, we cover the bare steam lines with some form of insulation to reduce the heat loss to the surrounding space. These insulating materials may be magnesia or foam glass, for steam service.

D. Total Heat of Steam

As we said in the beginning, steam is a good conveyor of heat and pressure energy. For the time being, let us think only of the heat in it. We have seen that steam generated in our boiler has two kinds of heat in it, sensible and latent, and that the quantities are explained as so many BTU per pound. The sensible heat in the water raises the temperature of the water to the boiling point, or its saturated temperature. Further addition of heat changes the water at the boiling point into steam at the same temperature. This heat is known as latent heat. The total heat in the steam is sensible heat plus the latent heat, and we show you this in picture form below. You will see from Fig. 2-2 that the quantity of latent heat is considerably greater than the quantity of sensible heat.

In every 1 lb weight of our steam that is at a temperature of 212°F and atmospheric pressure, there are 180 BTU of sensible heat and

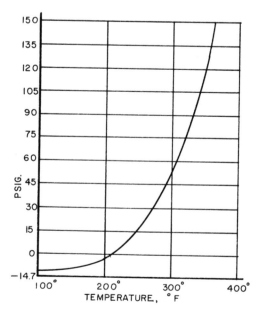

Fig. 2-1. Boiling point of water.

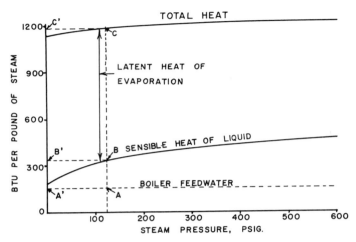

Fig. 2-2. Total heat in 1 lb of saturated steam.

970 BTU of latent heat, making the total heat about 1150 BTU. If, instead of 1 lb, we had a quantity of 100 lb, we take the figures given above and multiply them by 100. This means that we should have at our disposal a total heat of roughly 115,000 BTU. This would be made up of 18,000 BTU of sensible heat and approximately 97,000 BTU of latent heat. It is important to remember that by far the greater proportion of the total heat of the steam is latent heat.

In Chapter 1 we described briefly the operation of the total steam plant cycle. Using Fig. 2-2 we shall now show it in another form, which may help to clarify it. This discussion will be of a general nature, and the reader may wish to supply conditions existing in his own equipment to the points on Fig. 2-2.

Feedwater enters the boiler at point *A*, and heat is added to bring it up to the boiling point *B*. This heat is known as "sensible heat," for reasons already given, and the heat content at point *B* is known as the "sensible heat of the liquid."

Once the water is at boiling temperature, as shown in Fig. 2-1, any further addition of heat converts the water to steam at boiling temperature, bringing the steam to point *C*. The heat added to each pound of steam from point *B* to point *C* is known as the "latent heat of evaporation," as shown on Fig. 2-2.

The steam at point *C* is then sent out to the process in the plant, where it is condensed, giving up its latent heat of evaporation, and producing condensate at point *B*. This condensate is then sent back to the boiler plant for reuse in the boiler for another cycle. This is assuming that the steam is being used in some heat-exchange process, which is the case in most industrial plants, permitting recovery of the condensate for recycling through the boiler plant.

If the steam is being injected directly into the process stream in the plant, the steam is lost, which means the loss in water represented by that steam must be replaced in the boiler feed-water supply system.

If the steam is being used to drive a blower, pump, or similar unit in the plant, then it expands its energy in a turbine or piston, discharging into a condenser or into a low-pressure steam supply system for further use in the plant.

The primary purpose of the steam plant is to supply steam containing the heat from point *B* to point *C*. That is the sole purpose for the entire boiler plant, and the personnel required to operate it.

Regardless of how the steam is used in the plant, it can readily be seen that steam is simply a convenient method of distributing heat from a central source to the various locations in the plant where it is needed, as we stated at the beginning of Chapter 1.

In most cases, the heat in the boiler feedwater from point *A* to point *B* is carried along as excess baggage.

E. Use of the Steam Tables

Table 2-1 is a brief tabulation of the various important characteristics of saturated steam, which we shall now describe. Please note that this table may not be in the general form that you will find in other references, but it will illustrate our discourse very well. You should become familiar with using this table and the trends you will note when you read from one portion of the table to another.

The steam tables are based on the following premises and limitations:

1. All values are for 1 lb of saturated steam, or water.

2. Water at 32°F has no heat content, or the total heat content is 0.

3. The pressures are usually in psi absolute (psia), whereas Table 2-1 is in psi gauge (psig). This is done for convenience, although it should cause no difficulty either way, as you no doubt are aware of the relationship between the two systems. Remember, psia = psig + 14.7, at sea level.

4. The table is based on saturated steam that is dry, having no entrained water.

5. The steam is pure, having no contaminants such as air or any type of gas.

6. The specific volume (sp. vol.) of the steam is in cubic feet per pound of steam.

As the pressure of the steam increases from the vacuum range through the higher pressures, note that the following take place:

1. The steam temperature increases, as we have just explained in connection with Fig. 2-1.

2. The heat content of the liquid (water) increases, which is sensible heat of the liquid.

3. The latent heat of evaporation decreases with increase in pressure.

4. The total heat of steam is equal to the sum of the two previous columns (column #5 = column #3 + column #4).

5. The volume occupied by 1 lb of steam decreases rapidly with increase in steam pressure.

As we have previously stated, there are several different forms of this table, but the relationships are the same regardless of the form in which it is presented. See pages 393 and 394 in the Appendix for a more complete version.

Using Table 2-1 we shall now apply concrete examples to our discussion of Fig. 2-2. We shall take the case of a boiler producing saturated steam at 125 psig and 353°F, using boiler feedwater at 180°F.

At point A, the heat content is = 180 − 32 = 148 BTU/lb.

At point B, the heat content is = 325 BTU/lb, from Table 2-1.

Heat added from point A to Point B = 325 − 148 = 177 BTU/lb.

Heat added from point B to point C = 868 BTU/lb, from Table 2-1.

Table 2-1. Properties of Saturated Steam

Pressure (psig)	Temperature (°F)	Heat of Liquid (BTU/lb)	Latent Heat of Evaporation (BTU/lb)	Total Heat of Steam (BTU/lb)	Specific Volume of Steam (ft³/lb)
20 in. vacuum	162	130	1001	1131	75
10 in. vacuum	192	160	983	1143	39
0	212	180	970	1150	27
50	298	267	912	1179	6.7
100	338	309	881	1190	3.9
125	353	325	868	1193	3.2

Heat available for the process = 868 BTU/lb, less some losses from condensation in the transmission lines.

F. Steam under "Vacuum"

The term "vacuum" is often misused and misunderstood. It is generally used to denote pressures below atmospheric. And that is the way we will continue to use it throughout this book, although there are at least three ways in general use of denoting pressures that lie between absolute zero and atmospheric pressure. There is another term for it that is probably better than "vacuum" and that is "negative pressure."

Over the years of steam usage a belief has developed that some mysterious change takes place when steam enters the vacuum range, or pressures below atmospheric. However, if one were to take the time to study page 393 in the Appendix it would soon be evident that any change that takes place is of a gradual and orderly nature. The columns showing the pressure and temperature relations for saturated steam, if plotted as a graph, would be a smooth curve throughout its entire range, all the way down to 32°F. This is illustrated in Fig. 2-1.

There are probably only two things that have to be watched when steam goes into the vacuum range.

Steam spaces, as with any space under vacuum, would then have a tendency to leak inward, instead of outward. Thus, as long as the steam pressure inside is less than atmospheric, the system could be leaking continually, and it would not be readily detectable. To find such leaks would require a soap suds bath at the suspected points, while the system is put under a slight air pressure, or bathing the joints with water or oil while under the vacuum.

Also as shown in the Saturated Steam Tables, page 393 in the Appendix, when steam approach the lower absolute pressures, the volume occupied by 1 lb of steam increases tremendously. This means that any equipment or piping to handle this extremely

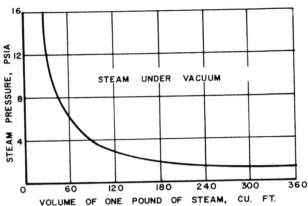

Fig. 2-3. Pressure vs specific volume for saturated steam under vacuum.

low-pressure steam must be fairly large in capacity. Such equipment must also be capable of withstanding vacuums without collapsing. See Fig. 2-3.

G. Superheated Steam

We have just discussed the generation of saturated steam, and you have probably noted in your plant that all of the heat input to produce that saturated steam is directed into the water from which the steam is generated. In the average industrial plant the saturated steam is then bled off from the boiler through pipes to the point of use. However, if we were to pass that saturated steam through pipes into another source of heat, we would find that the saturated steam will absorb more heat, and be changed in the process. This will result in "superheated steam." The definition for "superheated steam" is steam that is at a temperature higher than its saturation temperature. See Fig. 2-4.

Exposing saturated steam to an additional source of heat results first in the drying up of any entrained moisture that may be present in the steam. After all of the entrained moisture has been evaporated, producing dry and saturated steam, more heat is then absorbed by the steam, and its temperature is raised by the addition of sensible heat.

From this point, the steam no longer follows

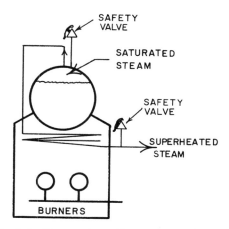

Fig. 2-4. Generation of superheated steam.

the relationships shown in the Saturated Steam Tables, but a separate table is required, known as "Superheated Steam Tables." See Table 2-2.

In the use of superheated steam there are some trends that may be important to you, such as:

1. As the steam temperature rises so does the total heat of the steam, as explained for Fig. 2-2

2. As the amount of superheat increases, the volume occupied by 1 lb of the steam increases also.

3. Therefore, there is a net decrease in BTU per cubic foot of superheated steam as the temperature increases.

The superheated steam table in Table 2-2 is in a form that you will not see elsewhere, since it is shown thus simply for illustration. There are many forms in use, and some of them are rather complicated and will require study to

be able to use them properly. The majority of the superheated steam tables found in use today will be in the form of a graph or tabulated groups of figures.

The first three columns of Table 2-2 are a brief version of Table 2-1. The superheat columns were chosen merely for purposes of demonstration, as the amount of superheat is not limited to 50° and 100°, but may be any amount from a few degrees to several hundred degrees. The final steam temperature is determined by its intended use. For instance, most steam turbines are designed for use on superheated steam, often as high as 1000°F. It will seldom be used in heat exchange operations, since the heat transfer is very low, relative to saturated steam.

As with any high-temperature invisible gas, superheated steam calls for care in its use and handling.

The fact that the steam is superheated does not indicate that it is definitely dangerous: it depends entirely upon its actual temperature. In short, it is the temperature that may be dangerous, not the fact that it is called "superheated." For instance, the air you breath contains superheated steam, but it is low enough in temperature that it is not only harmless, but may actually be beneficial to your health. But if you were to breathe superheated steam at 400°F, it would be your last breath!

Dealing with the temperatures expected in the steam plant using superheated steam, we should categorically consider it dangerous. It is dangerous for several reasons, as follows:

1. The temperatures are usually high enough to cook human flesh, or at least to burn it very badly.

Table 2-2. Properties of Superheated Steam

Pressure (psig)	Saturated Temperature (°F)	Total Heat (BTU/lb)	50° Superheat		100° Superheat	
			Temperature (°F)	Total Heat (BTU/lb)	Temperature (°F)	Total Heat (BTU/lb)
0	212	1150	262	1175	312	1194
25	267	1170	317	1196	367	1220
50	298	1179	348	1207	398	1232
100	338	1190	388	1213	438	1246
125	353	1193	403	1222	453	1249
150	366	1196	416	1225	466	1254
200	388	1199	438	1232	488	1260

2. Superheated steam is invisible.

3. It has a tremendous amount of internal heat energy, which carries it great distances when released into the air.

As you will note by studying Fig. 2-5, a leak in a superheated steam system may be very dangerous. It is a good rule to keep that in mind when entering a room or an area where it is known superheated steam is being used, *always* keep a sharp lookout for the cloud of swirling, silver-gray visible end of the jet, accompanied by a high-pitched shriek or roar. If you see one, do not move until you have studied it, and found the source of the leak. To cross through the invisible portion of the jet is inviting disaster, as you can see by studying Fig. 2-5. Be sure and notify all others in the area of the danger immediately. Shut down that portion of the system and repair it immediately.

Many plant operators who have superheated steam in the higher, dangerous temperature ranges, have adopted a simple safety precaution. When entering a room or area where the superheated steam is either generated or piped, a lightweight wooden wand, such as an ordinary yard stick, is held in front of them and waved up and down as they proceed cautiously through the area. A high-temperature invisible jet of the superheated steam hitting the stick will be detected immediately, permitting the operator to take appropriate action.

Of course, every part of the superheated steam system should be well insulated, not only to preserve heat, but also to protect personnel from accidental burns.

As an example of the internal energy in the jet of high-temperature superheated steam, we point out that pinhole leaks in boiler super-

heater tubes have often sliced through the walls of the adjoining boiler tubes! The appearance is similar to cuts made with an oxyacetylene torch.

It is important to remember that the safety valve shown on the outlet of the superheater section in Fig. 2-4 must be set to relieve at a pressure below that on the drum. This is to ensure that when an overpressure occurs, the steam being relieved to the atmosphere will flow through the superheater, thus keeping the superheater tubes cool enough to prevent them from becoming overheated and damaged.

For the same reason, when starting up a boiler with a superheater and getting it on the line, it is absolutely necessary that the drain on the superheater outlet be open until the boiler is sending steam into the plant system. This means that the discharge from the drain must be in a safe location.

The location of the superheater section is dependent on the design of the boiler and the desired temperature of the outlet steam. If all of the superheater tubes are in only one section of the furnace, then there will usually be a fairly wide swing in final steam temperature with changes in steaming rate, which is often the case with the smaller boilers.

In the larger boilers, the superheater tubes are often split into two sections, with part in the convection section, and the remainder in the radiant section of the furnace. The split between the two sections are then designed to produce a fairly constant outlet temperature with reasonable swings in steaming rates.

A few of the larger boilers will control the superheated steam temperature by controlling bypass dampers in the flue gas stream, diverting the flue gas flow between the various sections of the superheater tubes, with some installations using two separate superheater sections for the purpose. A few large boilers are arranged to swivel the burners vertically, thus obtaining a degree of final temperature control.

Another method is to produce the steam at an elevated temperature, then reduce its temperature to that required for the plant. Figure 2-6 is an example of this method, which gen-

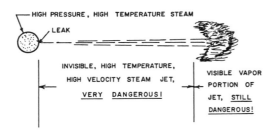

Fig. 2-5. Superheated steam leak.

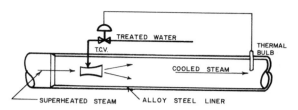

Fig. 2-6. Basic desuperheater principle.

erally results in a close control of the final temperature. It is important to remember that with this system, no attempt should be made to lower the superheated steam temperature to the saturation temperature, since once the temperature-control element has been adjusted to produce the saturation temperature, all control over the percentage of moisture in the steam is lost, and the result could be very wet steam. It is best to adjust the control to give at least

a few degrees of superheat on the outlet of the desuperheater.

A very large number of industrial boilers of the watertube design are arranged as shown in Fig. 3-15. This works very well for those plants desiring only a few degrees of superheat, such as when transmitting steam long distances and a minimum of line condensation is desired with fairly dry steam for the final process. For this purpose, usually only a few feet of superheating tubing in the furnace is required.

Superheater tubes are more often found in watertube boilers, as most firetube boilers have no location suitable for their insertion. They cannot be placed in areas that make them subject to direct impingement of the flame, which limits their application in some firetube boilers.

C H A P T E R 3
BOILERS

A. Heat Absorption in the Boiler

There are three types of heat-transfer methods, and all three are in general use in the average boiler. Some boiler designs use one system more than the others, depending on the type and size of boilers.

Figure 3-1 shows the primary type of heat transfer in the boiler, which is radiant heat. You have probably heard a great deal about radiant heat in the past, in regard to home heating, where the term has been misused considerably.

Radiant heat is similar to light in its behavior, in that it follows almost all of the laws that light rays obey. Radiant energy travels in a straight line from a high-temperature to a lower-temperature body, without heating the air or gas through which it travels. Radiant heat is given off by all bodies to others at lower temperatures, and is highly affected by the condition of the two surfaces involved; a black, dull body absorbs heat readily, while a shiny surface radiates heat readily.

The radiant rays that transport the heat energy are not too well understood, and we do not intend to add to the confusion here. We know how it behaves; therefore, we may make proper use of it without understanding the nature of its composition.

Radiant heat in a boiler is usually utilized by the placement of the tubes in a watertube boiler so that the glowing, high-temperature flame has a direct "view" of the dull black surfaces bearing the water to be heated. In this case, we say that the radiant heating surface of the boiler "sees" the flame.

The next type of heat transfer you should learn is shown in Fig. 3-2, which is "convection." This simply means the transportation of heat energy by moving streams of gas or liquid. In the boiler, this takes the form of heating the incoming air for combustion, producing hot gases as a result of the combustion, then guiding the high-temperature gases across the heat-transfer surfaces in the boiler. Of course, along with the heat contained in the gas we may find fine particles of waste matter that cause no end of problems. That which does not attach itself to the heat-transfer surfaces in the boiler or the stack is carried out to the atmosphere.

Now that we have covered the transportation of heat by carrying it along with streams of gas from the burners, it then finds its way into the water in the boiler by the process of "conduction." See Fig. 3-3, which illustrates the method by which heat finds its way into the tubes and into the water. Remember, we stated that heat always travels from a high temperature to a lower temperature. Now that the convection currents in the boiler and the radiant rays from hotter surfaces in the boiler have all deposited heat on the tube surface, the heat must be transferred into the water. This

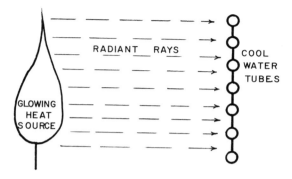

Fig. 3-1. Principle of radiant heat in the boiler.

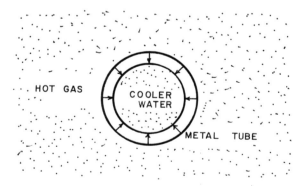

Fig. 3-2. Heating by convection in the boiler.

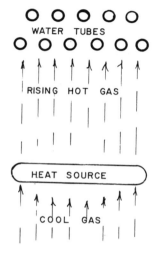

Fig. 3-3. Heating by conduction in the boiler.

is accomplished by passing heat energy along from molecule to molecule in the metal of the tube. This is the basic method of heat transfer in most of the heat exchangers found in industry today.

The larger watertube boilers, which usually have superheaters in them, rely on all three types of heat transfer described. Modern boiler design in all types and sizes is basically a matter of balancing the three types of heat-transfer methods to get the desired result. All boilers make use of the method of conduction. In some of the more compact "flash" boilers, it is difficult at times to determine which surfaces are radiant and which are convection.

In the case of the firetube boilers, it's not too difficult to determine that the majority of the surface is heated by the convection principle, since there are relatively long gas passages after the flame is produced in the furnace tube. The furnace or firing tube is the only radiant surface, the remainder being convection surfaces.

It is the relative proportion of radiant to convection surface in a boiler that determines its output characteristics at different firing rates. This is true not only of saturated steam boilers, but also of superheating surfaces.

Figures 3-4 and 3-5 illustrate the two general types of boiling that may exist across the tube wall of a watertube boiler. In both cases, the water temperature is the same, which is set by the pressure in the boiler. However, the temperature to which the tubes are subjected on the opposite side is determined by their position in the furnace and the firing rate of the burner.

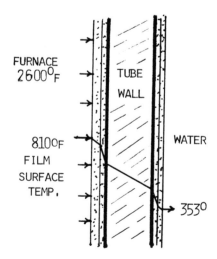

Fig. 3-4. Nucleate boiling temperature profile.

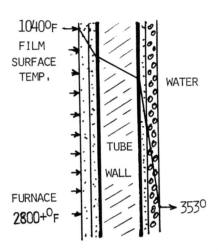

Fig. 3-5. Film boiling temperature profile.

Referring to Figure 3-15, you will note that there are generally two types of tubes in the saturated section of the boiler and one set of superheater tubes. In general, nucleate boiling takes place in the convection and the radiant sections of the boiler. The superheating section follows an entirely different type of heat-transfer characteristic, similar to a gas–to–gas exchange, since no boiling takes place in this section.

Nucleate boiling is the condition whereby the water sweeping past the surface of the tube carries the bubbles of steam away from the hot surface as fast as they are formed. These bubbles are carried along with the water until they are discharged into the steam drum. In this manner, there is always water on the surface of the tube, which increases the heat-transfer rate across the tube wall.

In the case of film boiling, the bubbles of steam being formed on the surface of the tube are not carried away fast enough, so they accumulate and cause a film of steam between the water and the hot tube surface. This condition causes the heat-transfer rate to be lower, since the film of steam acts as a barrier to the passage of heat. You will note from Fig. 3-5 that in order to obtain the desired water or steam temperature with film boiling, the opposite surface of the tube has to be subjected to much higher temperatures than in the case of nucleate boiling, as shown in Fig. 3-4. Obviously, tubes under the higher temperature will not last

as long in the furnace as will tubes under nucleate boiling.

How does this relate to your operations in the boiler plant?

The average watertube boiler contains a combination of radiant and convection tubes of the nucleate type, since they are the most efficient and give the longest service life without trouble. As long as the boiler is operated at or below its rating, and providing the tubes are kept fully operable, being regularly cleaned and checked, there should be no difficulty. However, should a condition arise where the circulation is reduced in the boiler, then in order to get the desired output, it may be necessary for the burner to produce higher combustion temperatures, and thus higher gas temperatures, and film boiling may result. This condition could also come about if you attempt to fire the boiler beyond its rating. This is possible with many boilers, since the burners are often rated at about 10% capacity over the actual boiler requirements.

Remember, when you overfire a boiler, or attempt to operate when some of the tubes are plugged, or the water circulation is impaired, you are only asking for trouble, and trouble you are sure to find.

A boiler is a form of heat exchanger in which fuel is burned to form heat and products of combustion. As much of the heat as possible is conducted through the heating surface, which causes the water to vaporize under controlled pressure. The hot gases of combustion, after having as much heat as economically feasible absorbed by the water, are then discharged to the atmosphere at elevated temperatures.

Figure 3-6 illustrates the basic principle of the steam boiler. The boiler is an energy conversion apparatus that takes in fuel, air, and water in the right proportions, and produces hot gases, steam, and radiant heat. The hot gases are allowed to escape up the stack, because they are waste products in most cases. The radiant heat, which escapes from the hot surfaces of the boiler, is usually kept to a minimum by the application of insulation. The only usable portion of the process is the steam, which is the part we are primarily interested in here.

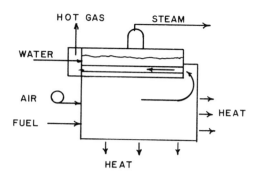

Fig. 3-6. Basic boiler principle.

Boilers are classed generally by the relative arrangement of the hot gas passages. If the hot gases go through the boiler tubes, with the water outside of the tubes, it is known as a firetube boiler. If the water is inside the tubes and the hot gases outside the tubes, it is called a watertube boiler. Each one has its own distinctive styles and characteristics.

B. Firetube Boilers

Firetube boilers generally have large diameter tubes, from 2 in. diameter and up, and they are usually straight and relatively short, so that the hot gases will have a low pressure loss in passing through them. The water space surrounding the tubes is usually contained by a large-diameter cylindrical vessel. For this reason, firetube boilers are seldom designed for more than 250 psig, since the required wall thickness would be excessive. See Fig. 3-7.

Firetube boilers will have a fairly large volume of contained water relative to the steaming rate, so that there is a considerable amount

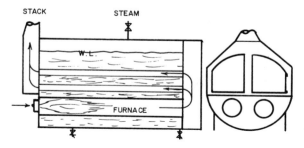

Fig. 3-7. Common two-pass firetube boiler arrangement.

of stored heat energy in the boiler. This makes for steady steaming rates and smooth operation, usually. It also increases the inherent danger in case of a tube or furnace wall rupture. These boilers produce a great deal of damage when they burst, and often fly some distance from their setting, owing to the tremendous amount of released energy in the water.

There are many styles and designs of firetube boilers in use today, since it is the oldest class in existence. Firetube boilers have evolved from a simple drum mounted over a fire into the Scotch-type boiler, which probably represents the most advanced design in firetube boilers.

One of the old favorites for years in boiler plants was the horizontal-return-tube (HRT) boiler. It is basically a cylindrical drum on a brick furnace containing the combustion space directly under the bottom of the drum. The bottom of the drum receives the first heat from the furnace, and then the gases pass to the rear of the furnace, reverse direction, and come forward through straight tubes in the lower portion of the drum. They end at the uptake, which is at the front, or firing end, of the boiler. From the uptake the gases go into the breeching to the stack. Fig. 3-6 is an example.

There are many variations of the HRT design, which may still be found in some older plants. Some may have two passes of tubes, and some may have the tubes inclined upward in the direction of flow.

One of the earliest improvements of the firetube boiler was the Scotch-type boiler, first used in marine service. It is a large-diameter pressure vessel sitting horizontally, with the fuel being fired into smaller cylindrical furnaces submerged in the water in the boiler. The firing tubes or furnaces are considered as one pass of the flue gases. Fig. 3-7 is this style.

The gas passes to the rear end of the firing tubes, then is reversed and comes back through smaller-diameter tubes above the furnaces, to the front of the boiler, as in the HRT design just described. Most boilers are two-pass design, with the uptake in front of the boiler. Some have three and four total passes, which require

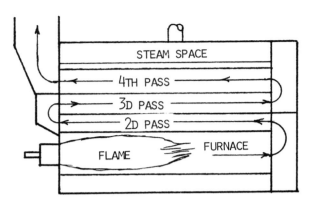

Fig. 3-8. Four-pass firetube boiler.

more draft to obtain good combustion. See Fig. 3-8.

The modern version of this boiler, which is enjoying a great deal of popularity, is the package style, made in sizes up to about 600 horsepower. The entire boiler—burners, controls, draft fan, and sometimes the feed pumps—are all contained on one base, skidded into place, and connected. Many companies have concentrated on this style in recent years, and, as a result, it has been developed into a very reliable and efficient unit. Competition, insurance companies, and local codes have combined to produce a class of boiler that is fairly standard

from one brand to another, with only a few differences. The main differences in design of the boiler itself are in the number of flue gas passes, forced or induced draft, and wet or dry back style. Figure 3-9 illustrates a typical modern packaged firetube boiler.

C. Watertube Boilers

Watertube boilers are characterized by smaller-diameter tubes, which are longer, often bent or curved, with the water inside the tubes. Many boilers are made today having watertubes that are serpentined or even coiled in a spiral. Furthermore, they are often headered in parallel to make up a complete solid wall or panel of heat-absorbing watertubes. For this reason, the heat absorption rate per unit of furnace volume or furnace wall area is relatively high, as compared to the firetube boilers. See Fig. 3-10 for an example of this style of boiler.

All high-pressure boilers are of the watertube design, because tubing of a given diameter and wall thickness will withstand a much higher internal pressure than external pressure. In the firetube boiler, the tubes have to

Fig. 3-9. Typical packaged firetube boiler. (Courtesy of Cleaver-Brooks Co.)

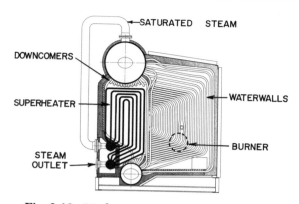

Fig. 3-10. Modern watertube boiler section.

withstand the water pressure from the outside of the tube.

Most of the failures in watertube boilers come from split or burned tubes, which result in a smothering of the fire and collapsing of the watertube walls at worst. Very seldom does a drum in the watertube boiler burst; but when it does, the result can be as disastrous as with a firetube boiler.

As most watertube boilers have a very high steaming rate, with a low contained volume of water, they are very sensitive to changes in demand and water level. Therefore, they require very close attendance, a better and more so-

phisticated control system, and water of high purity.

The larger steam generators found in central-station power plants are an excellent example of the present peak in design of watertube boilers, with high pressures, high temperatures, and combustion efficiencies as high as 91%.

Some larger industrial plants have smaller versions of this same design, with from one to three drums, and numerous bent tubes. This style is now available in factory-assembled package units requiring more installation expense than is the case for the Scotch firetube mentioned above. However, in the size ranges above 600 horsepower, the watertube boiler is the only packaged industrial type now available, unless multiple units are considered. Figure 3-11 shows such a design quite popular today.

One of the chief differences between firetube and watertube boilers is the condition of the water required to maintain proper and efficient operation. Generally speaking, the watertube boilers will require more care in the treatment of the feedwater, since they are more sensitive to scaling up of the tubes, and thus

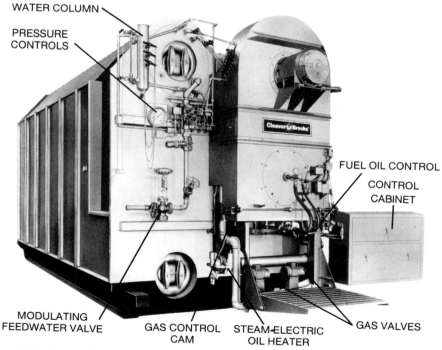

Fig. 3-11. Typical packaged watertube boiler. (Courtesy of Cleaver-Brooks Co.)

require more care in blowing down and in maintaining the right internal water conditions.

One of the most popular styles of small package watertube boiler now in use is the "flash"-type, forced circulation units being manufactured by several firms. They have a high rate of generation for the space they occupy, but this high rate of generation is usually accompanied by more maintenance requirements and a shorter life of the boiler. In addition, the steam produced is generally wetter than with other types having more conservative steaming rates. See Fig. 3-12.

The "flash"-type boiler uses multiple small-diameter coils, through which saturated water is circulated by a pump. The pump takes suction from the bottom of the return tank, into which the make-up water has been injected. This water is pumped under pressure into the heating coils, where it absorbs heat from the burner in its high velocity passage through the tubes. The water reaches a highly saturated condition and is then released into the top of a separation vessel, where flashing takes place, and the steam formed is then passed through the top of the separator into the steam main. The remaining water falls to the bottom of the separator for recirculation through the system. A level controller on the make-up tank supplies sufficient hot treated make-up water to maintain proper operation.

The rate of circulation of the water around the circuit is as much as three to four times the steam generation rate in some boilers. We refer you to Fig. 3-13 for a view of a typical flash-type boiler.

In "flash" boilers, blowdown is a constant

Fig. 3-13. Typical packaged flash boiler. (Courtesy of Clayton Manufacturing Co.)

problem, and is usually taken care of by some type of automatic adjustable blowdown device that either bleeds off sludge continuously or bleeds off sludge in pulses timed and regulated by the feed pump.

One of the best uses of the "flash"-type boiler is for remote locations and intermittent operation, where it is not too practical to supply a steam line from a central source. This type of boiler can be started from a cold start very rapidly, and can be shut down easily when there is no further demand.

D. Electric Boilers

There are several makes of boilers on the market that derive their heat from electric heaters immersed in the water. Most of them are made in package units from around 200 kW input up to 3000 kW input. Translated into boiler horsepower, this is the equivalent of about 20–300 bhp, since 10 kW/hr is the approximate equivalent of 1 bhp.

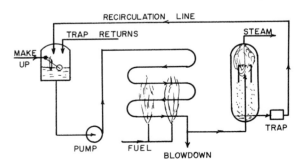

Fig. 3-12. Flash boiler principle.

Electric boilers have been around for some time, over 40 years, in one form or another. Like the other types of boilers on today's market, they have undergone considerable development and refinement. They have gained some popularity for certain classes of service, so that there are now several manufacturers who have them available.

Electric boilers work on either of two different principles. The simplest one is similar to a horizontal Scotch package boiler, with the substitution of an electric resistance element in place of the oil burner and tubular furnace, and without the flue gas passages. This style of boiler uses fairly conventional controls, except that the steam pressure switch controls the immersion heater through suitable control relays.

The other style of operation for electric boilers makes use of the electrical resistance of the water to produce the heat from the flow of electricity between two electrodes. The water forms the electrolyte, and the heat input to the water is in direct proportion to the amount of electrode exposed to the water. The steam pressure control is accomplished by altering the immersion of the electrodes: the larger the steam load, the lower the electrodes in the water, and the greater the electrical flow into the water. When the load drops to zero, the water level is below the electrodes, so no current flows, and no steam is generated. This is the same principle as the ordinary household vaporizer used in a child's sick room.

Figure 3-14 shows one version of this electric boiler. It is a typical package boiler, containing everything essential to its operation within one housing. It is this packaging that has helped considerably in making it attractive to many users. Also of importance are the ease of installation, quick steaming ability, cleanliness, safety, high degree of flexibility and turn-down ratio, and portability of some models. They are finding a great deal of acceptance for isolated spots where steam is desired for only a portion of the time, and only in one small location. Such uses include portable cleaning applications, small laundries and dry cleaning firms, research laboratories, hospitals, cafeterias, restaurants, and many industrial plants requiring only small quantities of steam for short periods of time.

The electric boiler utilizes the electricity at about 98% effficiency, and this factor has been used as a selling point by the makers of this type of boiler. One must keep in mind, however, that the electricity was probably generated in central generating stations at about 35% thermal efficiency from oil-, coal-, or gas-fired boilers and steam turbine generators. It reaches the industrial plant at about 25% thermal efficiency. So from an overall cost standpoint, they may not be as cheap as if oil or gas were to be used directly as fuel for the boiler instead of electricity.

There are no air pollution or flue gas problems with the electric boiler, of course. However, they suffer from all the water condition problems that conventional boilers are faced with. The electrodes and the immersion heaters have a tendency to become coated with deposited salts from the water, which calls for water treatment and blowdown, as with conventional boilers.

Fig. 3-14. Typical packaged electric boiler. (Courtesy of Automatic Steam Products Corp.)

All these boilers now on the market are being made to ASME code and are approved by the Underwriter's Laboratories. Most of them are made in pressures to 150 psig, with a few available at higher pressures for laboratory and research work.

E. Watertube Boiler Arrangement

In the last chapter we briefly covered the generation of superheated steam, mentioning that superheated steam is generated from saturated steam. Figure 3-15 will show how this is accomplished in the industrial watertube boiler.

The key to proper design of the watertube boiler is to arrange the size, number, length, and arrangement of the various tube sections to give the required output capacity, pressure, and temperature of the steam. The saturated steam is generated in the waterwall, convection, and radiant sections, and in these tubes the water is circulated by convection, always from the bottom of the boiler up to the steam drum, which is placed on or near the top of the boiler.

Starting with the hottest tubes, those in the radiant section, closest to the flame, are the most critical for quality of material and manufacturing, since they are usually the first to fail, owing to the very high temperatures on the outer surfaces.

Next, the hot gases of combustion are guided past the convection sections, where heat is absorbed by the cooler tubewall from the hot gases. There are many more convection tubes than radiant tubes, as the heat transfer is less in BTU per square foot of surface.

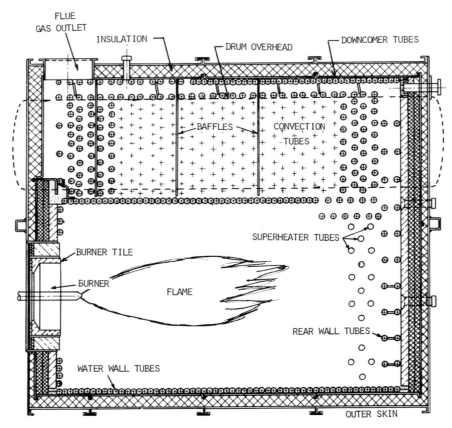

Fig. 3-15. Floor plan, typical packaged watertube boiler. (Courtesy of Cleaver-Brooks Co.)

The waterwalls are classed as either radiant or convection surface, depending on how well they are screened by the radiant tubes.

All of the tubes described so far carry water from headers or drums in the bottom of the boiler, up to the top drum, where water and steam are discharged into the drum. The water must be fed down from the top drum to the lower drum or headers, and this is accomplished by the downcomers, which are usually the tubes furthest from the flame and in the coolest section of the boiler. They are necessary to complete the convective circulation of the boiler water. If anything interrupts this circulation, tube failure usually results at some place in the boiler.

The steam being discharged from the tubes into the top drum is saturated, and carries large quantities of moisture with it. This steam is passed through internal baffles, or scrubbers in the top of the drum, then out into the steam main or nozzle. If superheated steam is desired, this steam is then passed back into the boiler through the superheating section, as shown in the rear of the furnace in Fig. 3-15. These superheater tubes are very critical as to circulation and soot accumulation. It is essential that while the boiler is being fired, steam be passed through the tubes, or they will burn out very rapidly.

The arrangement shown in Fig. 3-15 is typical, and not the only one on the market. The principles are the same in any case, regardless of make or arrangement. Not shown are the staggering of the tubes or the baffling, which help direct the flow of gases across the tube surfaces in the most efficient manner.

F. Boiler Rating and Sizing

Possibly you have become confused in past years over the various claims and counterclaims concerning the proper method of rating a boiler. Perhaps by covering some of the history of boiler rating methods we can dispel a small fraction of that confusion.

Years ago, it was ascertained that the boilers of that day, when coupled with a steam engine of the right size, would create a system that would have a definite, or fair approximation, equivalent of a certain number of horses in work performed. This was at a time when most of the work was easily measured in output, and could be compared with similar systems where horses were doing the work. It was then the practice to rate the boilers by the number of horses they replaced.

As the years went by, it was determined that on the average the boilers were employing about 10 ft² of heat absorption surface for each boiler horsepower rating, and that became the common method for rating boilers. One boiler horsepower was considered the equivalent of one horsepower of work for a long time, and is still used in some cases. Of course, it wasn't long before efficiencies of the steam generating equipment rose higher and higher, and much more than one horsepower was being obtained from each 10 ft² of boiler heating surface. The rating system still persists in many quarters, however.

Then came the high-steaming-rate industrial boilers, most of which came into prominence after World War II, and the old rating method began to be pushed aside. It was found that by careful design, and sacrificing some boiler life for high steaming rates, it was possible to get the same steam output with much smaller heating surface areas. Some boilers have been claimed to put out one boiler horsepower equivalent for each $2\frac{1}{2}$ ft² of heating surface.

As is usual with some trends, the pendulum of change swung first to an extreme, and then started to settle back at a more conservative midpoint. Today, when boiler horsepower is used as a rating, it often is based on 5 ft² of heating surface per boiler horsepower.

Along with the advent of the high-steam-rate boilers, there came into use another system of rating boilers, still using the term "boiler horsepower." Today it is more accepted practice to consider a boiler horsepower as being the evaporation of 34.5 lb/hr of steam, from and at 212°F. Now a look at the steam tables will show you that this simply means that one boiler horsepower equals the heat output of

33,475 BTU/hr. This gives you a strong clue as to how to convert the rating for any boiler from its normal operating conditions. Thus,

Boiler horsepower (bhp)

$$= \frac{\text{Hourly steam rate} \times \text{heat increase per pound}}{970 \times 34.5}$$

$$= \frac{(\text{Lb/hr}) \times (\text{heat in steam} - \text{heat in feedwater})}{33,475}$$

However, we give you here a table of values, Table 3-1, all worked out for you, so you can easily find the boiler horsepower rating of your boiler, providing you know the pounds of steam per hour the boiler will produce, and the temperature of the feedwater to the boiler.

Of course, if you have the BTU/hr output rating of the boiler, it is then simply a matter of dividing the BTU/hr output rating by 33,475 to give you the boiler horsepower rating.

Above about 600 hp, boilers are usually rated in BTU/hr output, or in lb/hr of steam produced. In the latter case, the steam conditions must be specified. In both cases, the fuel used must also be given, to properly tie down the true capability of the unit. As a result, any given boiler in any size or design may have several output ratings available. Close attention to the published data produced by the manufacturer must be given, in order to ensure that the unit chosen will perform as required when installed and operated under the conditions existing in your particular plant. This is very important, as to overlook this factor can, and often does, contribute to much ill-will between plant personnel and suppliers.

So much for rating methods. Let us now concern ourselves with the proper use of the ratings, and determine how to size or select the right boiler for the steam demand we wish to make upon it. We shall use the last described rating method, where one boiler horsepower equals 33,475 BTU/hr output.

Referring to the following calculations, we have a brief table of the steps required to determine the correct size of the boiler. We shall assume that it is required to heat up four tanks of solution requiring 432 lb/hr steam each, and four tanks requiring 216 lb/hr steam each. This gives us the total steam demand in 1 hr that the boiler must produce, at the process location. This is present peak demand, which must be reliable and steady at the point of use.

Starting with the basic demand, we shall now

Table 3-1. Boiler Horsepower Equivalents in Pounds of Dry Saturated Steam per Boiler Horsepower

Feedwater Temperature (°F)	Steam Gauge Pressure (psig)								
	0	15	50	100	125	150	200	250	300
60	29.8	29.5	29.1	28.8	28.8	28.7	28.6	28.5	28.4
70	30.1	29.8	29.4	29.1	29.0	28.9	28.8	28.7	28.7
80	30.4	30.0	29.6	29.3	29.2	29.2	29.1	29.0	28.9
90	30.6	30.3	29.9	29.6	29.5	29.4	29.3	29.2	29.1
100	30.9	30.6	30.2	29.8	29.8	29.7	29.6	29.5	29.4
110	31.2	30.8	30.4	30.0	30.0	30.0	29.9	29.7	29.7
120	31.5	31.2	30.7	30.4	30.3	30.2	30.1	30.0	29.9
130	31.8	31.4	31.0	30.7	30.6	30.5	30.4	30.3	30.2
140	32.1	31.7	31.3	31.0	30.9	30.8	30.7	30.6	30.4
150	32.4	32.0	31.6	31.2	31.2	31.1	31.0	30.8	30.7
160	32.7	32.4	31.9	31.5	31.4	31.4	31.3	31.1	31.1
170	33.0	32.6	32.2	31.8	31.7	31.7	31.6	31.4	31.3
180	33.4	33.0	32.5	32.2	32.1	32.0	31.9	31.7	31.7
190	33.8	33.3	32.8	32.5	32.4	32.3	32.2	32.1	32.0
200	34.1	33.6	33.1	32.8	32.7	32.6	32.5	32.4	32.3
212	34.5	34.1	33.5	33.2	33.1	33.0	32.9	32.8	32.6
220	—	34.3	33.9	33.5	33.4	33.4	33.3	33.3	33.2
230	—	34.7	34.2	33.8	33.8	33.7	33.6	33.6	33.5

develop the proper method of sizing the boiler for the process, using 125 psig steam.

Four tanks at 432 lb/hr = 1728 lb/hr

Four tanks at 216 lb/hr = 864 lb/hr

Maximum process load = 2592 lb/hr

Warm-up allowance, 10% = 260 lb/hr

Aging allowance, 10% = 260 lb/hr

Cycling allowance, 20% = 520 lb/hr

Future load =

Total steam load = 3632 lb/hr

Assume,

Feedwater temperature = 180°F

Boiler pressure = 125 psig

$$\text{Boiler horsepower} = \frac{\text{Total lb/hr}}{32.1 \text{ lb/hp/hr}}$$

$$= \frac{3632}{32.1} = 113 \text{ bhp}$$

The warm-up allowance is to handle the initial heating required for the cold piping system. This must be considered when starting from cold. The allowance of 10% is a minimum, and you must use your own good judgment here. If there is a long steam main feeding the equipment, then perhaps you will wish to increase the allowance to 15%, but seldom is it necessary to exceed that figure.

As the boiler accumulates soot and scale, it loses efficiency, unless it receives exceptional maintenance. The output drops steadily between cleanings, and an allowance should be included for this. The 10% included here actually allows for much more than 10% drop in efficiency. It means simply that of the total output of the boiler, which we shall find is 125 boiler horsepower, the boiler can drop by about 17% in efficiency without impairing the plant operation. The extra margin of safety is due to the difference between the 113 bhp actually required, and the 125 bhp finally selected.

The next item on the list is the cycling al-

lowance. This is to permit the boiler to cycle, or reduce firing rate at times even when the top demand is being met. It means that even at maximum load, there is still a reserve in the boiler.

We have not included any allowance for a future load in our problem, but we wish to discuss the possibility here. This is a subject that plant operators and maintenance people quite often are at a loss to calculate. Too often, even the top personnel in a company cannot shed much light on the expected future requirements. If they could predict the future of a company that accurately, they would be in demand by many other companies.

However, it is not a good idea to ignore the subject entirely. Something should probably be included for future possibilities, if there is the remotest chance of an increased future steam demand occurring. Perhaps there are certain physical limitations, such as the size or the layout of the plant, that will permit only a certain increase, or limit the possibilities in other ways. Truly, the plant operator is called upon to use his best judgment at a time like this.

One word of caution. It is not economically practical to consider future load increases occurring more than two or three years in the future. The longer ahead you attempt to guess the load demands, the less possibility that you will be accurate. Also, today's plants seldom allow for a capital recovery period longer than two years. This means that to size your equipment for longer periods increases the immediate capital expenditure to an unnecessary amount. It could conceivably result in making the project uneconomical.

Returning to our example, we now are in a position to total up the steam demand upon the boiler. We find that for our example we need a boiler output rate of 3,632 lb/hr steam. Table 3-1 tells the remainder of the sizing story. We are assuming here that the boiler is to operate at 125 psig, and that the boiler feedwater enters the boiler at a temperature of 180°F. The boiler horsepower equivalent in the equation shown is 32.1 lb/hp/hr. This is taken from Table 3-1.

The actual computed horsepower required is

113. However, most boiler manufacturers size their boilers in steps, and one of the common steps is from 100 to 125 horsepower, which we have assumed in this example. If you find one that is closer to the actual computed requirement, then you should choose it, but never go below your computed size without good reason.

When selecting a boiler from a manufacturer's catalog, remember to check the conditions under which his boilers are rated. They will probably be different than the conditions under which the boiler will have to operate once it is installed in your plant. It is best to give your operating conditions to the boiler representative and let him make the necessary corrections to bring his selection in line with your plant conditions.

The preceding method of sizing the boiler will result in a conservatively sized unit. Perhaps to some plant operators it will be too conservative. At least, you will know where the pitfalls are should you decide to compromise from the method given here. That will be much better than groping blindly, which is done too frequently, we are sorry to relate.

Another factor to be kept in mind when sizing and selecting a boiler, or a battery of them, is the best operating point on the efficiency curve. Figure 3-16 shows the efficiencies to expect from a typical small boiler of the industrial type. The one represented by this curve would be large enough to have some auxiliary heat reclaiming apparatus, such as an economizer in the stack, to obtain an efficiency of 80% or better. Some boilers will attain 80% without heat reclaiming, but they are probably the exception and not the rule.

The point to note here is that the boiler rating is usually given at a point on the efficiency curve below the peak, and beyond the highest point. You will note that in this instance the curve peaks at about 80% of rating. You will find this is generally true of most industrial boilers. Therefore, it is good practice to attempt to choose a boiler that will perform most of the time at about 75–80% of its rating. This helps to ensure long life for the boiler and to obtain maximum fuel economy.

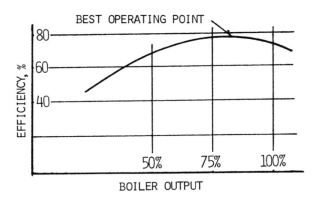

Fig. 3-16. Typical efficiency curve, industrial boilers.

In making the final selection of the boiler supplier, it is well at this point to analyze the actual present operating load required and see where it will fall on the efficiency curve for each boiler offered. In doing this we select only the actual process load and the aging allowance, since they are the ones that will make up the long-term duty for the new boiler, excluding the future load requirement at this time. The procedure is as follows;

$$\text{Actual process load} = 2592 \text{ lb/hr}$$

$$\text{Aging allowance} = \underline{260}$$

$$\text{Normal steady load} = 2852 \text{ lb/hr}$$

$$\text{Normal boiler horsepower} = \frac{2852}{32.1} = 88.8 \text{ bhp}$$

To find what percentage of the proposed new boiler rating this 88.8 bhp represents, use the following:

$$\text{Percentage of rating} = \frac{88.8 \times 100}{\text{New boiler rating}}$$

$$= \frac{88.8 \times 100}{125}$$

$$= 71\% \text{ of rating}$$

This procedure is performed for each boiler offered, and the final selection should be based on where the 71% rating falls on the efficiency curve for the new boiler. It should be fairly close to the peak of the boiler efficiency curve submitted with the bid. This selection should be done with care, as a difference of 1 or 2% in normal operating efficiency will make a substantial difference in the annual fuel bill, as we shall illustrate in later chapters.

When installing more than one boiler on a common steam header, it is best to have all of them of the same size whenever possible, since the problems of load balancing, maintenance, and spare parts will be much simpler.

If it is not possible to have all of the boilers the same size, then the problem of balancing the flows so that each boiler is operating at its peak efficiency point must be solved. The best way to do this, if the size of the plant warrants it, is to install a master boiler controller within the combustion control system, which is designed to balance or assign loads to each boiler according to its capability. This is a type of control that is usually found in the larger plants, and we will not go into it any further here. Chapter 27 covers this subject in more detail.

The other, and more simple, method is to place flow restrictors in the steam headers from the smaller boilers to limit the output from them to an amount approximating their point of maximum efficiency. We will go into this later in this book, however.

G. Boiler Efficiencies and Fuel Selection

We have already mentioned that large watertube boilers, of the type found in central station power plants, may run at efficiencies as high as 91%. This is when the personnel operating the system are highly trained in the best way to operate the plant for peak efficiencies, and with added heat-reclaiming equipment.

The boilers found in the average industrial plant, using steam at pressures up to around 250 psi, will not have the equipment needed for such careful tuning. Most boilers of the smaller package variety are designed for an efficiency in the range of 75–80% at best, and that is to be expected only when the boilers are first installed and are still in good condition.

After a few months of good, hard usage, with indifferent maintenance, you can expect the efficiency to drop, sometimes to as low as 65%, depending on many factors.

The fuel has a great deal of influence upon the type of maintenance required to maintain good efficiencies in a boiler. Gas is probably the all-time preferred fuel in this respect, being the cleanest fuel available for the boiler. Therefore, we could naturally expect that the boiler will maintain its peak efficiency relatively longer than with other fuels. This is assuming that all other factors are equal. Gas-fired boilers may be abused, also, with a resultant loss in efficiency.

Fuel oil runs a very close second in favor for today's industrial boilers. It has the advantage of being relatively easy to store in large quantities, which is difficult to do with gas. We find oil and gas competing very actively in most areas, the deciding factor being in many cases the cost of transportation to the plant, and the availability of gas on an uninterruptible basis.

If we were to rate the two fuels on a relative basis, we would do so in this manner: Gas is the cleanest and the quietest fuel, requires the least air to burn, and is probably the easiest to handle. Its combustion efficiency is slightly lower than oil.

From a safety standpoint, much depends on the condition of the controls and the burners, the grade of fuel oil, and the training and care of the operators.

We will be probing deeper into the subjects of fuels, efficiency, and the best methods in their utilization in later chapters.

H. Hot-Water Boilers

This text is written for the most part for the industrial steam boiler operator. However, much of what we discuss is applicable to hot-water boilers, also. In this section we shall cover the differences and the points of similarity, in an attempt to help those operators who have a hot-water boiler needing attention.

Boilers designed for hot-water service are

often adapted from a steam boiler design. The only real differences are usually in the connections, the controls for the initiation of the burner management system, the safety relieving equipment, and the water treatment. The similarities are in the basic circuiting of the boiler, the design of the hot gas passages, and the combustion equipment.

Section A in the Appendix contains most of the material published by the Cleaver-Brooks Co. in their hot-water-boiler manual.

First let us discuss a few pointers on the design of hot-water systems in general that have an affect on the operation and maintenance.

The purpose of the expansion tank is to maintain a flooded boiler and a positive suction head on the circulating pump, to provide a convenient location for feeding make-up water into the system, and to provide a point where air in the system may accumulate and be vented. If these points are kept in mind, it is often possible to analyze troubles a little easier.

If the expansion tank is on the discharge side of the pump, the system may become starved for water, and this may not be discovered until the boiler is damaged, or the pump cavitates and damages the impeller. Remember, in a closed-loop forced-circulation system, the lowest pressure in the system is in the suction eye of the pump impeller, and the highest pressure is at the discharge of the pump.

Whether or not the expansion tank is under pressure or open to the atmosphere depends on its position relative to the remainder of the system. Only if the return lines are under high pressure, such as when the boiler is in the lower floors of a high-rise building, will the expansion tank be under pressure. The general rule is that the expansion tank to be open to the atmosphere must be the highest point in the system. Otherwise, it must be under pressure. If it is under pressure, there must be a method of supplying air under pressure to the top of the tank, of sufficient pressure to balance the water in the portion of the system above the tank. This calls for a continual supply of air under pressure to be fed into the top of the tank. If this air supply fails, the result will be pressure surges caused by slamming check valves and other sources of water hammer. The air cushion smooths out much of the pressure surges that would otherwise be present for any reason.

Many hot-water systems are divided into zones, each zone being designed for its own temperature or required use time. For this reason each zone should be provided with its own circulating pump under its own control. Attempting to operate two or more zones on one circulating pump can lead to difficulties in control of the desired temperatures at the point of use, and may also lead to excess fuel consumption. The control valves used to control the circulation of water through the various heaters are *usually not* of the tight shut-off design, and may permit water to circulate through circuits that are supposed to be turned off.

If the boiler is installed in one of the lower floors of a high-rise building, and there are water hammer problems, the result may be frequent springing of leaks around the tube ends owing to the pressure surges caused by the water hammer. Slamming check valves often cause this, and the problem may often be cured by installing nonslam check valves, or two check valves in series.

Operationally, hot-water boilers are usually easier to operate than steam boilers, except that the combustion system and the combustion air supply problems are identical to those with steam boilers. With hot-water boilers there are fewer readings to be logged, as the temperature controls are simpler than the pressure control systems on steam boilers.

Hot-water systems must be kept clear of air pockets in the water side of the system. For this reason, all lines should slope upward in the direction of flow, and all high points where this air may collect must be vented with float-type vents. Air in the system can cause cold spots in the heating equipment, impaired circulation throughout the system, and erratic behavior of the control system. By all means, air must be kept out of the water side of the boiler.

Maintenance of hot-water boilers is usually less of a problem overall than on steam boilers, especially on the waterside. The properly designed hot-water system, with good maintenance and operation, will experience very little water treatment problems, as long as the system is maintained in tight and leak-free condition. As the system is inherently closed, once the corrosion-causing elements in the water have deposited a thin initial layer of scale on the heat exchange surfaces, including the boiler, the water no longer should be capable of further scaling, and the problem stops there. However, should too many leaks develop, requiring excessive make-up of raw water, then the corrosion problems increase in direct proportion to the amount of make-up required.

The water for make-up generally needs to be reasonably soft, and this may be further treated by means of slugs of the proper chemicals injected into the system by means of a shot-type feeder. One of these is shown in Fig. 10-1.

It goes without saying that in general, the maintenance problems will increase also with the increase in operating temperature, above approximately 170°F, and with temperatures below that point also. Below 170°F, corrosion on the hot gas surfaces increases due to condensation of moisture in the flue gas.

All portions of this text covering shut-down, wet lay-up, preparation for inspection, and placing back into service as described for steam boilers also apply to hot-water boilers. There may be some slight differences from those described in this material; but for the most part, the hot-water boiler operator will find very little difference in those aspects. Chapter 16 contains some helpful advise from the Cleaver-Brooks company on hot-water-boiler maintenance, and we refer you to Section A in the Appendix for additional material regarding hot-water-boiler operation and maintenance.

From a safety angle, hot-water boilers are equally dangerous from the possibility of explosions in the furnace, since the combustion system is usually the same for hot-water boilers as for steam boilers. The heat-release problems are the same in both cases, and, consequently, the control and safety equipment for the firing systems are usually the same.

Danger from the hot water increases with the increase in stored energy in the water. Therefore, those high-temperature hot-water systems where the high temperature is obtained by means of high pressures maintained with steam or high-pressure gas will approach steam boilers in possibility of damage from explosions.

Relief valves for hot-water systems are sized by the BTU rating method. This means that the relief valve is sized to relieve water at a rate sufficient to remove as much heat from the system as the combustion system is capable of supplying to the boiler. The relief valves supplied with the boiler should be rated by this method, as stipulated by the ASME Code, and they should not be tampered with, or changed, unless replaced by valves with the approved capacity rating. All other rules governing code rated and supplied relief valves must be applied to hot-water boilers.

C H A P T E R 4
AIR FOR COMBUSTION

A. How Much Air?

We mentioned earlier that a boiler requires air, fuel, and water in order to make steam, radiant heat, and hot flue gas. We have discussed the properties of steam very briefly, and we shall now discuss the air requirements for the boiler. Later, we shall cover the subject of the various fuels.

Figure 4-1 illustrates the point very well, as it shows the four basic requirements for combustion to occur in the furnace. All four must be satisified, and if there is any alteration in the balance of the four elements, then there is danger of trouble developing in the furnace, which could result in an explosion.

FUEL PARTICLES, FINELY DIVIDED

+ SUFFICIENT AIR

+ TURBULENT MIXING

+ IGNITION SOURCE
_____

= FLAME

Fig. 4-1. Requirements for combustion.

The same principles exist in the boiler combustion chamber as exist in the piston of the automobile engine. The fuel must be finely divided and mixed thoroughly with sufficient air, and the temperature must be elevated sufficiently to support continuous combustion. This sounds easy, especially the part about mixing

thoroughly with the fuel, as air being the elusive gas that it is, it should be an easy matter to mix the fuel and the air thoroughly enough for our purposes. However, this is not the case, for the simple reason that the whole process must be done in a very short time. There is but a fraction of a second between the mixing of the air and the fuel and the point at which combustion commences at the discharge from the burner. For this reason, it is necessary to provide more air than the amount theoretically needed to support perfect combustion.

In the case of natural gas, the amount of excess air is about 10–20%. In the case of fuel oil of the #5 or #6 grades, it is normal to provide about 15% excess air. Table 4-1 shows the approximate amount of air required in weight and cubic feet, for each million BTU burned in the boiler.

The values in Table 4-1 are based on average conditions, and may vary slightly with local conditions and composition of the fuels being used. For more accurate figures, it is best to consult the fuel supplier and the burner manufacturer.

Where does this air come from? In most cases, it is taken right from the room in which the boiler is located. Another look at Table 4-1 will show you that this could be quite an enormous draw from the boiler room air. Therefore, the first problem is to see that the boiler room has sufficient air intake to supply the combustion air, unless the boiler is supplied with separate

Table 4-1. Air for Combustion

	Pounds of Air per Million BTU	Cubic Feet of Air per Million BTU
Oil	780	11,000
Natural gas	750	10,500

ductwork to the outside air supply. This is the best method, and is recommended where it is convenient. However, if the boiler is rather small, and the air requirement is not too large, then all that is necessary is to be sure there is at least 1 ft^2 of outside air opening into the boiler room for each million BTU per hour burned in the boiler.

Actually, there are several methods of arriving at the proper amount of air to bring into the boiler room, as well as the amount required for the boiler furnace to support combustion. The method used will depend mainly on the size of the boiler and its location: inside or outside. Obviously, if the boiler is outside, then no allowance must be made for ventilating the "boiler room." If the boiler is inside, then we must bring into the boiler room enough air to ventilate the room and keep it fairly comfortable, as well as provide the air for the boiler furnace to support combustion.

There is one rule of thumb that works very well for gas- and oil-fired small boilers, and it is the first approach. It may be the only approach you need to take.

For every 100 BTU released in the furnace, bring 1 ft^3 of air into the boiler room. This will be sufficient for both ventilation and for combustion.

One of the boiler manufacturers recommends another method, which will work very well, also, as follows:

For combustion air, supply 8 cfm for each boiler horsepower rating.

For ventilation, supply 2 cfm for each boiler horsepower rating.

Whichever method you use, there will be sufficient air for the burners and for personal comfort.

Bringing the air into the boiler room, releasing it into the room, then allowing it to be

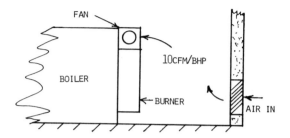

Fig. 4-2. Boiler air supply, warm climates.

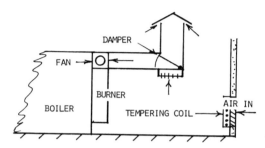

Fig. 4-3. Boiler air supply, cold climates.

drawn into the boiler as in Fig. 4-2, works very well in moderate climates. However, in cold climates it is not the best method, since the boiler room can get very cold and drafty. Therefore, the method illustrated in Fig. 4-3 is much better.

In Fig. 4-3 the air is pulled into the boiler room through louvers directly from the outside by the draft fan. During cold weather this air is heated as needed by a heating coil. The operator has the option of adjusting the damper to allow the boiler to take air from the boiler room floor during mild weather, thus providing ventilation at the same time. The air intake at the top of the vertical stack will be explained subsequently.

B. Boiler "Draft"

Air for combustion does not automatically find its way into the burner and the boiler; it must be either forced in or pulled in. The method by which this is accomplished is known as the "draft system," from the term "draft." There is an official definition of the term, but for our purposes it will mean the method or the system for getting the combustion air into the combustion chamber and out the stack.

The draft of a boiler is also the pressure of the combustion air and gases as they pass through the boiler and stack system. This pressure is measured in very small amounts, since it does not take very large pressure differences throughout the boiler and stack system to produce the flow required. This pressure is measured in inches of water column, abbreviated as "wc," and refers to the static pressure in the system at any point we wish to take a measurement. This pressure may be above or below the pressure of the atmosphere, and it may pass from the negative to the positive side of the atmospheric pressure existing outside of the boiler and stack system.

To properly explain the principle of the draft gauge, we refer you to Fig. 4-4, where we show a column of water in a vertical tube being balanced by the pressure in the pipe. This is another method of measuring pressures, and is used for those cases where the usual style of Bourdon tube gauge is not sensitive enough to give us an accurate reading of the pressure.

As shown here, for every pound existing in the pipe or vessel on the left, that pressure will support a column of water 27.7 in., or 2.31 ft, above the level of the water in the left leg of the U tube, providing that the top of the right-hand column is open to the atmosphere. This is the principle also of the manometer, where the pressure is measured by the height of one

level of the fluid column above the level in the opposite leg.

As small boiler drafts are usually measured in very small amounts, at times less than 1 in. wc, we must use some method of increasing the sensitivity of the gauge or device for measuring it. The ordinary inclined-tube draft gauge, as shown in simplified form in Fig. 4-5, serves the purpose very well.

Instead of the balanced water column extending vertically upward, we incline it as shown in Fig. 4-5. Now, if a vertical scale in inches were laid off on a rubber scale placed along the line *B* vertically, and if we were to pivot that rubber scale about its base where *A* and *B* meet, and stretch the top of the scale out along the line *A*, we would be increasing the sensitivity and accuracy of that scale. Every inch along vertical line *B* then becomes several times greater along line *A*, and results in a scale that can be read to much smaller divisions of an inch.

We have assumed water as the medium in the example, but in practice other liquids are often used, and are usually colored to make them much easier to read. In this case, it is absolutely necessary that when replacing the liquid, only the specified liquid should be used, since the scale is calibrated for only that liquid.

Figure 4-5 shows the arrangement for a positive draft reading. If the expected draft is to be negative, or below atmospheric pressure, then the connection to the breeching or boiler would be at the top of the inclined scale.

In the discussions which follow, covering the various types of systems for producing the required combustion air flows, or drafts, there is a horizontal line under the schematic drawing of the system, representing atmospheric pres-

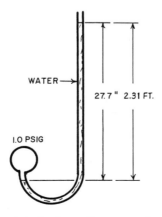

Fig. 4-4. Inches of water column.

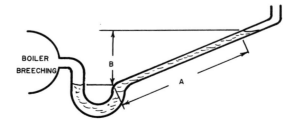

Fig. 4-5. Basic boiler draft gauge.

sure. On the left side of each diagram, there is a vertical line, representing the pressure in the system, with those pressures above the atmospheric line being positive (above atmospheric pressure) and those pressures below the line being negative (below atmospheric pressure). The sloped line lying either above or below the atmospheric pressure line represents a graph, or a trace, of the pressures in the boiler draft system directly above it, generally from left to right. Thus, it is possible to follow the changes in pressure inside the boiler draft system as the air and gases pass through the boiler from the air intake on the left to the stack discharge on the right.

A few words concerning the importance of maintaining the proper amount of draft while firing the boiler will be very appropriate at this point.

The boiler in your plant was designed for a specific purpose, firing rate, and set of firing conditions. This includes the size and arrangement of all of the various metal portions of the furnace, tubes, baffles, and casing. Each was designed to perform its function under reasonably predictable and uniform conditions. Consequently, to alter these conditions is inviting trouble.

One of the prime rules in designing and operating a boiler is that at no time should the flame impinge directly upon any of the metal surfaces not specifically designed for that condition.

If the draft is not properly adjusted and controlled, it is possible to cause the flame to be drawn up into the breeching or stack in some boiler installations. Or this may happen when the boiler is being pushed beyond its designed rating. The result may be to cause the tubes in the radiant portion of the furnace to be subjected to extremely high temperatures, beyond their capability to withstand the temperatures and pressures. Should there be an appreciable layer of scale on the water side of the tube, the flame side of the tube will rise to an abnormal temperature, which would cause the tube to fail.

On the other side of the coin, if the draft is insufficient, then the flame may be starved for air, and the result could very well be a sooting

of the heat-transfer surfaces, a reduction in boiler output, or an explosion. Should this condition exist for very long, any sudden increase in the draft or firing rate in the boiler controls could cause the layer of soot in the stack or furnace surfaces to ignite, overheating the surfaces, or releasing sparks to the surrounding area, or both.

The preceding reasons for maintaining the proper draft are based solely on safety considerations. We shall cover later one other very good reason for keeping a critical eye on the draft in your boiler, and that is the matter of economy.

How much draft do boilers require? That depends entirely on the size, the style, and the fuel. The average small industrial boiler of the package type, burning gas or light oil, requires in most cases less than 1 in. wc of draft. As might be expected, the longer the gas passages, the higher the draft required. Some Scotch-type package boilers come in a wide range of sizes and gas passage arrangements. This very popular style of boiler will require anywhere from 0.10 to 2.0 in. wc draft. Some firetube boilers are equipped with spinners inside the gas tubes to increase the turbulence and the heat transfer into the water, which will require slightly higher drafts.

C. Natural Draft

Figure 4-6 illustrates one of the oldest known methods for supplying air to the boiler, known as "natural draft." The principle upon which it works is the production of a negative pressure in the boiler owing to the column of hot gas contained in the stack, as illustrated in Fig. 4-6. The result is an inflow of cold air to the boiler through the air registers around the burner provided for that purpose. This inflow is maintained as long as the stack "draws," and as long as the remainder of the system is kept in the proper operating condition.

The basic principle of the natural draft stack is that air becomes lighter when heated, and tends to rise. See Fig. 4-6. In doing so, there

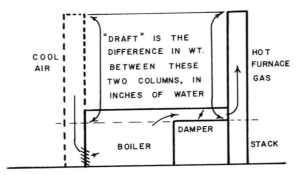

Fig. 4-6. Natural draft principle.

must be a replacement of the air with cooler air from below. In the natural draft system, we simply guide the inflowing cooler air so that it enters the boiler at the desired place to supply the burner. The pressure difference—draft—causing this inflow of air is the difference in weight between the stack of hot air and an imaginary equal column of cold outside air, as shown in Fig. 4-6.

Figure 4-7 represents the draft condition existing in most of the smaller boilers and other apparatus fired by gas, and requiring a very low draft to produce the combustion air for efficient operation. The gas pressures found in this type and size of boiler are usually a few inches wc only, and the draft required is about 0.01–0.50 in. wc. The gas passages are very direct, sometimes forming a direct vertical path from the burners up to the draft hood, or with only a few baffles in the way up.

The burners are usually of the inspirating type, utilizing the gas flow through a venturi to pull in the combustion air, in one of the more quietly burning arrangements.

The draft hood, sometimes called a draft diverter, is simply a method of limiting the draft to prevent the flame being pulled up into the tube nest and impinging directly on the tubes. The draft hood also prevents a sudden down draft from the stack blowing out the flame. This is always a danger with low-pressure burners.

The damper controls the amount of draft in conjunction with the air inlet at the base of the stack. As the draft pressure is below atmospheric pressure at all points of the system, the inflow of air helps to keep the draft under control, as well as serving as a relief from a sudden down draft.

The damper, which is not always present, should be automatically controlled from the combustion control system, and manually controlled only on the smaller boilers. Also, it must not shut off 100%, but should have about a 15% opening even when in closed position.

The danger from down drafts is very real, as a very violent one could force the flame outside of the boiler casing, through the air intakes, and possibly ignite any combustible surrounding the boiler, or giving anyone standing close a bad burn.

Referring again to Fig. 4-7, we present a typical method of explaining the draft system that we include in this discussion. The left side of the figure contains a diagrammatic view of the boiler and draft system, while the right side of the figure contains a graph. This graph illustrates the draft existing at the corresponding point in the boiler and flue system. In this case, all portions of the draft line which lie to the left of the atmospheric pressure line are above atmospheric pressure, while all portions of the draft line to the right of the atmospheric pressure line are below atmospheric pressure.

Following the passage of the air and flue gas through the boiler, we see that the action of

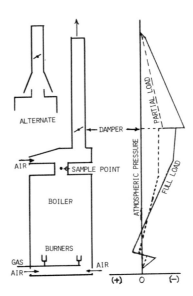

Fig. 4-7. Draft hoods.

the burners pulls combustion air into the burner throat through the openings in the bottom of the boiler. This air mixes with the fuel gas and is ignited, and the products of combustion pass up through the boiler with the pressure dropping further into the negative side, as shown by the "Full Load" line on the graph.

When the flue gases reach the air intake duct at the base of the stack, air is pulled into the draft system, which limits the amount of the draft at this point by diluting and cooling the flue gas. The damper just above the air intake further restricts the flue gas flow up the stack, and limits the draft still more. From the damper, the flue gas passes up the stack and out the top at atmospheric pressure.

The dotted line on the draft graph labeled "Partial Load" merely shows the effect of reduced firing and the simultaneous further restriction imposed by the damper when the combustion control system operates to limit or reduce the load.

This system of showing the draft throughout the boiler and stack arrangement will be used throughout this chapter, with the exception that the remaining ones will have the graph running horizontal below the boiler, as the remaining boilers in this chapter are arranged horizontally, and not vertically as is this one.

Remember, on a natural draft boiler, the furnace, uptake, and stack are always under negative pressure (below atmospheric pressure).

The damper in Figs. 4-6 and 4-8 may be either motor operated or of the barometric type, which will be explained next. One or the other will be required to control the air inflow, and provide the proper amounts of air throughout the operating range of the burner.

As explained in the previous section, too much draft will pull the flame up into the breeching, and may burn out portions of the boiler that were not designed to take direct impingement of the flame. Many boilers, especially the uptakes, have been ruined from improperly adjusted draft dampers.

And, we repeat here, insufficient draft will result in poor combustion, sooting of the tubes and uptakes, reduced output from the boiler, and possibly an explosion. Deposition of black soot over the surrounding real estate may also result, and most of you are probably well aware of what trouble that can cause.

Figure 4-9 illustrates the principle of the barometric damper in its application. The damper is hinged and has an adjustable weight on an arm at an angle to the damper. By changing the position of the weight on the arm, it is possible to alter the differential pressure required to open it. When the stack is drawing at its fullest amount, there is an excess draft available, as shown by the dotted line in Fig. 4-9, below the atmospheric pressure line. This reduction in pressure in the base of the stack, permits the barometric damper to open due to the differential pressure between the atmosphere and the stack. When the damper opens, air enters the stack, cooling the gases and raising the pressure inside the stack, thus altering the draft inside the boiler, as shown by the solid line in the diagram. In this manner, some degree of control is obtained over the amount of inlet air to the burners. As the firing rate is lowered in the burners, there may be some fluctuation in the pressure inside the boiler, and

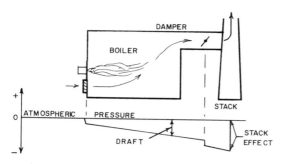

Fig. 4-8. Boiler natural draft.

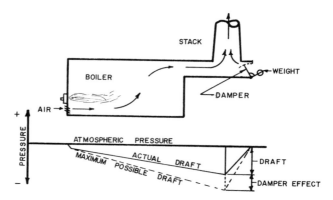

Fig. 4-9. The barometric damper.

some reduction in burner efficiency. The damper should be mounted as close as possible to the boiler for best control, as it maintains a constant negative pressure only at the point where it is mounted. Therefore, they work best on burners that fire at a constant rate, or on burners that have a very low turn-down ratio. When the burner shuts down, the barometric damper closes slowly as the stack cools and stops drawing. Being a simple butterfly valve with a counterweight, they are relatively cheap and easy to install and maintain, and are safe to use. However, they increase the amount of air drawn into the boiler room, over that required for proper combustion, as shown in Table 4-1, which may or may not be desirable. Very seldom will you find these dampers used in sizes over 36 in. in diameter, since they become expensive over that size.

On larger installations, it is more common to use the motorized damper, which is usually contained in a section of the boiler ducting, and consists of a series of parallel leaf dampers, all operated from one shaft. See Fig. 11-13. The motor operator is controlled from the fuel or burner control in some manner so as to vary the amount of inlet air with the demands of the burner. This style is calibrated upon installation, and should never need further adjustment, as long as it is properly maintained. Burner efficiency is maintained at the highest possible value with this method, for larger turn-down ratios.

Motorized dampers are less expensive in the larger sizes, and require no additional air to be admitted to the boiler room. However, motorized dampers require maintenance, are relatively complicated, and increase hazards from a furnace explosion. When an explosion occurs, the relief ordinarily offered to the atmosphere through the stack is partially blocked. Another place where you will find the motorized damper used is in the breechings from several boilers installed on a common duct to the stack, which will be explained later in this chapter.

Regardless of which damper is used, there is one basic rule to keep in mind: it should never close off the stack passage completely, but should allow at least a 15% opening when in the closed position. This is usually a requirement of the insuring agency, as the 15% opening allows venting of any gas or oil vapor that may leak into the furnace. Fuel valves do leak at times, and some very bad explosions have resulted from not paying attention to this factor.

You will find natural draft is used on older installations, and in some recent ones of smaller firing rates. Natural draft is particularly adapted to the low-pressure gas burners, which require very small drafts, on the order of 0.5 in. wc or under. If you will remember our discussion in the previous section, 1 psi is equal to 27.7 in. wc.

In our discussion on the various types of draft systems, when we refer to "damper," we are assuming the use of the motorized damper, unless stated otherwise.

D. Induced Draft

Figure 4-10 illustrates another common type of draft system, the induced draft arrangement. This consists of a fan located between the boiler and the stack, pulling the air around the burner registers through the boiler, and then discharging the hot gases up the stack. The graph in Fig. 4-10 indicates that the boiler is under a negative pressure, which is a characteristic of induced draft systems. You will note, also, that the amount of pressure boost which the fan must provide depends partly on how much assistance the stack is able to give. On the graph, the upward vertical portion of the

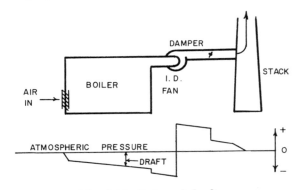

Fig. 4-10. Induced draft.

draft line indicates, to scale, the increases in pressure through the fan, where it starts at the lowest point and rises vertically up to the highest point, directly under the location of the fan in the ductwork.

Induced draft fans are used on many of today's package boilers, which have to be designed to take the negative furnace pressures without excessive in-leakage of air where it is not desired. The fan must handle large volumes of hot, corrosive flue gases, requiring special materials in its construction.

E. Forced Draft

Figure 4-11 illustrates another type of mechanical draft system, the forced draft arrangement. Here you will note that the fan is on the inlet to the burner, placing the furnace under a positive pressure throughout. Here again a damper is used to regulate the draft with the demands on the burner firing rate. This is common with the smaller boilers, in both the induced fan and the forced draft systems. On larger boilers, it is sometimes found more feasible to regulate the draft by varying the speed of the fan. However, the equipment to do this is rather expensive.

Here, as with the induced draft fan, the boiler must be constructed tight enough to prevent hot gases of combustion from leaking through the furnace seams into the boiler room. However, this is not too difficult to do in the mod-

ern boiler, and we find this type of draft used to an increasing extent on boilers, with ever higher draft pressures. Some large central station power boilers are now being built with forced draft fans designed for as much as 1 psi, or around 28 in. wc. This has been made possible since the completely enclosed waterwall style of watertube boiler has become popular in the larger sizes.

The forced draft fan handles cool, dense air, with the use of ordinary steel and aluminum construction. Consequently, forced draft fans require lower horsepower motors, and less maintenance than do the induced draft fans.

F. Balanced Draft

Finally we come to a combination of the forced and the induced draft system, known as "balanced draft," and shown in Fig. 4-12. This is one combination that you may never see, because it is not used too often in industrial boiler plants, and then only in the larger plants. For some time it was not popular, for several reasons. The total power costs are higher than either of the other systems, with higher maintenance than either of them. Also, the furnaces must be designed to withstand both internal and external forces, since the point of "balance" or atmospheric pressure inside the furnace changes with the firing rate. It was an attempt to split up the total draft required into two portions, and thus permit higher total resistances in the gas passages without placing

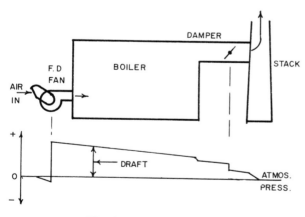

Fig. 4-11. Forced draft.

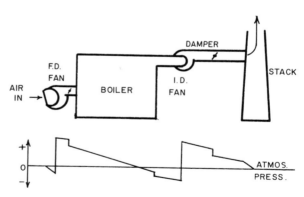

Fig. 4-12. Balanced draft system.

too high a stress in either direction upon the furnace walls. With the advent of the pressure-tight welded waterwalls in the watertube boilers, in many cases it no longer is necessary to employ the balanced draft system.

We will not go through the graph of this system, as it is easy to follow it from our previous discussions on the two mechanical draft systems from which it is derived.

There are two very prominent trends in recent years that have resulted in a resurgence of interest in the balanced draft concept. One of these trends has been the increased cost of fuel, forcing higher efficiencies in the boiler combustion process, which in turn requires heat recovery apparatus on the boiler flue gas outlet system. This, at least in the case of retrofitting existing installations, often calls for another fan on the outlet system, to compensate for the increased pressure drop across the heat-recovery equipment. We shall cover this subject more in later chapters.

The second trend operating in this same manner has been the requirement to reduce stack gas emissions. Scrubbers, filters, or similar equipment are being installed in many existing boiler plants on the outlet gas systems to meet local air pollution limitations. These items added to the boiler draft system will usually require increased pressures in the draft system.

The net effect of these two trends in added equipment is to produce a draft system graph very similar to that shown for the balanced draft system in Fig. 4-12. Should you ever be called on to add equipment to your boiler draft system, be sure and keep in mind that the effect may change the pressure conditions throughout your boiler draft system, and added strength may be needed at some locations, such as in the duct work, owing to the original design not having been based on the new pressure conditions. For instance, if a duct is designed to withstand an internal pressure of 1.0 in. wc, and a redesign places it under a negative static pressure of, for instance, 0.5 in. wc, the duct may collapse, or at least, vibrations could be set up from the tendency to collapse.

G. Boiler Breechings

Now that we have covered boiler draft, we must provide a proper method of getting the flue gases out of the boiler and into the stack. One common method for connecting the smaller packaged boilers to the stack is to simply install the stack directly above the boiler flue gas nozzle, and the stack may or may not be supported separately from the roof. The boiler supplier's literature should be consulted in this respect, as there is a definite limitation as to how much weight the boiler flue gas nozzle will take. Also, the matter of side sway from winds, earth shocks, etc., has to be taken into consideration. When in doubt, it is best to support the stack from the roof or use some other method to take the stress off the flue gas nozzle. In addition, guy wires will be required if the stack is of any appreciable height.

The recommended method of connecting the boiler to the stack is shown in Fig. 4-13. The stack in this case is supported separately to one side of the boiler, and it has a clean-out connection and drain. The drain is for the purpose of removing any condensate that may collect in the bottom of the stack and also to check for the presence of water at that point. If you find water in the base of the stack, you may expect trouble in the lining and life of the stack.

In Fig. 4-13, the boiler is connected to the stack by ducting as shown, with its own clean-out on the end of the horizontal breeching section. The ductwork will have to be properly supported to prevent sagging as well as to re-

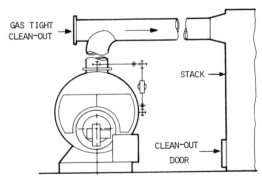

Fig. 4-13. Recommended stack arrangement, single boiler. (Courtesy of Cleaver-Brooks Co.)

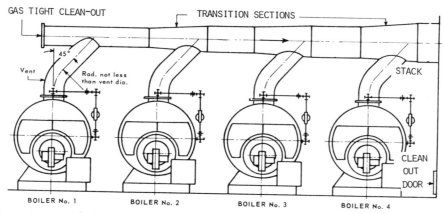

Fig. 4-14. Breeching for several boilers on one stack. (Courtesy of Cleaver-Brooks Co.)

move the stress from the boiler flue gas nozzle.

When connecting two or more boilers to one stack, the method shown in Fig. 4-14 is recommended. Note that each boiler outlet has its own damper to control the draft. Without this feature, each boiler would not be able to produce its proper share of the total demand, as the draft requirement for all boilers will very seldom be equal.

Not shown in this case are the duct flanges at each boiler flue gas nozzle for insertion of blanking disks when it is desired to work on a boiler when the remaining boilers on the system are in operation. It is necessary to isolate the cold boiler from the others in all instances where there is danger of either flue gas, steam, hot water, or blowdown from any of the hot boilers finding its way back into the cold boiler, if work is being performed on the cold boiler.

Figure 4-15 shows the accepted method of attaching a side outlet boiler to the stack. In this case the usual arrangement for multiple-boiler installations is to have one stack serve two boilers, with each boiler being a mirror image of the other one on the common stack, with the ducting entering the stack from the two boilers at different elevations. In any case, means for isolating each boiler from the remaining ones on the common stack is still required for servicing a cold boiler while the other one in the system is being fired.

We shall return to this matter of servicing cold boilers on common header systems when we discuss the shut down of boilers.

Figure 4-16 illustrates an accepted method of bringing a boiler stack through the roof of the boiler room. Not shown is the flashing around the sleeve through the roof, required to waterproof the sleeve. You should be aware that local codes often govern the method of taking a boiler stack through a roof or shelter structure.

The size of the flue gas ducting required to pass the hot gases will usually be recom-

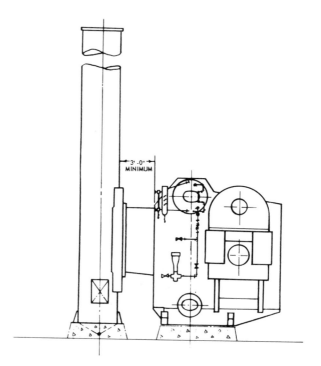

Fig. 4-15. Common boiler and stack arrangement. (Courtesy of Cleaver-Brooks Co.)

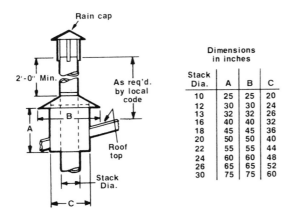

Fig. 4-16. Typical stack detail. (Courtesy of Cleaver-Brooks Co.)

Dimensions in inches			
Stack Dia.	A	B	C
10	25	25	20
12	30	30	24
13	32	32	26
16	40	40	32
18	45	45	36
20	50	50	40
22	55	55	44
24	60	60	48
26	65	65	52
30	75	75	60

mended by the boiler manufacturer. The literature provided with the boiler will usually carry tables giving the sizes and length limitations and they should be followed very carefully. Any attempt to reduce the size or increase the length limitations may lead to a reduction in output, at the least.

The hot flue gas ducting should also be insulated, not only to reduce heat loss, but also to protect boiler plant personnel from the hot surfaces of the ducting. Most specifications for hot surface insulation require that sufficient insulation be installed to reduce the surface temperature to about 140°F. Here, again, the boiler supplier may be able to help you in case you are contemplating a change in arrangement or improvement of the boiler plant equipment.

H. Stacks

We shall now devote our attention to that necessary and distinctive feature of all power plants using common or "fossil" fuels—the smoke stack. Such a distinguishing feature it has become, that it is very easy to orient yourself as soon as you step foot on the grounds of another plant, if you have had any experience in one of today's steam plants.

We have already discussed the effects of the stack on the boiler operation, now what about its construction?

First, we shall take up the matter of height required to obtain the desired draft, and other factors associated with the height. The major factors determining the height of the stack are:

1. Type of fuel and burner used.

2. Flue glas temperatures.

3. Prevailing wind directions.

4. Interferences from surrounding structures and land.

5. Type of draft system used and required draft.

6. Effects on neighboring population.

7. Air traffic hazards.

8. Cost of construction and maintenance.

It goes without discussion, of course, that the height of the stack should be just high enough to satisfy as many of the above factors as possible, and no higher.

The type of fuel, burners, and draft systems are all very much related, and are usually dictated by the boiler and burner supplier, as a minimum requirement. This provides the starting point in most cases. As stated in an earlier chapter, it is well to provide an extra margin of safety when designing any feature of the steam plant, and the draft system is no exception. It is well to provide an extra 15% of draft over the minimum specified, and reduce it to the required amount with one of the damper systems described. This is to take care of those "transient conditions" mentioned earlier—such as extreme variations of atmospheric pressure and temperature, and firing rates, or changes in fuel characteristics.

In the case of the flue gas temperature, we are forced here to decide on as good a compromise as possible. For high steam plant efficiency we would like to discharge the flue gases as cool as possible, since the heat going up the stack is a total loss. However, there is another factor we must take into consideration here, and that is the condensation of the water vapor in the flue gas as the gas cools down. If the gas is discharged at too low a temperature, as it cools on its way out the stack, it may cool to its "dew point," or point where condensation occurs, which can cause trouble. The flue gas contains a considerable amount of gases, which have a tendency to dissolve into any water of condensation formed, with the result that a highly corrosive solution is formed. And as the water vapor contained in the flue gas is a very high proportion of the total flow, the result on the stack walls may be disastrous. Many stacks are constructed of special steels for this reason, or lined with special concrete or brick and mortar, to prevent this attack.

Even if the fine droplets of corrosive moisture do find their way up the stack without attacking the lining, the effect on the surrounding buildings may be equally expensive. Which brings us to another listed factor—wind direction.

Actually, three of the above listed factors should be considered together: wind direction, interferences from surroundings, and effect on the population, for they are all part of the same general problem. To solve this part of the problem requires some detailed study and planning, since to fail may result in hard feelings upon the part of the neighbors.

The droplets of corrosive moisture are part of the general subject of "fallout" to be expected from stacks. Tables are available from stack designers and suppliers that show the dispersion effect of various heights and wind speeds. This is important, since one of the chief methods so far used to solve this problem of fallout is to carry the pollutants high enough so the scattering effect of the winds is sufficient to reduce complaints from the populace. The amount of fallout is not reduced, it is merely spread over a wider area.

The air traffic hazard is very real in most areas today, with the increased use of private, commercial, and military aircraft. In fact, there may be definite rules in your area concerning your stack, set down by our government agencies regulating air traffic. At the least they will require warning lights on the top of the higher stacks, and may even require distinctive coloring, such as neon red. Another method is to paint the stack in a bright color, and have

searchlights directed upon it at night. These remarks apply only to the larger industrial installations, of course.

Still on the subject of height, we find that many of today's extremely high stacks are built that way simply to reduce the possibility of complaints from fallout and not to produce the required draft for the boiler. To produce high drafts with high stack heights requires higher temperatures as the stack height goes up, with increasing resistances to flow from the stack walls. The point of diminishing returns is reached very soon. Stacks which are around 500–600 ft high are for the purpose of fallout dispersion, mostly, unless connected with one of the older steam plants.

On the subject of cost, it is easy to see from the previous discussion that the stack may represent a very high investment, especially in the larger sizes. It is useless to discuss the actual cost to expect in this discussion, since there are so many factors affecting the cost as to constitute a complete subject in itself.

Maintenance is something that you will no doubt be highly interested in considering. It is no small matter in the older installations.

When high stacks are designed, they are usually provided with staging provisions around the rim at the top, from which working platforms may be suspended for servicing the stack. Servicing consists mostly of examining the coating and the lining, repainting, and occasionally relining the stack. There are types of concrete and mortar available that are designed to resist the attack from the wet flue gases, which have helped to reduce the maintenance on modern stacks.

Let us now consider the other extremes in stacks—the short, steel, stub type of stack. This is the type found in many of the average smaller steam plants.

Package boiler suppliers often advertise that their boilers need only enough of a stack to discharge the gases to a safe height, as the draft is furnished usually by the draft fans supplied with the boiler. As a result, there have been some ridiculously short "stub" stacks installed, often to the chagrin of the plant operator or owner. It is best to use some discretion when reading boiler maker's literature on this subject.

The average small to medium boiler in an industrial plant will probably have a metal stack, extending through the roof, high enough to carry the gases into a clean sweep of wind. As the total cost of such a stack is nominal, it is best to invest in one of the better grades of steel for these stacks, and lower your maintenance costs.

There is one type of patent stack that we should mention here, and that is the venturi stack. There are several designs on the market, all designed with the idea of obtaining a higher throw of the flue gases with a shorter stack. This is done by increasing the velocity of discharge. However, they may still be affected by local wind conditions, so use them with care, and do not try to select one without professional help.

A few words on the trend in design of stacks, and then we shall go on to another subject.

In the smaller stacks for the average plant, we have already mentioned the venturi stack design. In addition to that, there is a trend toward more corrosive resistant steel, or special linings to prolong the life of the stack and reduce the maintenance costs.

In the larger plants, requiring the higher stacks, the trend is toward more attention to appearance and public acceptance. This has led to some fairly exotic proposals, from architecturally designed configurations, to artistically blended simulations. Tomorrow's plants may vie with each other in seeing which one can build the best disguised smoke stack. We may see trianons, triangles, pylons, space needles, sweeping vertical curves, and even giant flowers or trees!

CHAPTER 5
FUELS AND COMBUSTION PRINCIPLES

A. The Combustion Process

We have listed briefly the requirements for proper combustion in the boiler, and have also covered the subject of one of the main components, the air requirements. It is now time to look at the specific conditions to attain the most efficient flame possible in the boiler. First, we must look into the actual combustion process in sufficient detail to permit the average boiler operator to grasp a feeling for what is taking place in the boiler furnace.

Figure 5-1 is a diagram of the average oil burner flame produced in a typical boiler furnace, greatly simplified for our purposes.

Fuel, finely divided into atom-sized particles if it is oil, or in its original gaseous state if one of the gas fuels is being burned, is forced through the burner port under pressure. This pressure will depend on the characteristics of the fuel and the required firing rate.

The pressure behind the burner port forces the fuel into the furnace at a certain velocity, where it mixes with the air entering the furnace around the burner tip. This air is "secondary" air, if "primary" air has already been mixed with the fuel before it enters the burner tip. If no air was added to the fuel ahead of the tip, then all of the air for combustion must enter the furnace around the tip, and this is then the "primary" air for combustion.

Assuming that the flame has been established and is stable, there is sufficient radiant heat being released by the flame to cause the fuel/air mixture to increase in temperature from the burner tip outward to the flame, until a temperature is reached that will cause the mixture to ignite. This takes place at a location known as the flame front, and its distance from the burner tip will depend on two factors.

The flame propagation along the fuel/air path in the furnace progresses at a certain definite velocity, both outward from the burner tip and back toward the burner tip. The fuel/air mixture leaves the burner tip and travels outward at a certain velocity, as previously mentioned, and this velocity depends on the pressure of the mixture ahead of the burner tip. This combination of the two velocities establishes the flame front, since the velocity of the fuel/air mixture decreases as its distance from the burner tip increases owing to the increased area into which the mixture is expanding. Where the fuel/air mixture velocity matches the velocity of flame propagation toward the burner tip, the flame front is established.

As we shall see, the burner firing rate is controlled by the pressure of the fuel entering the burner tip. This variation in burner tip pressure causes variations in the velocity and the amount of fuel entering the burner, which in turn causes the flame front to seek its new position to match the change in fuel/air velocity. As the burner firing rate decreases, the flame front moves closer to the tip; as the firing rate increases, the flame front moves further away

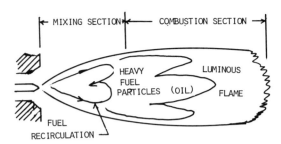

Fig. 5-1. Typical flame pattern.

from the burner tip. And herein lies two of the potential sources of furnace explosions.

It is quite possible under certain conditions actually to blow out the flame by forcing the flame front out too far. This could happen if the controls fail to limit the fuel pressure entering the tip. When it happens, the fuel/air mixture enters the furnace at a very high rate, and the hot surfaces in the furnace could ignite the mixture under unfavorable conditions, causing a violent furnace explosion.

If, under the preceding conditions, the explosion does not take place, the result could be to carry the flame out to the point where actual impingement takes place, which could be almost as disastrous, as we have mentioned.

In the other direction, if the fuel pressure entering the tip is reduced too much, the flame front will be established at the burner tip, and the result could be the flame becoming very inefficient owing to improper mixing of the fuel and air. This could cause the flame to "pop," which is caused by the flame going out, then reigniting with the popping noise. This generally takes place in cycles, with the noise becoming very noticeable. This is also a potential cause of a violent furnace explosion.

This discussion has laid the groundwork for the later discussions concerning the problems of controlling the combustion process.

To get back to the established flame burning in the furnace, note that in Fig. 5-1 there is a recirculation taking place within the invisible portion of the fuel/air mixture between the burner tip and the flame front. This turbulence is highly desirable and necessary to produce the complete mixing of the fuel and air. The higher the degree of turbulence, the more efficient the flame will be.

One other basic fact of combustion we wish to bring out here is that the lighter liquid fuels do not burn directly. It is the combustible vapors from these fuels that burn, and in order to burn them efficiently, it is necessary to break the fuel up into very minute particles, then heat these particles to a temperature which will drive out the combustible vapors, then heat the vapors to the ignition point.

If these vapors are not heated sufficiently, or if the finely divided solid fuel particles, such as from pulverized coal and heavier fuel oils, are not heated sufficiently to drive off all of the vapor, then the remaining particles may leave the furnace in the form of soot, which is simply unburned fuel particles. We shall see later what troubles this can cause, if you do not know already.

Another limitation in the combustion process is a condition known as the flammability limits, which means if the fuel/air mixture is too lean, the mixture will not ignite, and if the mixture is too rich in fuel, it will not ignite, either. The boiler operator should keep this fact in mind, and remember that it is usually impossible to look at a mixture that has not yet been ignited and tell whether it is on the lean side or on the rich side. Table 5-1 gives representative mixture limits for several common fuels and fuel constituents.

The discussion so far has been of a general nature, and it is not possible here to delve further into the various flame patterns and characteristics that are in use throughout the boiler industry. Each burner design has its own shape, flame pattern, color combinations, and patterns, so that it is the duty of the boiler operator to learn what to watch for through the observation ports provided in the burner front.

Table 5-1. Fuel/Air Flammability Limits

Fuel	Percentage of Fuel Gas by Volume	
	Lower	Upper
Butane	1.86	8.41
Hydrogen	4.0	74.2
Methane	5.0	15.0
Natural gas	4.3	15.0
Propane	2.4	9.5

Also, it is next to impossible to determine from direct observation of the flame if the correct or optimum fuel/air mixture exists. There are an infinite number of combinations of fuel compositions, firing rates, furnace and burner configurations, and draft conditions that make it next to impossible to list or classify them all. The boiler operator is forced to rely on instruments in today's energy-conscience economy to ensure that the particular installation is operating at the best available efficiency.

Probably the most that the operator can hope to gain from experience in operating the particular boiler/burner system is to detect drastic changes in operating conditions by looking through the observation ports.

However, page 356 in the Appendix gives some guidance in troubleshooting with oil burner flames, which may be of help.

B. Fuels

This discussion is based on boilers found in the average industrial boiler plant. Such boilers are usually fired with either oil or gas, and occasionally with coal, or one of the waste fuels now being burned in the larger boilers. Consequently, we shall limit our discussion here to those boilers firing only gas or oil. The gas could be natural gas, or one of the liquified petroleum gases, such as propane or butane, or one of the air mixes of them. The oil usually burned in this type of burner ranges from diesel oil, which is commercially classed as #2, up through the heaviest oil, #6. Very rarely will these small boilers be expected to handle the extremes in the lighter grades, such as kerosene, gasoline, or naphtha, and the heavier extremes such as bunker C or waste tars.

However, with the increased trends toward burning almost any waste that can be consumed in the boiler, it is possible that the readers may find themselves using one of these fringe fuels. Let us assure them that the principles are the same, and this discussion will not be in vain. The temperatures and the burner equipment will be different, possibly, but that is all.

The fuel has a great deal of influence on the effort required to maintain good efficiencies in a boiler. Gas is probably the all-time preferred fuel in this respect, being the cleanest fuel available for the boiler. Therefore, we could naturally expect that the boiler will maintain its peak efficiency relatively longer than is the case with other fuels. This is assuming that all other factors are equal. Gas-fired boilers may be abused, also, with a resultant loss in efficiency.

Fuel oil runs a very close second in favor for today's industrial boilers. It has the advantage of being relatively easy to store in large quantities, which is difficult to do with gas. We find oil and gas competing very actively in most areas, the deciding factor being in most cases the cost of transportation to the plant and the availability of gas on an uninterruptible basis.

If we were to rate the two fuels on a relative basis, we would do so in this manner: Gas is the cleanest, the quietest, and probably the easiest to handle. However, its combustion efficiency is 2–3% lower than for oil.

From a safety standpoint, much depends on the condition of the controls and the burners, the grade of fuel oil being considered, and the training and skill of the operators.

The heating value of fuels is one of the important things you should know, along with other basic characteristics. It is measured simply by burning a measured quantity of the fuel, and absorbing the heat produced in a water bath. The process may be either a continuous flow process or a batch process using a known quantity of fuel and water.

Two heating values are usually stated, but the one used most is the "higher heating value," since it is the simplest one; it is used in this country extensively.

Natural gas varies in composition from place to place, and sometimes from day to day. In those areas that obtain their natural gas from various sources or that pull on a stored supply of gas during part of the day, the composition may vary considerably. Most of the storage of natural gas is done in abandoned oil wells, deep beneath the surface of the earth. This permits the gas to be stored at extremely high pres-

Table 5-2. Fuel Oil Data

Grade	Common Name	HHV[a] (BTU/gal)	Sp. Gr.	Pumping Temperature, Minimum (°F)	Atomizing Temperature (°F)	Air (ft³/gal)
1	Kerosene	137,000	0.81	Ambient	Ambient	1370
2	Distillate	141,000	0.865	Ambient	Ambient	1410
4	Very light residual	146,000	0.90	15	60–110	1460
5	Light residual	148,000	0.94	40–60	120–170	1480
6	Residual	150,000	0.96	60–140	140–260	1500

[a]HHV = higher heating value.

sures during the off periods and withdrawn at reduced pressures during peak periods. This is usually known as "peak storage." Underground storage chambers often have passages to other oil and gas pockets nearby, which cause leaks, both in and out of the storage chamber. The result may be mixing of gases, and variations in heating values from day to day.

For all practical purposes, however, we may assume that most natural gases have a heating value of from 1000 to 1100 BTU/ft³ at atmospheric pressure and normal ambient temperatures. If in your calculations you use a value of 1050 BTU/ft³, it will be accurate enough for most applications you will be handling.

Fuel oils come in several standard grades, from #1 to #6. Table 5-2 summarizes the important data for all of them in which you will be most interested. The temperatures given are the minimum in all cases. It is safest to allow yourself some margin when following the values given for handling temperatures. Usually it is best to follow the advice of the company supplying the oil.

Table 5-3 shows similar data for the three major gas fuels used in small boilers. Here again, we recommend the boiler plant operator check with his source of supply for the actual heating value data. We have given the average values for propane and butane, but be aware that in the case of air mixes of these two fuels, various heating values are possible, and the final value will depend on the equipment and process used in the air mixing system.

Propane and butane are part of what is commonly known as "LPG," which stands for liquefied petroleum gases. They are both heavier than air, and thus require certain precautions in their use, as we shall explain in a later chapter.

Table 5-4 contains the normal flue gas composition produced by the combustion of natural gas, when being fired with different tabulated amounts of excess air. The values given are representative only. As we explain elsewhere, it is the last two compounds, water and nitrogen, which are the highest potential source

Table 5-3. Fuel Gas Data

Type	Sp. Gr.	HHV[a] (BTU/ft³)	Cubic Feet of Air per Cubic Foot of Gas
Cal. natural gas	0.64	1060	11
Butane, vapor	1.95	3370	34
Propane, vapor	1.52	2560	26

[a]HHV = higher heating value.

Table 5-4. Typical Flue Gas Composition for Natural Gas in Percent by Volume

Element	Composition (%)	Percentage Excess Air to Burner				
		0[a]	10	20	30	40
O_2	0–4	0	1.7	3.2	4.3	5.6
CO_2	10–12	10	9.1	8.3	7.5	6.8
H_2	0–1	0	0	0	0	0
H_2O	18–21	18	16.7	15.0	14.0	13.1
N_2	31	72	72.5	73.5	74.0	74.5

[a]Assumes perfect combustion.

of atmospheric pollution, especially if there is any trace of sulfur in the gas.

C. Pilots

During the discussion concerning the requirements for producing a flame, one item was the source of ignition, which we shall cover at this point.

The majority of today's industrial boilers are ignited by means of a pilot flame, and the simplest one is the continuous burning pilot with a flame detector. This pilot remains lit for the entire period that the boiler is in service, even when the burner is off, waiting for the start-up signal from the control system. This pilot must usually be lit off by some manual external source of flame, such as a flaming torch. It is usually found on the smaller boilers with on-and-off burner controls, since the continuous pilot is then needed at all times. It is usually gas fired.

The next type of pilot found on industrial boilers is the intermittent pilot, which is ignited at the start of the light-off process. It remains in service during the time that the boiler is in use, and shuts down when the main burner shuts down. It must have a method of ignition if the boiler is on an automatic cycle control, such as a program controller. This auxiliary method of ignition is usually from an electric spark, and the pilot must have a flame detector wired into the control circuit to permit the programmer to continue the cycle once the pilot has been proven. This type of pilot is also usually gas fired.

And finally there is the interrupted pilot system, which uses the pilot only during the ignition cycle for the main burner. Once the main burner is on, the pilot goes out, and if the main burner goes out, then the entire cycle, as controlled by the program panel, must be restarted. This requires an electric spark to ignite the pilot, with a flame detector to prove it has ignited, and the usual programmer control circuit, which will be described subsequently.

In the case of burning the heavier fuel oils, there is often an intermediate step between the ignition of the pilot and the heavy fuel oil. Often the initial burner light-off is performed on one of the lighter grades of fuel oil, under pressure atomization. Once the flame has been established, the heavier fuel oil is then turned into the burner, and the lighter fuel oil is phased out gradually. Thus, there may be four steps to lighting off a heavy fuel oil burner, consisting of spark ignition, gas pilot light-off, light oil light-off, and then the heavy oil light-off.

D. Pollution

We have already mentioned the matter of air pollution several times. Here we shall cover the subject in somewhat more detail, but not so much as to frighten the reader. We shall not go into the deeper realms of the chemistry of pollution, but shall merely attempt to acquaint you with the causes, and what you can, and are expected to do about it.

The major items causing concern are what are termed oxides of nitrogen and sulfur. These are caused by burning sulfur and oxygen in the presence of nitrogen in this manner; the nitrogen in the combustion air and the nitrogen in the fuel combine with oxygen during the combustion process to form various oxides of nitrogen, while the sulfur in the fuel combines with oxygen to form oxides of sulfur. This generally comes about from improper combustion, which in turn is usually caused from poor burner adjustment. The composition of the fuel plays a large part, also.

If the combustion process is not complete, then there is excess oxygen that can form oxides in the flame. These oxides pass out in one of several forms from the furnace into the stack, and out into the surrounding atmosphere, to the consternation of nearly everyone.

The oxides of nitrogen escaping to the air are one of the chief causes of what is termed "smog," as the sunlight acting on the oxides of nitrogen produce the harmful chemicals in the atmosphere.

The oxides of sulfur produce one of several

forms of acid when they come in contact with water vapor in the air. These acids are highly corrosive and harmful to our health. They also produce, in some areas, what is now termed "acid rain" where the concentration is sufficient and the water vapor in the air is plentiful.

Obviously, it is to our mutual advantage to see that such substances are kept under reasonable control. To do this requires a concerted effort on the part of boiler plant operators, fuel suppliers, and equipment suppliers. In short, the steps to reasonably clean stack emissions are:

1. Burn only clean fuel, low in sulfur.

2. Keep the excess air for combustion to a minimum, which will be covered subsequently.

3. Maintain the combustion controls and the burners in proper adjustment at all times.

4. Keep a constant watch on the stack output.

5. Keep the heat-transfer surfaces on the hot gas side clean of soot.

Later in this book we shall go into more detail concerning several aspects of boiler plant operation that will help you reduce the stack emissions from your boiler. By following these procedures, and by applying some diligence and common sense, you should have little fear of ever being cited for polluting the air. The operators of relatively small boilers will have a much easier job of it than will those who operate our large industrial and utility boilers.

We mentioned that the sulfur and part of the nitrogen comes from the fuel. The amount of sulfur in the fuel is the most important element to watch, since the nitrogen in the combustion air overshadows the amounts of it in the oil and gas being burned in the boilers. No doubt you are already aware of the great emphasis being placed on removing as much of the sulfur from fuel oil as possible in order to make it acceptable for boiler fuel. The amount of sulfur inherently in the fuel oil depends

mainly on its source. Normally, without any attempt to remove it, sulfur in fuel oil would be between 1% and 5%, which is much too high for health reasons. Of course, the more effort being expended to remove the sulfur, the higher the price of the fuel oil. Some large utilities install their own sulfur-removal apparatus, because their consumption is sufficient to make it economical.

In the Southern California area, fuel oil must not have over 0.5% sulfur by weight. Natural gas shall have a maximum permissible sulfur content of 80 parts per million (ppm). These limitations are subject to several qualifications, and we will not attempt to interpret the rules. For those of you who live outside that area, we strongly urge you to contact your local governing agency in this matter and obtain a copy of their rules and regulations. Section H in the Appendix contains some data pertinent to pollution control.

Now that we have explained what is generally expected of you to help us have clean air to breathe, there still remains the problem of how you are to monitor the results. Of course, you could wait for the local governing agency to do it for you, but that may be a little late, and you should be at least one step ahead of them, and know what your equipment is producing in the line of air pollution.

You have probably noted that the point of detection for pollution is usually the stack from the boiler. And the chances are very good that your present stack does not have any opening in an accessible location to permit you to insert a stack gas probe. It may not even be convenient for you to keep a watch on the discharge from your stack.

The stack discharge may not be in a convenient location for your observation. This may be solved by a mirror placed in the proper place, to make it possible for you to observe the discharge without you even having to leave the boiler room.

Of course, if your budget allows, there are any number of more sophisticated pieces of equipment on the market to make life easier for you in monitoring your stack conditions. But, they are really not necessary, and we believe the methods we shall describe later will prove sufficient in most cases.

Later, when we go into the matter of regulating the combustion controls to obtain the best operation from your burners, you will find that the key to the whole procedure is being able to sample the stack gas. Therefore, if you do not already have a sampling point in either the stack or the boiler uptake, then we suggest you begin thinking about installing a $1/2$ or $3/4$ in. pipe coupling in the side of the stack, with a pipe plug installed for easy removal and insertion of a probe. This plug should have Teflon tape wrapped around the threads to prevent freezing or rusting in place. It should be only hand tight in the coupling.

CHAPTER 6
FUEL OIL SYSTEMS

A. Oil Burners

There are two types of oil–flame combustion, and they are known as vaporization and atomization.

The light oils, Nos. 1 and 2, can be completely vaporized by heat, leaving little or no residue, and burned as a vapor. The flame produced is nonluminous, and blue or purple in color.

A common design, Fig. 6-1, consists of a pressure-atomizing oil nozzle that discharges with the combustion air into a venturi. After a brief firing interval the venturi will draw hot gases back through the gasifier line, the oil is vaporized by the returned hot gases, and the mixture is drawn into the venturi. The atomizing nozzle is then shut down, and combustion continues.

The heavier oils, Nos. 4, 5, and 6, cannot be satisfactorily burned in vaporizing burners because when heated they break down into light oils and residual material. Most of the latter is carbon, which does not burn as readily, may eventually obstruct the burner, and cause air pollution problems.

In simple atomizing burners, light oil is forced through a spray nozzle at high pressure and discharged as a fine mist. These burners usually have a tendency to clog because of the small port sizes used.

In a gun-type burner oil is pumped to the burner by a rotary pump and discharged through an atomizing nozzle into the furnace. A relief valve regulates the pump discharge pressure at about 100 psi by returning excess oil to the tank or pump suction.

In a heavy-duty mechanical-type burner, preheated oil at 150–300 psi is discharged from the burner nozzle in a fine mist, Fig. 6-2. Combustion air enters through a register and mixes with the oil spray after passing a diffusing plate.

The mechanical atomizers are limited to less than 2 to 1 turndown, since a drop in pressure drastically affects the atomization. We shall describe this in more detail later.

In the low-pressure air-atomizing burner, oil at approximately 35 psi is picked up by the air stream from a low-pressure blower system and carried, at 1–2 psi, into the combustion chamber. In the medium-pressure type the atomizing air is proportional to the fuel flow rate. These features produce a type of burner that is extremely flexible, since its spray angle can be readily changed. Maximum turndown can be obtained with this design by using a minimum airflow.

Steam-atomizing burners resemble the low-pressure air-atomizing type to a certain extent in method of operation. Oils up through No. 6 grade can be used and good atomization is obtained, owing to the rapid expansion of steam as it leaves the nozzle. Air is supplied by natural or forced draft around the oil nozzle, in the usual manner.

The plugging potential is greatly reduced as

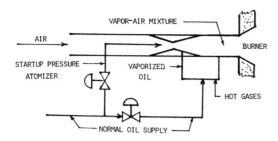

Fig. 6-1. Oil-vaporizing burner.

Fig. 6-2. Pressure-atomizing oil burner nozzle. (Courtesy of North American Manufacturing Co.)

orifice sizes are larger with this type burner. Approximately 1–2 lb/gal of steam are used to atomize light oil and 2–3 lb/gal for heavy oil, which is seldom missed from the boiler's output. See Fig. 6-3.

Combination burners are usually the nozzle mix type with the gas in the center and air in the space around the gas tube. An oil lance is then installed through the center of the gas tube. Either gas or atomizing air can be passed through the gas tube, or the gas burner may be a ring burner, with the oil gun inserted through the center, as shown in Fig. 7-10.

Special combination burners can be arranged and ordered, depending on the user's particular requirements, with separate connections for gas and atomizing medium. An existing low-pressure air-atomized burner can be converted to a combination simultaneous burner by substituting gas for atomizing air, mixing gas and atomizing air as they enter the

burner, or inserting an aspirator mixer in the main air line. Be sure you check with the burner manufacturer before attempting any of these exotic combinations.

B. Fuel Oil Storage

We stated earlier that one of the advantages of using fuel oils is its storage capabilities. Under certain ideal conditions it may be stored for very long periods without losing its ability to provide a ready and convenient source of heat for the boiler.

The storage of fuel oil is relatively easy, providing certain local and national codes are followed. Chief among these are probably the Factory Mutual and the Underwriter's Laboratories organizations, and both of them will cooperate with anyone who has a need for advice on the proper storage arrangements for fuel oil or one of the LPG fuels mentioned earlier.

Table 6-1 lists the more common standard sizes and dimensions of storage tanks for fuel oils. The weights are for steel tanks, but the trend now is for more use of fiberglass tanks. This is due to the lasting qualities of fiberglass under corrosive conditions, as they do not have to be given special coatings to prevent corrosive attack from the weather or the soil. Also, they do not need cathodic protection when buried.

The required size of the storage tank depends on each individual installation. Things to be taken into consideration are

1. Availability of commercial suppliers of fuel oil of the desired specifications and quantities, including advance notice to get delivery.

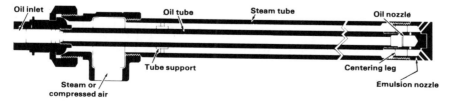

Fig. 6-3. Steam-atomizing oil burner gun. (Courtesy of North American Manufacturing Co.)

Table 6-1. Dimensions of Underwriters Fuel Oil Storage Tanks[a]

Capacity (gal)	Diameter (in.)	Length (ft-in.)	Shell Thickness (in.)	Head Thickness (in.)	Weight (lb)
550	48	6-0	$^3/_{16}$	$^3/_{16}$	800
1,000	48	10-10	$^3/_{16}$	$^3/_{16}$	1,300
1,100	48	11-10	$^3/_{16}$	$^3/_{16}$	1,400
1,500	48	15-8	$^3/_{16}$	$^3/_{16}$	1,650
	65	9-0	$^3/_{16}$	$^3/_{16}$	1,500
2,000	65	11-10	$^3/_{16}$	$^3/_{16}$	2,050
2,500	65	14-10	$^3/_{16}$	$^3/_{16}$	2,275
3,000	65	17-8	$^3/_{16}$	$^3/_{16}$	2,940
4,000	65	23-8	$^3/_{16}$	$^3/_{16}$	3,600
5,000	72	23-8	$^1/_4$	$^1/_4$	5,800
	84	17-8	$^1/_4$	$^1/_4$	5,400
7,500	84	26-6	$^1/_4$	$^1/_4$	7,150
	96	19-8	$^1/_4$	$^1/_4$	6,400
8,500	108	18-0	$^1/_4$	$^5/_{16}$	7,250
10,000	96	26-6	$^1/_4$	$^5/_{16}$	8,540
	120	17-0	$^1/_4$	$^5/_{16}$	8,100
12,000	96	31-6	$^1/_4$	$^5/_{16}$	10,500
	120	20-8	$^1/_4$	$^5/_{16}$	9,500
15,000	108	31-6	$^5/_{16}$	$^5/_{16}$	13,300
	120	25-6	$^5/_{16}$	$^5/_{16}$	12,150
20,000	120	34-6	$^5/_{16}$	$^5/_{16}$	15,500
25,000	120	42-6	$^3/_8$	$^3/_8$	22,300
30,000	120	51-3	$^3/_8$	$^3/_8$	28,000

[a]Actual dimensions will vary slightly with respective manufacturers.

Courtesy of North American Manufacturing Co.

2. Size of the delivery trucks available.

3. Amount of fuel oil expected to be used for each 24 hr operation.

4. Price break for various quantities ordered at one time.

5. Space available for the tanks.

To determine the amount of fuel oil you expect to consume during each day's operation, it is best to contact the boiler supplier for this information, or the burner supplier, if it is being furnished separate from the boiler. If this is not convenient, then we suggest you turn to page 308, where there is a list of the boiler sizes available from one boiler supplier, with the oil consumption rates.

Figure 6-4 is of a typical aboveground storage tank, mounted on concrete supports, with an unloading pump installed in case the delivery truck does not have one, which is quite rare. However, if there is more than one tank installed, it may be desirable to be able to transfer the oil from one tank to another, in which case the one pump may be piped to do double duty.

If the tanks are aboveground, they should be in an isolated location safe from damage from trucks and other traffic, but they should still be readily accessible for the delivery truck. It is good practice to have a fence or guard posts around it. Often, you will find that local codes will dictate restrictions as to the tank location.

Figure 6-5 shows the same tank installed underground. Basically, the same tank may be

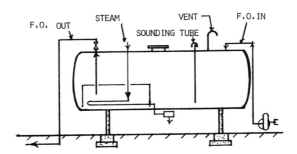

Fig. 6-4. Aboveground oil tank installation.

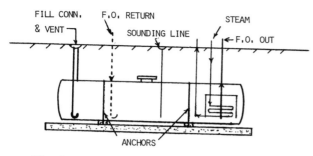

Fig. 6-5. Underground oil tank installation.

used, but if it is made of steel, then it must be coated and may have to be protected from galvanic attack by a cathodic protection system, as we mentioned earlier.

Underground tanks should be installed to prevent their floating out of the ground if the ground in the vicinity is capable of collecting and holding water. If there is any doubt, it is good practice to install a concrete pad as shown in the illustration, and strap the tank to the pad. The combined weight of the tank and the pad should be greater than the buoyancy of the tank when empty.

The trend recently is to substitute a belowground concrete vault in place of the buried tank. This is to prevent contamination of the underground water table in case of an oil leak from the tank. This is simply an adaptation of the aboveground installation, with provisions for removing any water that finds its way into the vault, along with any oil that may overflow through the tank vent.

The vault must be large enough to permit easy access to all parts of the tank to inspect for leaks, and it is good practice to have the top of the vault covered against the weather. This removes one of the advantages to be gained from a strictly buried tank, in that the area encompassed by the vault is completely dedicated to the use of the tank, and cannot be used for any other purpose, in most cases.

For aboveground tanks, local codes may require that the tank be contained in a diked area, capable of holding the entire tank volume in case of a major accident or spill. The same may also be true for a tank holding one of the LPGs, such as propane or butane, which is a safety feature of excellent merit.

The drawings of the tanks are shown with a device called a "Suction Box," which surrounds the heating coil and should be installed in all tanks. This is merely to help contain the heated oil around the inlet to the oil suction pipe, and prevents the entire tank contents from being heated. If this happens, the oil may deteriorate or separate into some of its components, which is undesirable. It may also cause the formation of sludge on the bottom of the tank.

Whether aboveground or belowground, the fuel oil tank should be sloped downward to the fill or vent connection, to permit water in the oil to flow to that location. It can then be pumped out with a self-priming pump.

The other items in the tank are self-explanatory and will not be discussed any further.

The heavy fuel oils, usually Nos. 5 and 6 grades, often require additives to improve the qualities of the oil, and to prolong the oil's storage abilities. There are additives to prevent sludge formation and settling, and additives to improve the combustion qualities, as well as additives to prevent water from settling out.

To illustrate the method of adding these additives to the oil, we refer you to Fig. 6-6, which illustrates the method recommended by one of the major suppliers of this material. Also, we will include here the most important portions of the sales literature describing two of the materials on the market now being sold by that same firm:

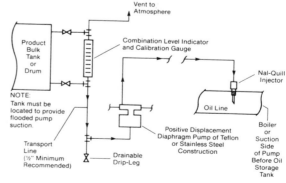

Fig. 6-6. Nalco 7260 feed system. (Courtesy of Nalco Chemical Co.)

NALCO 7260

* Unique water-based formula provides total dispersibility in oil.

* Ionic-state metal ensures high reaction efficiency.

* Waterlike viscosities decrease potential for application problems.

* Absence of suspended solids eliminates solids settling, reduces abrasion to the system metals.

* Can reduce slag deposition with its high-melting-point magnesium component.

* Can inhibit cold-end corrosion by efficiently reacting with sulfur trioxide (SO_3) in the flue gas stream.

NALCO 7260 is a magnesium-based treatment in a readily pumpable liquid form. NALCO 7260 helps control problems caused by sulfur, vanadium, sodium, and other residual-oil contaminants.

NALCO 7260 helps control slag deposition by means of a high-melting-point component that inhibits slag particles from fusing to each other or to fireside surfaces.

NALCO 7260 is designed to control corrosion by reducing the formation of aggressive molten sodium vanadate and alkali slags and neutralizing acidic sulfatic deposits.

NALCO 7260 dosage varies with concentration of sulfur, vanadium, and sodium in fuel and the operating conditions of the combustion system. Typical dosage can range from $1/2$ to 3 pints per 1000 gallons of oil.

NALCO 7260, when applied to a system burning oil, is continuously metered to the fuel oil line. A positive-displacement pump with stainless steel or Teflon-brand parts is suitable for this purpose. Pump directly from the feed system into the oil line supplying the burners through the NAL-QUILL® injector (NALCO Part P4610) to ensure uniform dispersion.

If the active storage tank has a turnaround time of less than 30 days, NALCO 7260 can be added to the oil as the storage tank is being filled. A FEEDPAC® 9B feed system should be used for this application.

Do not get it in the eyes, on skin, or on clothing. Wear goggles or face shield when handling. Avoid prolonged or repeated breathing of vapor. Do not take internally. NALCO 7260 has a maximum recommended in-plant storage life of six months. Store away from heat and open flame. Keep container closed when not in use.

Pumps

* Any positive-displacement pump with stainless steel or PVC construction.

* Specify double-ball check valves on all pumps, both suction and discharge ports.

* Size pump so average feed rate is near 75–80% of pump capacity.

Orientation

* Ensure flooded pump suction.

* Use of a drum rack is recommended with chemical feed pump located below the discharge bung of the drum.

Miscellaneous

* Eliminate restrictions in suction line. Keep suction line short. A filter in the suction line is suggested.

* Installation of fluid level indicator will help determine when full drum should be installed. The pump and lines should not be allowed to run dry.

* Maintain temperature between 60 and 80°F if possible.

NALCO FIRE · PREP 8256

Principal Uses

FIRE · PREP 8256 is a concentrated formulation designed to reduce sludge formation, to disperse existing sludge, and to inhibit corrosion. Typical applications include:

* Utilities with diesel or turbine peaking units using distillate oil and maintaining it on standby.

* Industrial plants with small package boilers, thermal fluid systems, and hot water systems using distillate oil. Often permits use of less stable fuels and "incompatible" fuels.

Feeding

FIRE · PREP 8256 is applied either by batch treating, adding to suction of transfer pump or to storage tank during fuel delivery to assure complete mixing or by continuous proportional pumping to fuel line at a point as far back in the system as possible for maximum protection.

Dosage

For fuel oil stabilization—Effective at a dosage of $\frac{1}{2}$ pint/1000 gal or less, depending on characteristics of fuel. Start at $\frac{1}{2}$ pint/1000 gal and adjust as needed. Where practical, tests described in NALCO Technifax 9 can be employed to periodically adjust product dosage.

For bacterial and fungal microorganism control—Effective for control of filter-plugging bacteria and fungi growth at dosages of 0.3–0.6 pint/1000 gal (40–80 ppm). Higher dosages may be required to completely inhibit some bacteria not considered to be factors in filter plugging.

Handling

Danger: Harmful if swallowed, inhaled, or absorbed through the skin. Causes severe eye and skin damage. Do not get in eyes, on skin, or on clothing. Wear goggles or face shield and rubber gloves when handling. Keep away from heat and open flame. Keep container closed when not in use. Avoid breathing of vapor. Use with adequate ventilation. Do not take internally.

C. Fuel Oil Piping

Getting the fuel oil from the storage tanks into the burner tips requires a pumping unit and interconnecting piping. Each boiler supplier will usually furnish recommended layouts and specifications for this portion of the plant. As an example, we have called upon two major suppliers of package boilers to give typical examples, and these are duplicated in the Appendix. Also included are specifications for steam and condensate piping, valves, and fittings.

Neither of these two offerings is intended to be final or binding on your particular situation, but they do serve as an example of what to watch for and will at least give you a starting point. If your proposed pumping and piping system will not fit these situations, then we suggest you call the boiler supplier or the pump supplier.

Pumping of heavy fuel oil is similar to pumping light fuel oil, with the exceptions of the pumping temperatures and the wear and tear on the pump. Heavy fuel oil must be heated to a certain minimum temperature for pumping, as shown in Table 5-2. Even then, the heavier fuel oils will usually result in a shorter life for the pump than if it were handling the lighter grades of oil. The heavy grades often contain sludges that are abrasive or that may pick up abrasive dirt in the piping and carry it into the pump parts. Fuel oil pumps are usually positive-displacement style, with very close clearances between the pumping surfaces in contact with the oil. Any grit in the oil will shorten the life of the pump between overhauls. Of course, a strainer ahead of the pump will catch much of the larger particles, but not all of them.

Note that a recirculating system is recommended in both cases. This is recommended for the following reasons:

1. It tends to keep the oil lines flushed out.

2. It helps to keep the oil in the tank stirred up and of a constant composition.

3. It ensures the presence of a fairly con-

stant oil pressure at the burner gun, meaning a smoother operating combustion control system.

The design of the fuel oil supply and return systems depends on many factors, such as type and grade of fuel oil, distance being piped, number and design of burners, type of atomization, ambient conditions, codes, and insurance requirements. Therefore, we shall not go into the subject here, other than referring you to the Appendix, Section D for two typical systems. Also, the North American Manufacturing Co. publishes a very good book titled, *North American Combustion Handbook*, which covers the subject in fairly comprehensive detail.

D. Soot Blowers

If the boiler has been designed to burn No. 6 fuel oil, coal, wood, bagasse, etc., it should have been provided with soot blowers, properly placed to ensure that all soot accumulations can be eliminated. These are simple devices, which have been in use for years, especially for those boilers burning coal. They employ steam jets to blast the soot off of the heat-transfer surfaces, and thus maintain the efficiency of the boiler.

Most of them use a rotary header that contains many steam nozzles in such a pattern as to give complete coverage of the boiler tubes when the high-pressure steam is turned onto the header, causing the header to rotate. The soot being washed off drains to the bottom of the boiler, and must then be drained out through the bottom drains provided for the purpose. It is a messy job, and often there are considerable quantities of soot blown out the stack, covering the surrounding area, and resulting in much unpleasantness. This is particularly true of soot blowers installed in economizers and stack-mounted air preheaters.

If steam is not available, as, for instance, when the boiler is down for maintenance, high-pressure air may be used as a substitute. When this happens, the resultant soot and air mixture tends to go up the stack, rather than down to the drain. For this reason, it is often more desirable to water wash the heat-transfer surfaces when the boiler is down for maintenance.

Obviously, soot blowers find their use in watertube boilers, as it is then easy for the steam jets to cover the gas-side surfaces. In the case of firetube boilers, it is necessary to shut the boiler down, and either blow or wash the soot from the insides of the individual tubes, manually.

Either way, soot removal is a messy job, but a necessary one.

CHAPTER 7
FUEL GAS SYSTEMS

A. Gas Burners

As discussed in Chap. 4, when gas is used for fuel, a certain amount of air must be mixed with the fuel to provide the oxygen required for combustion. This gas and air mixing can be accomplished in several ways.

There are three general types of gas burners, based on the method of mixing the gas and the combustion air, and they are premix, nozzle mix, and seminozzle mix. It is not always easy to classify them in this manner, since, in some cases, some mixing of secondary air takes place after the mixture has left the burner tip. Also, subclasses are often given for the method of mixing, but we shall confine ourselves here to the three basic types just given.

PREMIX

Figure 7-1 shows the basic principle of the premix burner, where the gas and combustion air are mixed together in the proper ratio ahead of the burner, then fed into the burner gun. The most common form of this is the ordinary low-pressure inspirator burner, as found in the smaller boilers and many gas appliances, and shown in Fig. 7-2. It is also used on the higher-pressure gas supply systems, up to about 4 psi.

As shown in the illustration, the gas, being higher in pressure than the atmospheric air surrounding the burner base, is injected into the inlet of a venturi, which produces a suction

and pulls in air at atmospheric pressure, raises its pressure, and mixes it with the gas. The mixture is then fed into the nozzle. The air is regulated by means of adjustable ports surrounding the induced air intake. The venturi principle involved is well known and will not be explained here.

The main advantages of this burner are its simplicity and its quietness of operation, and that no electrical power is required to operate it. Its main limitations are its relatively small size and its requirement for a fairly constant firing rate, limiting its output range. It is often found in an upshot arrangement and in multiples.

Another premix style of burner is shown in Fig. 7-3, where the aspirator burner is depicted. In this case, the air and gas are interchanged from that arrangement shown for the inspirator burner, just described. The gas is reduced in pressure and fed into the suction chamber of the venturi, and the combustion air is used as the inducing force. The air must be raised in pressure from a few inches of water column to as much as 4 psi, which requires a blower. This burner is used mostly on low-pressure gases, where a variable firing rate is required.

A very common low-gas-pressure type of burner in the larger sizes is shown in Fig. 7-4, known as the radiant burner. The one shown here is firing horizontally, but they can also be arranged for firing vertically.

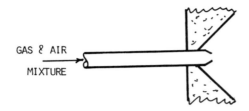

Fig. 7-1. Premix burner.

In this burner, the principle of mixing is again the premix, either the inspirator or the aspirator type as previously described. But here the fuel/air mixture is broken up into multiple flames, which are enclosed by refractory blocks. The gas flames heat the refractory blocks to a radiant glowing temperature. This eliminates the possibility of flame impingement on the tube surfaces, shortens the furnace required, and increases the firing range of the burner, depending on the fuel gas or air pressures available.

Another premix burner, more sophisticated, and often found on burner systems requiring a higher turndown ratio (described later in this book) is that shown in Fig. 7-5, the variable-pressure-mixing burner. In this arrangement, the sensing line between the gas-pressure-reg-

ulating valve and the air line ahead of the air orifice keeps the two volumes in correct proportion at the different firing rates.

The blower-mixer style of premix burner shown in Fig. 7-6 is a more complicated arrangement, which accomplishes the same thing. Here, the mixing takes place inside of the blower, and the mixture is increased in pressure as required for the firing rate.

In the variable-pressure burner, the firing rate is adjusted by means of the air valve position, while in the blower-mixer burner, the firing rate is changed by positioning the ratio control lever.

NOZZLE MIX

Figure 7-7 shows another basic style of burner, the nozzle-mix style, in which the gas and air are kept separated until they reach the burner tip. This eliminates the possibility of flashback, as there is no flammable mixture in the air or gas supply systems. This produces a burner in which more flame patterns and higher turndown ratios and smaller furnaces are possible. The combustion air is supplied by a low-

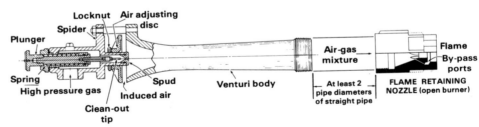

Fig. 7-2. Inspirator-type gas burner. (Courtesy of North American Manufacturing Co.)

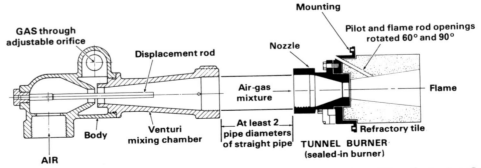

Fig. 7-3. Aspirator-type gas burner. (Courtesy of North American Manufacturing Co.)

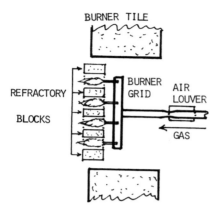

Fig. 7-4. Radiant gas burner.

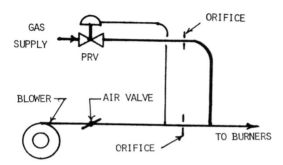

Fig. 7-5. Variable-pressure-mixing burner.

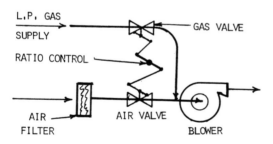

Fig. 7-6. Blower-mixer style burner.

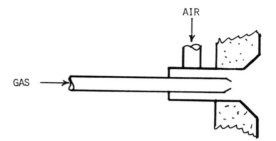

Fig. 7-7. Nozzle-mix burner.

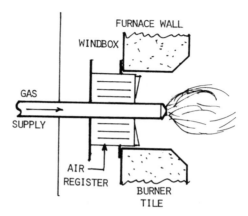

Fig. 7-8. High-pressure-gun-type burner.

If the boiler has the induced draft system, the windbox is eliminated, and the subatmospheric pressure existing in the furnace pulls air in around the burner tip through the air register and the spinner vanes surrounding the burner gun.

SEMINOZZLE MIX

The seminozzle mix burners are a further development of the nozzle-mix style, and are often made by converting the nozzle-mix burner in an endeavor to increase the firing capabilities. As shown in Fig. 7-9, it is simply a nozzle-mix style with a secondary air supply into the burner tip area. This permits increasing the gas pressure and the firing rate without changing the primary air blower.

However, we warn you in attempting to make this conversion, be sure to contact the burner and boiler supplier for their approval and comments.

The combination oil and gas burner shown

pressure blower, either through ducting or into a plenum surrounding the burner gun.

The high-pressure-gun-type burner, nozzle-mix style, is shown in a typical arrangement in Fig. 7-8. The windbox is pressurized by a blower, and the air is introduced through a register having variable vanes controlled by linkages from the combustion control system. The air is usually introduced into the area surrounding the burner tip through directional louvers, which impart a twist to the air, thus improving the mixing of the fuel and air.

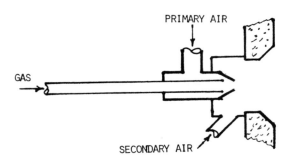

Fig. 7-9. Seminozzle-mix burner.

in Fig. 7-10 is quite often found on today's boilers, especially where there is competition between the suppliers of fuel oil and gas. The one shown is for the heavier grades of fuel oil, such as Nos. 4 through 6, requiring atomization of the oil to obtain proper combustion. When the atomization is performed with air as shown here, the burner becomes a type of seminozzle-mix burner.

We have shown only the more common and basic styles of burners, both gas and oil. There are other styles made to suit particular combinations of fuels.

B. Gas Piping

Supplying gas to the boiler is usually a matter of piping it from the supply line in the street to the point of connection of the gas piping train on the boiler. In most cases, the boiler is furnished with the required controls and fittings

to satisfy the burner and the governing codes. It must all be compatible with the type and pressure of the gas being supplied, of course.

Figure 7-11 is of a typical situation wherein the gas train piping comes with the boiler, and all that is needed is the piping and the fittings to connect the gas from the gas main in the street to the designated point of connection on the boiler gas train. Also shown, and highly recommended, is a meter in the line to the boiler to enable you to keep a check on the gas consumption when you initiate the energy-saving measures that will be explained to you later.

The gas train on the boiler consists of all of the necessary valves, controls, and fittings to satisfy the governing codes; this equipment has to be specified when the boiler is ordered. This subject will be covered in later chapters, so we will not go into it here.

In choosing the proper gas piping from the main to the boiler, we strongly urge you to call in your boiler supplier for help. The supplier will be willing to see that the proper gas piping is installed, since it will reduce the possibility of a trouble call later. The supplier will be glad to ensure the proper operation of the boiler.

If the boiler is to be fired on propane or butane, the actual burner and control system does not vary appreciably from that for natural gas. As propane and butane have much higher heating values for each cubic foot of gas burned, less of it is required. This means simply that the piping and valves may be smaller, and the

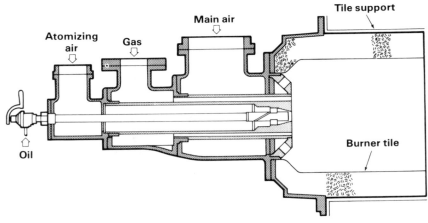

Fig. 7-10. Combination gas and oil burner. (Courtesy of North American Manufacturing Co.)

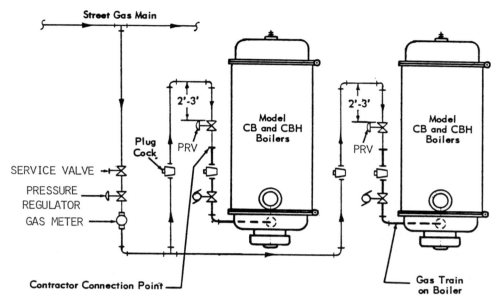

Fig. 7-11. Typical boiler gas piping. (Courtesy of Cleaver-Brooks Co.)

orifices that control the flow of gas entering the burner tips may be smaller than for natural gas.

The main difference between natural gas and propane or butane is that the latter two must be stored someplace on the premises, since they are not normally distributed long distances from a central location. The storage and vaporizing equipment must be located at the proper minimum distances from the boiler, plant, and railroad or truck routes. The National Fire Codes stipulate these distances, and they should be consulted when designing your facilities.

The storage, pressure control, and vaporizing equipment are often supplied by the distributors of propane or butane, and it is wise to first consult with them before making your final decision on the use of these gases. The distributors usually have years of experience behind them, and they will be very familiar with the local codes and the general system that will best suit your operation. They often will furnish the entire station equipment, either on a lease or sale basis.

There are several differences between natural gas and propane or butane that must be kept in mind at all times, and they are:

1. Both gases are invisible and odorless.

2. The LPG gases are heavier than air, and thus tend to flow along the ground and will settle in low spots, undetected until a fire results.

3. The reduction of LPG pressure at the control valve results in a drop in temperature, also. This can cause the valve to freeze up, especially if there is an appreciable amount of moisture in the gas. If the valve has external moving parts, rain, snow, or condensate may accumulate around the valve and cause the valve to hold up. Electric heat tracing around the reducing valve will usually solve this problem.

Sizing the propane- or butane-storage facilities requires much the same procedure as for fuel oil, which we described earlier. There are several other minor details that must be taken into account and that the gas distributor will be able to help you with. When propane or butane are converted from liquid to gas, heat must be absorbed by the gas. This heat may, under certain conditions, be supplied by transfer from the air through the wall of the tank into the liquid propane or butane. Otherwise, additional vaporizing equipment must be installed. It is best to let the gas distributor advise you on this, as we suggested earlier.

There may be situations where propane or butane are being used as a standby fuel, or as an alternate fuel, to natural gas, or other gas. Therefore, it would not be very convenient to change the burners and control equipment every time the switch is made. In this case, it is wise to consider the installation of one of the air-mixing systems, which will mix air with the propane or butane, to produce a mixture with approximately the same heating value per cubic foot of mix as the gas for which the burners were designed. Figures 7-12 and 7-13 illustrate two of the better known methods, as supplied by one of the major suppliers of this type of equipment, the C. M. Kemp Manufacturing Co.

There are two general systems used, and Fig. 7-12 shows the one in which the gas from the vaporizer is mixed with atmospheric air at low pressure, then the mixture is fed into a pressure blower that raises the pressure of the mixture to that necessary to satisfy the burner and

controls. As shown in both figures, the heart of the system is the proportioner, which mixes the air and gas in the right combination to give a resultant heating value as required.

In the system shown in Fig. 7-13, the gas and air are of the correct pressure before entering the proportioner, and therefore the pressure blower is not required. This is assuming that the air is available from an outside pressure system, such as the plant air supply. In this case, the entire system will not cost as much as for the premix system, since the pressure blower is a fairly expensive piece of equipment.

As we mentioned a few paragraphs earlier, the use of propane and butane often referred to as LP gas, require a few precautions.

General regulations for location and safeguarding of gas/air-dilution equipment are covered in Factory Mutual's *Handbook of In-*

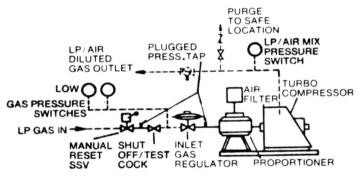

Fig. 7-12. Premix LPG/air mixing systems. (Courtesy of C. M. Kemp Manufacturing Co.)

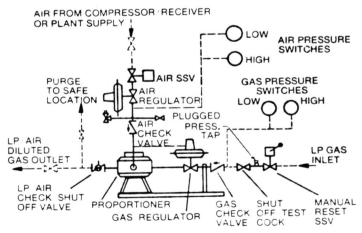

Fig. 7-13. Prepressured LPG/air mixing system. (Courtesy of C. M. Kemp Manufacturing Co.)

dustrial Loss Prevention, Chapter 43 and in National Fire Protection Association Chapter 58. IRI requirements are generally in accordance with one or both of these, but certain exceptions may be taken by IRI.

In addition, state and local codes may add specific safety regulations for LP/air gas plants. Safety requirements immediately related to the LP/Air Dilution Unit (Mixing Station) call for a Manual Reset Safety Valve in the entering LP gas line. This valve is actuated by pressure switches responding to abnormally high or low gas pressure. Where a gas/air turbo (premix type) or air pump is used, a third pressure switch is incorporated. Any abnormal pressure shuts down this turbo or pump. Where a unit air compressor/receiver, or where plant air is utilized, a safety shutoff valve in the air line is required, actuated from pressure switches responding to high or low air pressure (and gas safety valve is "triggered" also by these).

Any electrical device used in or at the mixer station is to be suitable for Class 1, Group D, Division 2 location provided required clearances from relief valves, etc., are maintained. Depending on the device, it may be suitably housed in a general purpose (NEMA-1) enclosure or may require an explosionproof (NEMA-7) enclosure.

Should the reader desire to delve further into the matter of burners, fuels, combustion, and controls, then we suggest the purchase of the book *North American Combustion Handbook*, available from North American Manufacturing Co., 4456 East 71st Street, Cleveland, OH 44105.

For complete fuel data on LPG fuels, we refer you to page 315 in the Appendix.

C H A P T E R 8
WATER FOR THE BOILER

This chapter is meant to be but a brief summary of some of the problems encountered in the average industrial steam plant. If any of these situations should prove to be a problem in your plant, then it is best to call in professional help. The material given here will help guide you to the right trade, service, or supplier. Or it may intrigue you sufficiently to inspire you into further study.

A. Water

Hot water and steel are compatible only under a very limited range of operating conditions. These limits are the ideal that all steam plants attempt to attain, and the degree to which they succeed depends primarily on the diligence of the operators and maintenance personnel. Very seldom is the ultimate ever reached and maintained for any length of time. Once the proper combination of operating conditions is reached, constant attention and diligence are required to maintain it.

Water is considered to be about the most universal solvent attainable. This means that there are very few materials on the face of the earth that will not go into solution in water in some degree. It also means that the raw water available for the boiler plant is seldom in a pure state, but will have dissolved materials in it. The problem for the boiler plant operator then

becomes one of controlling the dissolved materials in the water to the extent that any troubles resulting can be kept in safe limits. This is the basis upon which the entire field of water treatment operates.

It is now possible to remove completely all impurities or minerals from water, with the use of a complicated and expensive system of treatment. This is done in some central station power generating plants, in some space exploration programs, and in some food and beverage industries. However, in the case of water that has been completely "demineralized," the result is a water that is highly corrosive, due to its "hunger" for dissolving other materials.

Water technology is a separate field in itself, and is now becoming a major subject of investigation. For our purposes here, we are faced with the necessity of making the best use of ordinary water as obtained from the supply mains or from the plant's own wells or other source of supply.

Before we get any further into the subject, perhaps we had better define some of the terms used in water chemistry. It's not that we expect you to become experts from the material presented here, but it will help you to know what the outside water consultant is talking about when you receive a report.

pH: This is a numerical scale that denotes how acidic or how alkaline (which is the opposite of acidic) the water is. The numbers go from

0 to 14, with 1 being the strongest acid, 7 is neutral water, and 14 is the strongest alkali.

ppm (parts per million): The number of pounds dry weight of the substance under discussion, dissolved in one million pounds of water. Thus 50 ppm total dissolved solids means that there are 50 pounds of dry solids dissolved in one million pounds of water, or at an equivalent rate.

ppb (parts per billion): The same meaning as ppm, except here we are talking about one billion pounds of water, which is 1,000 times larger than one million. To convert a reading from ppb to ppm, divide the reading by 1,000.

Hardness: A term denoting the relative scale forming capabilities of water.

Total hardness: A term which implies that all of the scale-forming compounds have been included. Generally speaking, unless otherwise qualified, "total hardness" is meant when you see the term "hardness."

gpg (grains per gallon): Used considerably when stating the hardness of water; 1 gpg = 17.1 ppm, so to convert a reading in ppm to gpg, divide the reading by 17.1.

lb/Mgal (pounds per thousand gallons): Another convenient way of giving water treatment quantities, for it is easier to apply. To convert ppm to lb/Mgal divide the ppm by 120.

Another term used quite frequently in boiler water treatment is "calcium carbonate equivalent." This is a commonly accepted method of reporting the hardness and alkalinity of water by reducing all chemical components to one chemically equivalent compound, which is calcium carbonate. This is done mostly for convenience and simplicity, since calcium carbonate is one of the more common compounds to be found in water.

These are some of the basic terms. When we use a term in this discussion that is not given above, we shall define the term.

To give you an idea of the magnitude of the quantities we are dealing with in this discussion, let us assume that you have 1,000,000 lb

of pure distilled water, which is about 120,000 gal. If into this water you empty a 1-lb box of ordinary table salt, which is sodium chloride, you would then have water with 1 ppm of dissolved salt. You would not even be able to taste it, it would be so diluted.

Table 8-1 is a general classification listing of water by hardness, and is meant purely as a guide. Carrying our example further, you can see that to produce hard water you would have to dump at least 121 lb of scale-forming salts, such as magnesium or calcium carbonate, into your 120,000 gal of pure distilled water.

B. Water Composition

The degree of purity of the water that is intended for use in the modern industrial boiler depends primarily on the steam pressure and on the amount of make-up water required. Generally speaking, the higher either one of those two factors becomes, the more care should be taken to provide high-purity water for the boiler.

We usually consider the subject of water for boilers in two phases, "external" and "internal." "External" treatment is that applied to condition both the water for make-up and the returned condensate. "Internal" treatment is that applied to the water inside the boiler, although the actual treatment is often applied just outside the boiler, prior to feeding the water into the boiler. See Fig. 8-1 for an illustration of the breakdown.

For our purposes, we are going to assume that industrial boilers only are under consideration, and that steam pressures are all below 600 psig. This will cover most of the industrial ap-

Table 8-1. Water Hardness Scale

Total Hardness (ppm)	Classification
Less Than 15	Very soft water
15–60	Soft water
61–120	Medium hard water
121–180	Hard water
Over 180	Very hard water

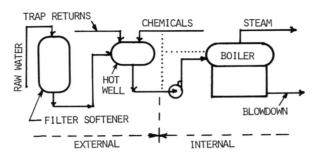

Fig. 8-1. Overall boiler water treatment approach.

plications. If pressures higher than this are being used in your plant, then we refer you to some of the more advanced books on boilers and steam generation.

As a guide to the quality of water required for our average industrial boiler, we refer you to Tables 8-2 and 8-3. These values are stated as a goal you are advised to follow in order to obtain the best service from your steam plant.

The values in Table 8-2 are maximum values. Most boiler manufacturers recommend lower values for safe operation in their boilers, especially in the total dissolved solids.

To complicate matters, quite frequently one or more of the components in the water-treatment chemicals may react with some chemical already in the water, and the result may be entirely unpredictable or, at least, it may cloud

Table 8-2. Recommended Boiler Water Condition #1

Boiler Pressure (psig)	Maximum Total Dissolved Solids (ppm)	Alkalinity, CaCO₃ (ppm)
0–300	3500	200–700
300–450	3000	160–600
450–600	2500	120–500

Table 8-3. Recommended Boiler Water Condition #2 (Regardless of Pressure)

Phenolphthalein Alkalinity	60–80% of Total Alkalinity
pH	9.5–11
Phosphates (PO_4)	30–60 ppm
Hardness ($CaCO_3$)	0
Sulfite	30–60 ppm
Condensate return pH	7.5–8.5

the desired result. This is one of the reasons why you will have to be patient with your water-treatment service, and be sure and follow all recommendations faithfully.

Note that we are assuming that you have engaged a service to come into your plant and regularly test the condensate, the boiler water, and the feedwater, and make recommendations for treatment for you to follow. The treatment will probably require you to feed prescribed batches of the service's own proprietary chemical solution or tablets at regular intervals. This will have to be followed until the next visit, at which time another prescription may be issued for you to follow. This battle will no doubt go on for as long as you are operating a boiler, so you may as well get used to it. We shall attempt to make the task a little easier by explaining what is taking place, so that you can at least be in sympathy with your treatment service.

C. Trouble from Water

Table 8-4 lists the major troubles to be encountered in most industrial boilers in the pressure ranges at which you will probably be operating. The table is simplified somewhat, but it will serve to illustrate what you can expect should your boiler water get out of control. There is some overlapping between the problems and between the causes, as well as between the solutions. One cause may enhance or reinforce another, so that the end result is that it may take some long-term experimenting for any boiler water-treatment service to stabilize the conditions in your plant. Coupled with that is the difficulty of predicting just what the composition of your water supply will be from day to day. It may change very rapidly, depending on where the supplier takes the pump suction or which reservoir is tapped.

We shall now cover the meaning of those requirements in Tables 8-2 and 8-3.

Total dissolved solids (TDS) refers to the dry weight, in pounds per million pounds of water, of mineral salts, acids, or alkalies that are dis-

Table 8-4. Major Boiler Troubles

Problem	Causes	Cure
Scale and sludge	Calcium and magnesium salts	Water softeners
Corrosion	Oxygen in water	Hydrazine treatment, Sodium sulfite, Deaeration
	Carbon dioxide	Deaeration
	Low pH (acid condition)	Add caustic soda, soda ash
Carryover	TDS[a] and pH too high	Increase blowdown
	Fast load changes	Install load limiter

[a]TDS = total dissolved solids.

solved in the water. They are measured by an electrical conductivity meter, since the ability of water to conduct electricity is determined by the dissolved chemicals in the water. The total dissolved solids may also be calculated.

Total alkalinity in ppm is the number of pounds of dry alkaline chemicals dissolved in one million pounds of water. The alkaline chemicals usually found in water are various carbonates, bicarbonates, and hydroxides, of calcium, magnesium, and sodium, for the most part. To measure them, it is necessary to make a chemical test that depends on a change in color after a measured amount of "indicator" has been added to a test sample. The result does not tell what is causing the alkalinity or how much of each chemical is present, but it does give the total alkalinity present. The result is usually given in terms of "equivalents" to calcium carbonate as a standard.

Suspended solids are those solids that are not in solution, but are in a finely divided form and that do not readily settle out. They may often be removed by treating with a coagulant, or precipitating agent, followed by settling and filtration.

Total hardness in ppm refers to the pounds of scale-forming chemicals in one million pounds of water. The principal compounds in this category are calcium and magnesium

compounds, with some aluminum, iron, manganese, and zinc also causing trouble. One of the common tests for the presence of these compounds is to take a measured quantity of the water to be tested, and add drops of a standard soap solution to the sample, shaking the sample after each drop, until a definite soap suds is formed on the surface of the water. The number of drops of the soap required indicates the hardness of the water in grains per gallon. To convert the results in grains per gallon to parts per million, multiply the gpg by 17.1 to arrive at the ppm. This test is not as accurate as one that the water chemists prefer to use, but it is simpler and easier to make. The aim, as shown here in Table 8-3, is to achieve zero hardness for the boiler, so that there are no scale-forming compounds in the boiler water. The extent to which this is achieved indicates how long the boiler may be operated between shut downs for descaling.

The iron found in boiler water will usually be in one of the combined forms, such as rust or one of the hydroxides. The amount of iron in the water depends mostly on the amount of oxygen in the water, which is the reason so much attention is given to keeping the oxygen out of the water as much as possible. The amount of oxygen permitted is very minute, being 7 ppb. Oxygen is removed by two processes, one of which, deaerating, is explained elsewhere in this book. The other method is to feed certain chemicals, such as sodium sulfite or hydrazine, into the feedwater that scavenge, or remove, any excess oxygen in the boiler water. This is one of the main purposes of the chemical feeders shown in Chapter 10.

The pH recommendations shown in Table 8-3 are usually maintained by adding caustics to the water at one or more stages in the process, as mentioned in Table 8-4.

Total organics refers mostly to the miscellaneous matter, such as decaying vegetable matter, sewage, algae, and similar material, often found in the nation's drinking water supplies. It is satisfactory to leave the total organics in minute quantities for use in our domestic applications, but not for boiler feedwater.

The preceding discussion has been brief, but

it should give the reader a fair idea of some of the requirements of water quality for use in the average well-maintained industrial boiler. At this point, it may well be worth your while to ask yourself how close to these ideals does your boiler make-up and feedwater come?

In most plants, the make-up water that is fed to the boiler to compensate for steam losses is mixed with the return condensate to form the boiler feedwater. Consequently, it is necessary to consider both the condensate and the make-up water to obtain the proper feedwater for the boiler. Naturally, of the two, the make-up water is usually the one requiring the most treatment, since it is usually supplied from the plant raw water system. The condensate, being purer, often only needs to be kept from absorbing oxygen before it is fed into the boiler.

Thus, we see that the total treatment problem consists of four areas of concern: make-up water, condensate, feedwater (which consists of the first two combined), and the water inside the boiler.

Before you can prescribe the proper medicine, you have to observe the symptoms of the disorder, much as a physician does when we are sick. In the case of the boiler plant, however, we go one step further; we attempt to analyze the condition of the water before it is used, then we try an initial treatment, observe the results, make corrections in the treatment, observe, and repeat the process. The ultimate aim, of course, is perfection, or as near to it as possible for the particular plant.

That is as far as we shall delve into the troubles caused by improper water treatment at this time. Later in this book, we shall cover the subject in greater detail, where we feel it more appropriate. At this point, you have been amply warned regarding the necessity of maintaining the correct water conditions, so we shall now proceed to assist you toward that goal.

C H A P T E R 9
WATER-TREATMENT METHODS

A. The Zeolite Treatment Method

There are many ways in which the water-treatment chemist may change the composition of a plant's water to make it suitable for use in the boiler. To do the best job, however, the chemist must know what to expect in the water analysis from one day to the next. When the water composition changes from day to day, or sometimes oftener, this makes the job of proper water treatment very difficult.

Some of the more common methods used in treating water for boiler make-up will now be discussed, very briefly, in order to give you a working knowledge of them, with a few comments on the pros and cons of each method.

The first one we will discuss is the old favorite of many plants, the zeolite softener, sometimes more scientifically called by the term of "ion exchange system." The name "zeolite" is given to the material used to alter the composition of the water. Originally, a natural mineral was used, but this has been replaced with an artificial form made of resin.

The water is passed through a bed of the zeolite, and the resin in the zeolite undergoes a chemical reaction with the calcium and magnesium elements in the scale-forming compounds in the water. In the process, an exchange takes place, so that the zeolite is changed, and the water leaves the softener with sodium compounds in place of the calcium and magnesium compounds that entered the softener. The sodium compounds do not form hard scale on the boiler tubes, and the water is known as "soft" water. This means that it will form soap suds very readily, and will work very well for washing and dyeing, and for the first stage in boiler make-up.

The resin soon is completely changed to calcium and magnesium zeolite, and must then be reverted, or regenerated, back to the original form so it may continue its job. To regenerate the softener, it is necessary to supply enough sodium to cause the resin to reverse its transformation, and again become sodium zeolite. This is done by washing with a solution of ordinary salt, or sodium chloride, which provides the necessary sodium for the zeolite.

After the regeneration is completed, the zeolite bed is rinsed with fresh hard water, before placing the softener unit back into service.

It is usual to have at least two tanks containing the zeolite, so that one may be in use while the other one is being regenerated. Also, it is usual to backwash the softener bed before commencing the regeneration, so that any foreign matter that the zeolite bed has caught will be flushed out the drain.

There are many variations of the zeolite softener, from simple little one tank units, to three or more, and with various means of controlling the cycles, and handling the brine. Figure 9-1 shows a typical zeolite system.

The most important thing to remember with

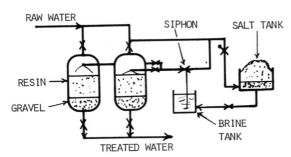

Fig. 9-1. Ion exchange softener.

Gallons of water treated

$$= \frac{\text{Grains capacity in tank}}{\text{Grains/gallon of hardness in water}}$$

For example, if the water being treated averages 15 grains per gallon hardness, then the unit will treat 100,000 gallons of the water between generating cycles, as per the following:

$$\text{Gallons treated} = \frac{1,500,000}{\text{Grains/gallon}}$$

$$= \frac{1,500,000}{15 \text{ gr/gal}} = 100,000 \text{ gal}$$

the zeolite softener is that it is not the complete treatment method to be used for boiler make-up. It is best to consider it only the first step in the process. The zeolite bed merely alters the composition of the chemicals in the water, it does not remove them. The resulting compounds may not form hard scale in the boiler, but, as we explained earlier, other troubles may be caused by its use. To get the best use from your steam plant, it is well to remember that zeolite-softened water should at least be heated and deaerated before it is fed into the boiler.

In addition, there are often other compounds in the raw water that are either untouched by the zeolite or that may be altered in an adverse way. In such cases, further treatment may be called for.

Many plants, it is true, have been using the zeolite softener alone for treating their make-up water for years and have not experienced too many troubles, so they report. But, when you look at the complete steam plant operation record, a different picture often emerges. We are not trying to discourage the use of the zeolite softener, but are merely attempting to place it in its proper perspective: at the entrance of the raw water into the plant, before the water line branches out to the various uses within the plant.

It is very easy to calculate the capacity of a zeolite softening unit. If a cell, or unit, has a capacity of 1,500,000 grains equivalent of zeolite, this means that 1,500,000 grains of water hardness will be removed by the zeolite before it is necessary to regenerate with brine. Thus,

B. More Advanced Treatment Methods

The cold lime and soda process is another method used to soften the make-up water and also to remove discoloration, manganese, and iron. Its operating costs are usually lower than with the zeolite system, but it does not always produce zero hardness as does the zeolite method. In fact, the cold lime and soda process is often followed by a zeolite system to give complete zero hardness.

The hot lime and zeolite softener, Fig. 9-2, is also used for softening the make-up water and for removing some of the dissolved gases, silica, and total solids. In this method, the water is heated by mixing it with steam, which drives off most of the dissolved gases. The hot water is then treated with a precipitating agent, such as lime and soda ash. This causes the scale-forming compounds to form a sludge that set-

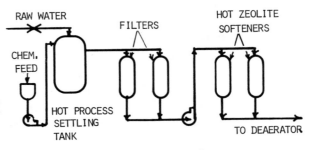

Fig. 9-2. Hot lime–zeolite softener system.

tles out at the bottom of the softener. The hot water is then forced down through this sludge bed, which acts as a filter, and removes most of the remaining sludge. The filtered water is bled off the top of the tank around the outside, into another bank of external filters, and into a bank of zeolite softeners for final removal of hardness. This method is often used where the hot water coming out is to be used in the boiler or in some other process very soon so as not to lose the heat that has been put into the water.

Demineralizers come in several types, and the final selection will depend on the desired result with the water available. The complete system may consist of from three to as many as six separate steps in the process, before the water is reduced to almost chemically pure water. It is possible to produce water that is as pure as triple-distilled water, and at a much lower cost than if the water had been produced by distillation.

In the demineralization process, Fig. 9-3, each step removes some element or compound, and in some steps, the result may be a completely different compound, which, in turn, must be removed by one of the succeeding steps. At times, the ingenuity of the water chemist may be taxed to the limit in order to produce the desired result.

Many of the troublesome compounds cannot be removed without first changing their chemical composition. In changing their composition, in turn, other undesirable results may be produced, which in turn must be handled in one of the succeeding steps. And throughout, the acidic or caustic condition of the water must be watched closely, so that the water emerging from the system has the desired pH range.

When the water has been completely demineralized, it still is a problem to handle, because in such a pure state it has a strong affinity for dissolving minerals and metals again. In some cases, it must be treated as if it were corrosive by using stainless steel, glass, or plastic pipe and tubing. In short, it must be handled almost as carefully as some of the acids or caustics.

As you can surmise, the demineralizing process is probably too expensive for the average small steam plant. However, the larger steam plants are finding that it is economically feasible and desirable from a long-term cost analysis.

Another system finding increased use for treating boiler make-up water is the reverse osmosis method. In this system, the water often receives an initial injection of either an acid or a coagulant, then it is pressurized by pumps to about 400 psig or more, then forced through a cell of closely packed membrane elements. The impurities in the water are left behind on the membranes as the pure water is passed through by what the chemists call "reverse osmosis." The impurities are bled off in a waste stream, which must then be disposed of in some acceptable manner.

There are several variations of the preceding sophisticated treatment systems, and they all must be tailored to the particular plant water conditions and requirements.

That covers the majority of the methods used in the treatment of make-up water in the average plant. We shall now list some of the other methods used from time to time, with a brief word or two on their application.

C. Miscellaneous Treatment Methods

Chlorination is often used to kill living organic matter, such as algae or plant slime, and also bacteria.

Aeration may be used to remove iron, which it accomplishes by oxidizing it so it may be re-

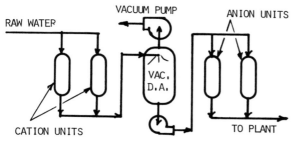

Fig. 9-3. Demineralization system.

moved by a subsequent filtering process. Some dissolved gases are also removed by this process. However, large amounts of air are added to the water, which must be removed prior to injecting the water into the boiler.

A dealkalizer reduces the alkalinity and sometimes will reduce silica in the water. It is usually applied following a zeolite softener, and operates similar to it.

The floc treator is similar in operation to the hot process softener described earlier, except that it is a cold process. The cold water is treated with a chemical, called a coagulant, which induces scale-forming compounds to form a sludge that will settle to the bottom of the tank. The chemicals used are compounds of lime, alum, and soda ash. The water is forced down through the bed of sludge, which acts as a filtering agent, and there the finer particles of sludge are removed, and the water is passed up to the top of the tank where it is drawn off for use.

There are also several variations of the zeolite softener, all of which are designed to serve a specific purpose. We will mention the major one in use, which is the manganese zeolite type. This is often employed for removing hydrogen sulfide, or what is known as "rotten egg gas," which may be found in waters adjoining sewage treatment or chemical plant discharges. It also removes manganese and iron compounds.

We mentioned previously the chlorination of water, which in turn may cause a problem if too much chlorine is added or if it is necessary to overdose to achieve the desired result. Often, it is then necessary to remove the remaining traces of chlorine. This is done by passing the water through carbon filters, or by adding sodium sulfite.

Filters used for removing sludge from water are of several types: sand, anthracite, settled sludge, carbon, or diatomaceous earth. They all have the same basic characteristic, with the exception of the carbon filter. We have already mentioned the carbon filter's ability to remove some of the organic matter or discoloration compounds in the water. In addition, it has the ability to remove odor.

The subject of filtering is a complete field of study in itself, and if the reader is so inclined, perhaps further knowledge may be gained by calling upon one of the better known filter suppliers for some educational material covering the products they supply.

That completes our discussion on the subject of treating make-up water for the steam boiler. Now, we shall cover the next step in the water cycle, which is the treatment and handling of the condensate and the mixture formed by the condensate and the make-up water. This is, as we have mentioned before, known as boiler feedwater.

D. Boiler Feedwater Treatment

The key to good boiler feedwater lies in the proper treatment of the make-up water, the careful handling of the condensate, and the final treatment of the mixture of the two, before it is fed into the boiler. Of course, in many of the smaller boiler plants, the process is not broken down in this manner. Some boilers take the make-up directly into the boiler, without being mixed with the condensate. Also, in a great many plants, absolutely nothing is done to the condensate before it is pumped into the boiler. It is all a matter of what is economically practical, or expedient (which is, too often, the case). For those of you who are operating plants falling into these last two categories, the material discussed here may serve as a guide should you decide to improve conditions in your plant. Of course, in some cases, the more complicated and expensive treatment methods previously described may not be necessary or economical.

Let us start the discussion by taking a hard look at the condensate right after it is formed, back in the equipment, or in the steam transmission main. Figure 9-4 shows what to expect when condensate is allowed to cool. As you can see, the amount of oxygen and carbon dioxide that the condensate will hold in solution is partially dependent on the temperature; the colder the temperature, the more gas it will

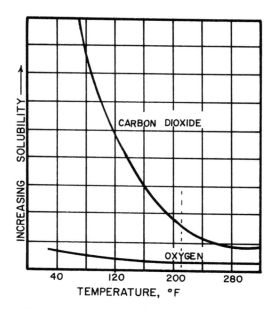

Fig. 9-4. Solubility of gases in feedwater.

dissolve. Also, we must keep in mind that hot condensate will absorb gases until it is completely saturated, if there is enough gas in contact with the condensate, and enough time for the absorption to take place.

Pressure also affects the amount of gas that will be absorbed by the water, since the absorption increases with pressure.

The last statements tell us directly what steps must be taken to keep the amount of dissolved gases in the condensate to a minimum. We must not allow the gases to accumulate over the surface of the condensate, we must keep the condensate as hot as possible, and the condensate should be put into the boiler as soon as possible after it is formed. The ideal case, of course, would be to blanket the surface of the condensate with an inert gas, such as nitrogen, as soon as it is discharged into the return main from the traps, but this is rather impractical. However, we can take some steps to keep the condensate hot until it is pumped into the boiler, and we can examine the entire condensate return system to make sure there is no undue hold-up at any point. This is where a pumped return system displays a distinct advantage, especially in the case of a long, sprawling return system. By using pumps, the condensate is returned rapidly, and relatively free from

contact with the air during its transit back to the feed tank.

Once the condensate is in the feed tank in the boiler room, steps may be taken to ensure that it will be kept hot, but not too hot for the boiler feed pump to handle it. It should, however, be kept as close to 212°F as is practical, even if a new boiler feed pump must be purchased. In this case, there are several new designs of pumps that have been brought on the market in recent years which will handle hot boiler feedwater, and these should be investigated. Or, it may be possible to elevate the feed tank to permit use of the existing pump.

There are still to be found in some older plants very large boiler feed tanks, which may, if the controls are not properly set, allow the condensate to sit in the tank long enough to cause an appreciable drop in temperature before it is pumped into the boiler. Such tanks are usually associated with sluggish condensate returns, such as those found in gravity return systems where the returns are slow in coming back to the boiler room.

Now that we have the condensate in the feed tank, and the treated make-up water is being supplied into the feed tank to keep a ready supply of boiler feedwater for the boiler demands, there is generally only one more treatment that may be given to the water before it is ready to be put into the boiler.

You will recall that in the chapter on boilers we mentioned that watertube boilers require a constant water feed and that they are very sensitive to water composition. The two requirements are partially solved by the use of a feedwater deaerator. This unit mixes the return condensate with cold make-up water, heats the mixture to a temperature high enough to drive off a large portion of the dissolved corrosive gases, and sprays or splashes the hot mixture into a deaeration chamber. Here the water falls to the storage chamber in the bottom of the unit, while the gases which are released pass through a small condenser, where any water vapor still remaining is condensed and recovered. The gases remaining are vented to the atmosphere, either in a throttled vent or through a thermostatic vent.

A glance at the graph, Fig. 9-4, will show the desirability of operating the deaerator at temperatures above 212°F. The final temperature required will only be determined by experience, and will depend on the make-up water composition and the purity of the return condensate. We suggest 10 psig as a good operating point.

Figure 9-5 shows a typical deaerator installation. As you can see, it is actually a combination of a deaerator, hotwell or return tank, and feedwater heater. In fact, it is often called a deaerating feedwater heater. The trap returns may come back by gravity, or they may be pumped back from condensate tanks at various places throughout the plant. The make-up water and the trap returns are mixed before entering the vent condenser, where the water passes through a shell and tube heat exchanger to condense the last remaining steam in the vent stream before the corrosive gases are discharged to the atmosphere through the vent. The water then drops down into the deaerator section, as described above. The make-up is controlled by a level controller in the storage section, or the hotwell, to maintain a constant level for the boiler feed pump.

The vent from the deaerator may be a thermostatically controlled vent, designed for the purpose, or a needle valve throttled down to produce the desired bleed rate from the vent condenser.

The thermostatic vent will discharge in puffs of vapor and air as the air accumulates ahead of it. The needle valve bleed rate, when used, should be adjusted to give a continual discharge of steam from the vent from 20–30 in. high. This is not to be taken as a waste of energy, since it is necessary for this steam to be discharged. It carries with it all of the corrosive air and gases being released from the feedwater, which would cause trouble in the steam and condensate system. If a plant official happens to mention that there appears to be a waste of steam coming from the deaerator, a brief explanation of the function and importance of the deaerator should be all that is necessary to calm any apprehension.

Figure 9-5 shows a system piped for pressures slightly above atmospheric pressure in the deaerator.

There are several styles and arrangement, of these deaerating heaters on the market, but they all follow the same basic concepts discussed here. We show one commercial unit in Fig. 9-6.

One more thing we should discuss before we leave Fig. 9-6 is the piping from the bottom of

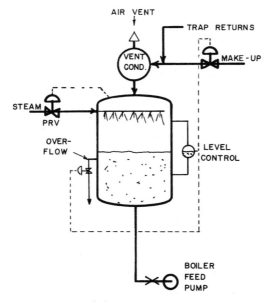

Fig. 9-5. Typical deaerator piping arrangement.

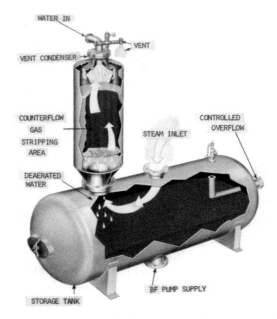

Fig. 9-6. Commercial deaerator and storage tank. (Courtesy of Cleaver-Brooks Co.)

the hotwell to the boiler feed pump. This piping should drop straight down as far as possible before any bends, fittings, or horizontal loops are put in. This is to ensure that the pump will have sufficient pressure at the suction to prevent flashing in the pump suction and in the eye of the impeller.

E. Warning!

The field of water treatment, like any area where there is a constant problem with money available to be spent in its solution, has its fair share of fast-buck artists. They seem to crop up from the woodwork every few years. At least, that has been the history so far, and there is no reason to expect any improvement in the future. The turnover in boiler plant operators is sufficient to produce a new crop of "marks" every few years for these peddlers of magical gadgets. And gadgets are usually all they are, often consisting of one or more special pipe fittings, made of undefined "patented alloys," or other such material. Or they may be a combination of two metals, similar to copper and zinc or magnesium, advertised as being the simplest answer to the water-treatment problems.

There are a few installations that can only be helped by such gadgets, but the range for their application is very narrow. Usually, after an inventor has installed them in a few test locations where they have proven to be of some help, the inventor engages in a promotional campaign to push them for all installations where water is being used. Then, the salesmen take over, and the market is flooded with advertising, claims of magical results, etc.

Then, the failures begin appearing, law suits result, the Federal Trade Commission is probably called in, and the truth comes out. By then, a lot of water-using equipment, including boilers, have been damaged, with expensive results.

There are some installations, as we stated, where a combination of copper anodes alternated with either zinc or magnesium anodes,

inserted in the proper location in a water system, will perform sufficiently to permit the operation to continue indefinitely. But the chances that they will work in your boiler are very remote. They have been known to perform in hotwells and make-up tanks to a limited extent, but this is not necessary if the overall water-treatment approach is properly handled by experts in the field. And the purveyors of these magical gadgets are usually not water-treatment experts, regardless of what claims they may make concerning their background.

For years, it has been well known that corrosion of steel in salt water may be forestalled by installing sacrificial zinc plates or plugs on the steel to take the full force of corrosive action, and thus save the steel from attack. These sacrificial units must be replaced regularly to remain effective.

To date, the only really lasting solution to the problem of impurities in the water has come from the chemical process industries. It is likely that at some future date, other professional disciplines, such as electrical or electronics engineers, may discover an esoteric solution that will revolutionize the water-treatment field. Such breakthroughs do happen from time to time, and we do not intend to leave you with the impression that all such progressive steps are to be ignored completely. It is just that the odds of it happening at this time are very slim.

What we are recommending is that you take the traditional conservative view of the average boiler plant operator, and watch any developments in your field very carefully, be skeptical of wild claims, and if you do decide to take a chance on any new process, be sure you have all of the data and references that are available. Then go into it with your eyes wide open, and be willing to take your lumps should they come your way.

The safest route to good water treatment for the average boiler plant, and the one most often pursued by the medium and smaller plants, is to engage the regular services of one of the local water-treatment-service firms. These are most prevalent in the more highly industrialized areas, such as the larger cities. Those plants in the rural areas may not have access to enough such firms to be sure of getting good and reliable service at a fair price, since there may be very little competition. Most of these firms feature personal service to the plant, with a more or less competent and complete laboratory service and personnel to back up the field representatives.

The success of these services depends on not only the integrity and ability of the service representative, but also on the cooperation of the boiler plant manager who is held responsible for proper operation and maintenance of the equipment. It must be a team operation, and as such, much depends on good communication between both parties. Should the operator not be satisfied with the results, as measured by the reports of the annual inspector, as well as daily observance of the boiler condition, then the best recourse is to inquire of other boiler plants in the area for their experiences and recommendations. The operator should not hesitate to obtain bids from other services, and make any changes necessary to get the proper results in the boiler. Remember that an experienced, diligent representative of a small, local water-treatment firm may be a safer bet than a poorly trained salesman for a large, nationally known firm.

CHAPTER 10
INTERNAL BOILER WATER TREATMENT

A. General Approach

We have just covered the methods and philosophy involved in preparing the boiler feedwater for entrance into the boiler. Now we shall go into the monitoring, treatment, and control of the condition of the water inside the boiler itself. This is the end use of the water, and the one most important aspect of the entire subject of our discussion.

The treatment methods for boiler feedwater and boiler water are subject to trends and fads, just as are many other portions of our existence. However, in the case of boiler water and feedwater, the trends or changes are usually due to a concentrated effort on the part of boiler manufacturers and the various chemical firms engaged in servicing the needs of the industrial and power boiler fields. In many parts of the country, the water available is of unknown composition from day to day, and in other areas, it may be of a fairly stable composition, but still very difficult to treat for our purposes. What is required for treatment in one area may be entirely wrong in other areas. Should you move from one area to another and take up boiler plant operation in your new location, one of the first moves on your part should be to investigate the common methods of water treatment being used in the new area. You may have to reeducate yourself before you can hope to get your boiler operating properly.

B. Feeding Treatment Chemicals

Assuming that you have engaged the services of a water-treatment firm, and are adequately supplied with their various tablets, liquids, powders, etc., and have a complete prescription to follow in treating the boiler feedwater and the boiler water, how do you proceed?

There are several methods in use for feeding chemicals for internal or boiler water treatment. The method depends on the size of the steam plant, the amount of chemicals that must be fed into the water, and the types of chemicals required. The methods vary from simple dump openings in the hotwell or receiver, all the way up through automatically controlled proportioning pumps, which vary the feed rate with the water flow, and may even record conditions at the same time.

We have shown one of the more common systems to be found in the smaller industrial steam plant in Fig. 10-1. This is a pressure pot that may be opened to permit pouring in a supply of chemicals—dry, liquid, or slurry. The pot is then closed off, the water under pressure that is being treated is then applied to the pot, and the outlet valve at the bottom of the pot is then opened to permit the chemical solution to flow into the water line by the difference in pressure between the suction and discharge sides of the pump.

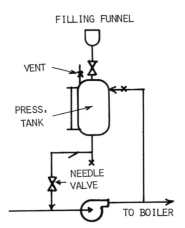

Fig. 10-1. Pot-type feeder.

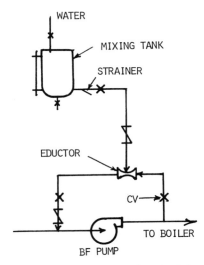

Fig. 10-2. Eductor-type chemical feeder.

There are some precautions to be taken when operating this system, such as operating the valves in correct sequence, and using reasonable care in the handling of the chemicals.

This system is to be installed wherever it is required, but one of the more common places is in the boiler feedwater line ahead of the feed pump as shown. It may also be placed in the make-up line. In either case, the solution must be thoroughly dissolved and mixed with the condensate before it enters the system. Placing the chemical feeder ahead of the boiler feed pump in the manner shown in Fig. 10-1 may cause trouble should the chemical not be thoroughly dissolved when it enters the pump. Lumps of treatment chemical can clog the pump suction, or the concentration may damage the materials in the seals of the pump, or even the pump itself.

Figures 10-2 and 10-3 show two other common methods of feeding water-treatment chemicals into the feedwater. Both systems use a mixing and feed tank that is at atmospheric pressure, and the chemical slurry or mixture is pumped from the mixing tank and then into the boiler. In contrast, the mixing pot in Fig. 10-1 must be capable of withstanding the full feed pump pressure. This usually is no problem, since the pressures encountered in the average industrial boiler plant are not high enough to cause a problem.

Figure 10-4 shows a typical package chemical mixing and feeding system, of a type very

common for this service. Its operation is shown diagrammatically in Fig. 10-3. To mix a batch of feed chemicals, a small amount of water is usually run into the tank from the water line overhead. The chemicals are then measured out and poured slowly into the tank while the proper amount of diluting water is running into the tank. The motor-driven mixer is then started and run until the mixing is complete, then it is shut off.

We should warn you, however, that many of the chemicals used may be hazardous to your health, and extreme caution should be followed in handling them. These materials will usually be accompanied with strict warnings by the supplier, with complete instructions on their use. Be sure and follow these instructions very closely.

It is well to size the mixing tank properly to last for several days between batches, in order to keep the chemical handling to a minimum.

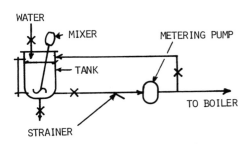

Fig. 10-3. Chemical mix and feed system.

Fig. 10-4. Package chemical feed unit. (Courtesy of Cleaver-Brooks Co.)

You must also keep in mind that the effective life of the chemical mixture may be limited, owing partly to impurities in the water and contamination from the air on the surface of the solution.

Figure 8-1 shows the internal chemical feed going into three possible places in the system. The one actually used should be dictated by the supplier of the chemicals, within certain logical limits. We have already warned you about feeding the mixture into the pump suction. You will no doubt find that the boiler manufacturer has provided a separate connection for injection of treatment chemicals into what the manufacturer has determined to be the best and most effective spot in the boiler. Follow this recommendation if you want to avoid trouble.

C. Treatment Chemicals

We shall give a brief description of a few of the more common chemical treatment compounds that are used to keep the boiler water under control and reduce internal corrosion to

a minimum. Keep in mind that these should be used only under the direction of a water chemist, since the proper balance of chemicals used with the particular boiler plant conditions varies from system to system. It is often very difficult to achieve the perfect result and when it is achieved, it is very easy to lose control of the situation. The battle against corrosion in the boiler and steam system is one of constant vigil and attention to minute details. Here, if any place in the boiler plant, the services of an expert is certainly to be recommended.

In spite of all attempts to remove dissolved oxygen from the feedwater, some usually manages to find its way into the boiler. To guard against this happening, it is recommended that a small surplus of an oxygen-hungry material be carried in the boiler. The most common material is sodium sulfite, and the usual quantity is between 30 and 50 ppm in the boiler water. It increases the total dissolved solids in the water, however.

Another internal treatment chemical used for control of oxygen is hydrazine, usually obtained in a 35% solution. It is similar to ammonia, and will form ammonia in the boiler under some conditions. Hydrazine is one of a family of treatment materials known as "neutralizing amines," since they form vapors which are carried over with the steam and which neutralize any corrosive acids in the condensate lines. Thus hydrazine serves as a double purpose in the boiler. See page 352 in the Appendix for a description of one of the commercial amines available.

Another type of amine is known as "filming amines," which form a protective film on the surfaces of the steam and condensate lines that helps to prevent corrosion.

Along with hydrazine quite often a phosphate compound will be injected into the boiler water. There are three forms available, and the selection depends on the particular conditions existing and the desired results. The two most often used are trisodium phosphate, commonly known as TSP, and disodium phosphate. The amount to maintain in the boiler is generally dependent on the pH of the water and the pressure. There are charts available that indicate the limits to be held in the boiler.

Another common treatment chemical is soda ash, which is sodium carbonate. It is not used as often as formerly, since there are several disadvantages inherent in its use. Its main uses have been to increase the alkalinity of the boiler water and to reduce the scale-forming compounds to a soluble sludge. This sludge is then easily removed by blowing down the boiler.

The disadvantages of the use of soda ash are the formation of carbonic acid from the carbon dioxide released and the possibility of caustic embrittlement of the boiler joints, and, in some boilers with poor circulation, the sludge that is formed causes trouble by reducing circulation still further.

Another addition to the water chemist's bag of tricks are complex compounds known as "chelants" (pronounced kee-lents). These serve much the same purpose as the soda ash, forming soluble sludges from the calcium, magnesium, and iron, which may easily be blown down from the boiler. The use of this material is very critical, however, since there must be no dissolved oxygen present in the water along with the chelants. It is also rather difficult to determine the proper feed rates until the results on the boiler metal surfaces have been observed.

Before we leave this subject, we should warn you again that the use of some of these chemicals, notably the filming amines, is extremely dangerous. They may be poisonous or they may form poisonous compounds in the boiler, and thus if the steam is used in conjunction with food or beverage production, local and state codes may prevent the use of such treatment. If not, common prudence will!

D. Sampling and Testing

One other subject we should take up before we leave the water-treatment field is the matter of testing to determine what results we are getting in our treatment methods.

The water chemists have over the years es-tablished several standard tests for checking the condition of the water at the various points in the steam plant system. The simplest one of all, and the least accurate, is visually investigating the water sample in a test tube and attempting to ascertain what the effects have been in your system. This is only useful for the detection of iron oxide (rust) in condensate, which may readily be detected from its characteristic reddish brown color. If your returns have this color, then it is wise to look further for evidence of oxygen and carbon dioxide.

The other chemicals usually found in the water at the various sampling points do not give any real color changes that permit such a simple test. In this case, one of the well-tried chemical tests will be required, as prescribed by your water chemist. Most of these require some practice in obtaining accurate results, but they are not beyond the average person's capabilities.

Of equal importance to the actual running of the tests, is the matter of where and when they are run. The sample should be typical of the water you are testing and taken from the main stream or pipe run, and not from stagnant pockets.

When sampling steam or hot condensate, care should be taken to see that the steam or condensate is cooled below 212°F before it is exposed to the air. This is to ensure that no steam, live or flashing, is lost when the sample is drawn. If some live or flashing steam is allowed to escape from the sample, the portion remaining will show a higher concentration of solids than actually exists in the original source. It will be almost impossible to accurately compensate for the loss of the steam when making your test.

The tendency in many plants is to take samples at the same time of the day or the same time of the week, continuously. This is most convenient, since it ensures that the samples are taken regularly; otherwise, it is very easy to forget. However, if this is the practice in your plant, then we suggest that you also duplicate some of the tests at different times, from those regularly scheduled. The purpose of this is in case your steam plant is undergoing periodic cycles of concentration and unloading of harmful materials at any point, then the regular sampling schedule may be missing the time of peak concentrations. Vary the sampling schedule frequently to insure against missing any tendency of your system to cycle periodically. By doing this, you may catch a source of trouble that you were previously unaware existed.

Where are these tests usually taken? There are several points that it is wise to consider when you set up the schedule for sampling. Following are some of them, but your water chemist will have something to say about the matter, and we suggest you follow the advice given:

1. Raw water at the plant supply.

2. Immediately downstream from any treatment system.

3. Condensate in the returns to the boiler hotwell.

4. Boiler feedwater before it is slugged with any further treatment chemicals.

5. Boiler water, taken at a point of normal circulation.

Also, if your plant is large enough to be using one of the larger boilers, then your water chemist will probably want you to take steam samples from one or more locations in the steam system. If so, remember to condense the steam before withdrawing the condensate for testing, as we explained a few paragraphs back.

E. Boiler Blowdown

Why is it so necessary to blow down a boiler regularly?

Perhaps the best illustration is to use one of the ordinary stills that are found quite frequently in the industry. Assume there is a distilling vessel into which a solution is fed, either in a continuous stream or in a batch process. This solution is then heated to boiling, and the vapor coming off is condensed in some manner, either for reclaiming or to dispose of it,

depending on the particular process. This vapor is some fairly pure component of the original solution, which is boiled off at a lower boiling temperature than the remainder of the solution. The whole process is one of concentration, as the solution left behind contains all of the thicker components brought in with the feed stock. As the process continues, the solution remaining in the still becomes thicker and thicker. This is because the feed stock is being separated, part of it being removed in the form of vapor, and part remaining behind, which is the thicker portion.

Now if the original feed stock has a concentration of heavy material dissolved in it in the proportion of 100 ppm, and the solution in the bottom of the still is at a concentration of 1,000 ppm, then we say there are 10 concentrations in the vessel.

The boiler is another form of distilling apparatus, in that feedwater is pumped into it, boiling takes place, and steam is produced that contains at the most probably 5 ppm of dissolved solids in it. This steam is piped away for use in the plant, leaving behind the remainder of the dissolved solids that came in with the feedwater. The remaining solids increase in concentration as the process continues, until the result may be troublesome, if something is not done about it.

What is the cure? It is necessary to purge the boiler from time to time in order to get rid of some of the dissolved solids. The thick solution in the boiler that is purged is then replaced by more feedwater at lower solids concentration.

There is a situation in nature quite similar to this process that may help to explain what we are attempting to convey.

Rivers flowing from the inland watersheds are considered to be freshwater, although they always contain considerable portions of dissolved solids. These rivers empty ultimately into the oceans, where a continual evaporation takes place into the atmosphere and this evaporation is pure vapor. The dissolved solids that came in with the river water are left behind in the ocean. This process has been taking place for centuries, so that today the seas are a rather high concentration of dissolved solids, approximately 32,000 ppm or higher. In the case of inland lakes that remain fresh, you will find that there is a continual purging of the lake, probably in the form of a stream or river, which maintains the balance so that the dissolved solids in the lake remain fairly constant. If there is no purging, or if there is insufficient purging, then there is produced a situation like the Great Salt Lake, Salton Sea, the Dead Sea, or, in extreme cases, our oceans.

How do we determine what amount to blowdown from the boiler each time it is done in order to maintain the desired results inside the boiler? Let us assume that we wish to maintain the boiler water at 3,500 ppm total dissolved solids, with the feedwater being supplied at 50 ppm total dissolved solids. We shall assume further that the steam leaves the boiler with sufficient carryover to contain a total of 5 ppm of dissolved solids. This means that in one million pounds of steam leaving the boiler there are 5 pounds of dry solids dissolved in it. Likewise, there are 50 pounds of dry solids dissolved in one million pounds of feedwater, and the total dissolved solids in the boiler water are the equivalent of 3,500 pounds of dry solids dissolved in 1,000,000 pounds of boiler water.

These figures in ppm of total dissolved solids are known also as "concentrations," and the formula for determining the amount of blowdown is as follows:

$$\text{Blowdown volume} = \frac{(\text{Feedwater concentration} \times \text{Feedwater volume})}{\text{Boiler water concentration}}$$

Figure 10-5 shows this example in another manner.

There are several things we wish to point out concerning this formula:

1. The amount of blowdown is based on the amount and concentration of the feedwater, over a definite period of time.

2. The total dissolved solids leaving with the steam is ignored, since the amount is so

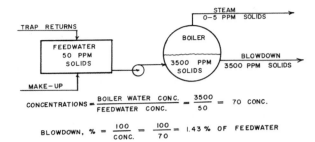

$$\text{CONCENTRATIONS} = \frac{\text{BOILER WATER CONC.}}{\text{FEEDWATER CONC.}} = \frac{3500}{50} = 70 \text{ CONC.}$$

$$\text{BLOWDOWN, \%} = \frac{100}{\text{CONC.}} = \frac{100}{70} = 1.43\% \text{ OF FEEDWATER}$$

Fig. 10-5. Principle of boiler blowdown.

small as to make practically no difference in the amount of blowdown.

3. The blowdown concentration is the same as the concentration in the boiler water, since it is boiler water that is being purged, or blown down.

4. The units of measurement must be consistent, so that if the feedwater is stated in gallons, then the blowdown is in gallons, over the same period of time taken for the measurement for the feedwater. If the feedwater is in pounds, then the blowdown required will be in pounds. If the feedwater is in inches on the gauge glass, then the blowdown must likewise be in inches on the gauge glass.

Study the preceding examples and equations for a few minutes. There are a few practical aspects that should be pointed out at this time.

This points out the desirability of maintaining a tight steam and condensate system, since the aim of all plants should be to reduce the amount of blowdown to a minimum. This is done by returning as much of the condensate as possible to the boiler, and treating the make-up to remove as much of the dissolved solids as economically reasonable.

The question now arises, how do we measure the amount of make-up water over a certain period of time? This varies, and in some cases it may not even be practical.

Actually in most cases, no attempt is made to measure it accurately, but the results are measured by testing the boiler water periodically, and then altering the blowdown program to make whatever corrections are indicated, until the correct results are obtained, or approached.

F. Intermittent Blowdown

We shall describe the most common style of blowing down the boiler, one that is used in the small to moderate sized plants. We refer you to Fig. 10-6, which illustrates the typical intermittent blowdown system used in the average steam plant, providing the system is properly installed.

The blow-off line is at least extra-heavy steel pipe, welded, and it leaves the boiler at the locations recommended by the boiler supplier. There are two valves in the line, as specified in the ASME Code.* The first valve may be a quick-opening valve designed for this purpose. The second valve is also designed for this purpose, and is the valve that controls the actual blowing down. The first valve is a safety precaution and it permits maintenance personnel to work on the blowdown valve while the boiler is in service.

Downstream from the blowdown valve, the piping may be standard weight pipe, since it contains no appreciable pressure. This line leads directly into the blowdown tank, which is an ASME Code tank, vented to the atmosphere as shown. As shown in Fig. 10-7, the inlet line is directed tangentially into the top portion, and the entrance to the outlet line to the sewer is

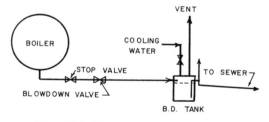

Fig. 10-6. Intermittent blowdown.

*This refers to the American Society of Mechanical Engineers, Boiler and Pressure Vessel Code, Sections I and IV. Whenever the term "Code" is used hereafter, it shall mean whichever of these two sections are applicable, unless otherwise qualified.

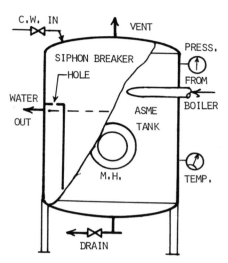

Fig. 10-7. Blowoff tank arrangement.

near the bottom of the tank. This means there is always a tank full of water in the vessel. The blowdown tank is sized so that this water that is retained in the tank is equal to at least one boiler blowdown, and preferably two. The reason for this is that, between blowdowns, the water retained in the tank cools down, and when the boiler is blown down again, this cool water is the water discharged to the sewer.

Most communities now have restrictions covering the maximum temperature of the water that may be discharged to the sewer, and this is usually around 120–140°F. In case the water is still too hot to discharge into the sewer, then it may be necessary to install a water cooling line as shown, which permits feeding cold water into the tank.

The procedure for blowing down the boiler is as follows:

1. Open the stop valve ahead of the blowdown valve.

2. Open the blowdown valve, fully keeping an eye on the level gauge glass.

3. When the correct amount of water has been blown down, as measured on the gauge glass, close the blowdown valve.

4. Close the stop valve.

There are some installations in which the preceding procedure may be modified, depending on the blowdown equipment furnished by the boiler manufacturer. The boiler operator should read the instructions furnished with the boiler and follow the recommendations given therein.

In the absence of any instructions by the boiler manufacturer, and if two blow-off valves in series are installed, then the method described here should be followed.

Warning! Never blow a hot boiler down so far as to uncover any of the tubes or surfaces exposed to hot gases. And never leave the blowdown valve unattended while it is open and blowing down. If an "emergency" occurs while blowing down, always take the time to close the blowdown valve before taking care of the "emergency" situation.

G. Continuous Blowdown

The intermittent method as just described has several disadvantages, especially in the larger steam plants and where the water conditions must be carefully controlled. The method described causes surges in the boiler conditions that may upset the steam plant operation, since the boiler water concentration varies considerably between blowing down and making up the amount blown down. Also, there may be considerable heat thrown away in the water blown down, if an appreciable amount has to be discarded over a period of time, as would be the case if there is a high percentage of make-up required. In addition, the blowdown water contains valuable treatment chemicals being discharged to waste.

In such cases, the continuous blowdown is usually installed. See Fig. 10-8, which we shall describe as being one of the more basic styles in use today. There are several variations available, depending on the degree of sophistication required for the size and application of the installation.

Basically, the blowdown is controlled to leave the boiler in a steady stream, throttled to obtain the proper amount of blowdown required. The water blown down is first piped into a flash

tank to reclaim some heat from the flash steam. The remaining water is then piped out into a heat exchanger, where it passes some of its heat to the boiler feed water, so that no heat is lost from the system owing to the amount of the blowdown.

Note the sample cooler, which follows the procedure described earlier in this chapter on taking samples properly.

The "flash" boilers, described earlier in the chapter on boilers, provide a particular problem when it comes to blowing down the coils. This problem was one of the chief obstacles which had to be overcome before this style of boiler could be considered commercially practical. Most of the designers of this style of boiler have settled on a variation of the constant blowdown, adjustable, and operating as a percentage of the feedwater flow. The percentage of blowdown is adjusted for each installation to suit the particular water and steaming rate conditions. Once set, it should require no further change until the operating conditions are changed.

One of the most common ways of providing the blowdown adjustment on the "flash" boiler is to have a valve operating off the boiler feed pump, which automatically changes its adjustment as the boiler feed pump varies the supply. In this manner, the blowdown rate is almost perfectly proportional to the steaming rate.

The location of the blowdown connection on any boiler is very important, and is always determined by the boiler designer. It is not a good idea for the plant operator or maintenance man to make any changes in that location without the advice of the designer.

Generally, in those boilers that contain sluggish spots where the sludge may accumulate, there will be a blowdown line from that point. This point is found most often at the bottom of firetube boilers, such as in the typical Scotch design of the package boiler.

In those cases of watertube boilers that have good circulation, it has been found that the point of highest dissolved solids concentration is near the center of the drum, a few inches below the waterline. The practice is to place a

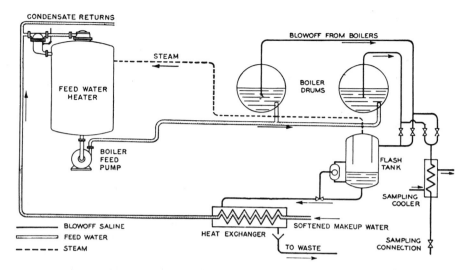

Fig. 10-8. Continuous blowdown. (Courtesy of Permutit Co.)

pan or perforated tube just beneath the surface of the water, with the blowdown connected to the pan. This is known as a surface blowdown system, and it is usually continuous.

All watertube boilers with continuous blowdown will also have a connection at some low point for blowing down sludge and dirt intermittently. In those plants where water condition is nearly perfect, this bottom blowdown may never be used.

Each design of boiler will have its own particular blowdown location, of course, and the preceding statements are meant to cover only the majority of the cases in use today.

In case the boiler operator is unsure which valves on the boiler are meant for blowdown service, remember:

Bottom, or intermittent, blowdowns will have two valves installed, back-to-back, in series, requiring both valves to be opened when blowing down.

If only one valve is installed, it is designed as a cold water drain for that portion of the boiler, and is to be used for maintenance only.

In Fig. 10-9, we have illustrated the method for calculating the amount of water per inch of depth in a horizontal boiler drum at any level. The dimension W is the actual width of the water level in the drum. Both dimensions in the equation are in inches, and the result will be in gallons of water for 1 in. of change in water level. The method is most accurate at the center line of the drum, and the accuracy reduces as the water level moves from the center line.

Figure 10-10 is the same approach for a much simpler case, with the boiler drum vertical. Here the amount of water represented by 1 in. change in level does not vary with the water level, as it does in the horizontal drum.

These two illustrations are for the purpose of assisting you in calculating the amount of water in the boiler for determining the amount of blowdown. We make no compensation for internals in the drum, but assume that the entire inside of the drum at the operating level is occupied with water. Some boiler manufacturers will provide in their literature the gallons of water per inch of difference in water level.

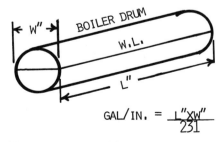

$$\text{GAL/IN.} = \frac{L'' \times W''}{231}$$

Fig. 10-9. Computing blowdown in horizontal drums.

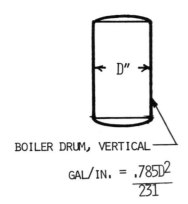

BOILER DRUM, VERTICAL

$$GAL/IN. = \frac{.785D^2}{231}$$

Fig. 10-10. Computing blowdowns in vertical drums.

H. Blowdown Piping

Most boiler codes state that any boiler operating at more than 100 psig shall be fitted with two slow-opening blowoff valves or one slow-opening and one quick-opening valve, with the quick-opening valve closest to the boiler. A slow-opening blowoff valve is one that requires five full turns to operate it from full open to full shut. The operation of the blowoff valves are explained in Section F.

Ordinary globe valves are not to be used as boiler blowoff valves at any pressure.

The intermittent blowoff piping shall be connected to the lowest point where there is water in the boiler. The minimum size is to be 1 in. and the maximum size shall be $2\frac{1}{2}$ in.

When the maximum allowable working pressure of the boiler exceeds 125 psig, the blowoff pipe from the boiler to the first valve shall be a minimum of extra-heavy steel pipe, or it shall be pressure rated to the working pressure of the boiler, whichever results in the heaviest walled pipe. This is in accordance with standard B31.1 of the American National Standards Institute (ANSI), which is the basic standard for piping in power boiler applications.

No boiler blowoff shall discharge directly into the sewer. When the blowoff discharge is ultimately to be bled to the sewer, local codes shall be consulted concerning the requirements for discharging hot products into the sewer, as mentioned earlier.

C H A P T E R 1 1
COMBUSTION CONTROLS

A. General Requirements

When considering the use of large amounts of concentrated power, it is essential that consideration be given first to the following elements in the use of that power:

1. Obtaining the energy in usable form.

2. Storing and handling it safely.

3. Adequate and safe controls over its use.

4. Safe disposal of the by-products.

In this chapter we shall explore the matter of safe and adequate controls over the use of power. The other three elements in the list are covered in other portions of this book, either in a direct manner or in allusions to them under other subjects.

The manner in which the concentrated power is released depends on the purpose to which it is to be put, and in our case it is the production of steam, vapor from some other liquid, such as Dowtherm, or heating some fluid such as water, or a combination of these. As the demands for the end product generally varies, as determined by a completely separate system of process control, it is necessary that the release of energy in the furnace be also controlled to follow the load demand as closely as reasonably possible. This was covered briefly in the first chapter, under the general description of the operation of the steam plant cycle.

Over the many years of the industrial revolution that started this vast use of steam power, there have been many incidents and accidents, some of them of disastrous proportions, which have caused the insuring agencies involved, as well as governmental agencies, to formulate regulations covering the use of boilers and their fuels. We learn by painful experience, and the more pain, the quicker the reaction and production of safer operating procedures. Today we have several relatively safe, but not foolproof, methods of controlling the combustion of many types of fuel in our modern boilers and furnaces. As you may have already noted, the number and complexity of the controls varies almost directly with the size of the equipment, from the small dry cleaning boiler to the large nuclear generating stations.

We shall start by listing a few definitions that will help you in absorbing the material in this chapter.

Burner management system: That portion of a boiler control system associated with the supply of fuel to the burners. This includes the complete fuel train, safety shut-off valves, fuel pressure and temperature limits, burner starting and sequencing logic, and annunciation of trouble signals.

Combustion control system: The portion of a boiler control system associated with the control and maintenance of air/fuel mixtures throughout the operating range of the burner and during changes in firing rate.

Igniters: A term used in industry to denote the device that provides the proven ignition energy required immediately to light off the pilot flame.

Interrupted pilot: A gas or oil pilot that burns during light-off and that is shut off during normal operation of the main burner.

B. Insurance and Code Requirements

All boiler and industrial furnaces or heaters are a source of potential damage and resultant lawsuits, and, consequently, all plants in which they are installed are today covered by insurance, or should be. There are several insuring agencies that serve the industries not only with liability coverage, but also with advice and counseling on the proper installation and use of potentially hazardous equipment and process. Although most of the insuring agency's requirements are not backed up by a government agency, the resultant insurance premiums levied, should a firm choose to ignore their recommendations, make the use of nonconforming installations prohibitive. In fact, most insuring agencies will refuse to carry liability insurance on equipment or installations that do not meet the standard set down by at least one of the major supervising agencies, as listed below:

ASME (American Society of Mechanical Engineers) Boiler and Pressure Vessel Code, Section IV, for steam boilers with design pressure not exceeding 15 psig; hot water boilers operating at pressures not exceeding 160 psig, and operating temperatures not exceeding 250°F.

ASME Boiler and Pressure Vessel Code, Section I, titled, "Power Boilers," for boilers with design pressures over 15 psig and hot water boilers for pressures over 160 psig or operating temperatures over 250°F.

Factory Mutual Engineering Corp. (FM).

National Fire Protective Association (NFPA).

Industrial Risk Insurers (IRI).

There are other similar agencies throughout the world, and some in various sections of the country, which may have a local jurisdictional following. It is up to the individual plant engineering department, through the firm's insurance carrier, to ascertain the one that must be consulted.

In addition, there are usually local code requirements covering the boiler plant operation, equipment, and accessories. These must be followed very closely, of course, if the firm expects to remain in operation for long periods of time without interruptions forced by noncompliance.

There are several minor differences between the preceding agencies, FM, IRI, NFPA, and others. Furthermore, the scope of coverage varies, also. However, they generally cover similar size ranges, for boilers having single burners.

There are many small boilers in use today, which are serving a definite purpose in industry, that are too small to be included in the general material covered here. Such units are usually covered under the NFPA bulletin 31, Installation of Gas Appliances–Gas Piping. The principles discussed in this book are equally applicable to these small boilers in most instances. The number and complexity of controls will be much simpler, of course, and some operations described here for the average boiler unit may be simplified, done manually instead of automatically, or may be omitted entirely.

Tables 11-1 and 11-2 following will give you a general idea of the requirements of burner and combustion controls for FM and NFPA, and along with those listings you will find a description of the IRI requirements.

The reader is advised, however, to check with the insurance carrier for the latest information, as this matter is a subject for continual monitoring and updating, and there may be recent changes in these lists. We give them here for basic information only, to be used as a guide and to illustrate their use.

Table 11-1

Insurance Requirements Automatic Gas and Oil Burners Single Burner Boiler-Furnaces	Factory Mutual Requirements												Industrial Risk Insurers Requirements						
	FM1 Burners under 10.5MM Btu/hr (250 BHP)							FM2 Burners 10.5MM Btu/hr (250 BHP) and Larger											
	Gas	Light Oil (Mech. Atomized)	Oil (Air or steam Atomized)	Heavy Oil	Gas and Light Oil (Mech. Atomized)	Gas and Light Oil (Air or steam Atomized)	Gas and Heavy Oil	Gas	Light Oil	Heavy Oil	Gas and Light Oil	Gas and Heavy Oil	Gas	Light Oil (Mech. Atomized)	Oil (Air or steam Atomized)	Heavy Oil	Gas and Light Oil (Mech. Atomized)	Gas and Light Oil (Air or steam Atomized)	Gas and Heavy Oil
Low water cut-off	•	•	•	•	•	•	•	•	•	•	•	•	•	•	•	•	•	•	•
Safety high input pressure or temperature control	•	•	•	•	•	•	•	•	•	•	•	•	•	•	•	•	•	•	•
Nonrecycling flame safeguard	•	•	•	•	•	•	•	•	•	•	•	•	•	•	•	•	•	•	•
Prepurge with four air changes	•	•	•	•	•	•	•	•	•	•	•	•	•	•	•	•	•	•	•
Blower air interlock (air flow and mag. starter)	•	•	•	•	•	•	•	•	•	•	•	•	•	•	•	•	•	•	•
Induced draft fan interlock													•	•	•	•	•	•	•
Enforced low fire start	•	•	•	•	•	•	•	•	•	•	•	•	•	•	•	•	•	•	•
Spark ignited interrupted gas pilot	•	•	•	•	•	•	•	•	•	•	•	•	•	•	•	•	•	•	•
Double pilot gas valves and N.O. vent[a]													•	•	•	•	•	•	•
10 sec pilot proving	•	•	•	•	•	•	•	•	•	•	•	•	•	•	•	•	•	•	•
10 sec main burner ignition	•	•	•	•	•	•	•	•	•	•	•								
10 sec main burner ignition												•	•	•	•	•	•	•	•
Low gas pressure switch	•				•	•	•	•			•	•					•	•	•
High gas pressure switch	•				•	•	•	•			•	•					•	•	•
Double gas valve with normally open vent valve												•					•	•	•
Main gas shutoff valve with "closure" switch								•			•	•							
Manual test valve in gas line	•				•	•	•	•			•	•					•	•	•
Bubble test between SSV and manual valve in gas line	•				•	•	•				•	•							
Low oil pressure switch		•	•	•	•	•	•		•	•	•	•		•	•	•	•	•	•
High oil temperature interlock																•			•
Low oil temperature interlock																•			•
Main oil shutoff valve with "closure" switch									•	•	•	•							
Atomizing medium interlock (steam or air)			•			•			•	•	•	•		•	•	•	•	•	•
Spring-loaded oil shutoff valve with 165°F fusible link		•	•	•	•	•	•		•	•	•	•							

[a] For pilots over 120,000 Btu/hr.

Source: Industrial Risk Insurers.

91

Table 11-2. Code Requirements—Factory Mutual Engineering Corp. (FM), Industrial Risk Insurers (IRI), National Fire Protection Association (NFPA)

	FM Automatic Lighted Boilers	IRI Automatic Single Burner	NFPA 85
Prepurge	• Air changes 50% avg. Air flow—supervised • Automatic dampers which close during downtime on natural draft boiler must be opened 90 sec prior to each start.	• Air changes @ 60% minimum. Air flow—supervised	• Eight air changes reaching not less than 70%. Air flow—supervised
Pilot Proving	10 sec	10 sec max	10 sec
Main Flame Trial for Ignition (Oil)	• #2 & 4–10 sec • #5 & 6–15 sec	• #2 & 4–15 sec • #5 & 6–30 sec	• #2 & 4–10 sec • #5 & 6–15 sec
Main Flame Trial for Ignition (Gas)	10 sec	15 sec max	10 sec
Flame Failure Response Time	2–4 sec	2–4 sec	2–4 sec
Interlocks Required	• Minimum air flow • Low oil temperature • Low fire start • Prove fuel valve closed prior to and during purge (over 250 hp) **Nonrecyling** • High & low gas press. • Low oil pressure • Atomizing medium	• High & low oil temperature • Low fire start **Nonrecylcing** • Supervised atomizing air or steam • High & low gas pressure • Low oil pressure • Minimum combustion air • Rotary cup—power outage to motor	• Low fire start • Manual start from cold start. • Prove safety shut-off valve closed. **Nonrecycle** • Power failure • Loss of control system activating energy • Low oil pressure and temperature • High and low gas pressure • Atomizing medium • Minimum air flow
Other	• Code applicable to gas-fired boilers 50 hp and over. Oil-fired boilers 100 hp and over		• Code applicable to single burner, oil or gas fired watertube boiler furnaces, 10,000 lb/hr steam and over

Source: Fireye Division, Allen-Bradley Co.

BURNER MANAGEMENT FOR SINGLE BURNER GAS FIRED AND OIL FIRED BOILERS AND FURNACES (BY INDUSTRIAL RISK INSURERS)

Purpose

These recommendations are for the guidance of Industrial Risk Insurers' personnel in determining risk evaluations for property insurance and are intended to supplement, not supplant, current federal, state, and local regulations.

The objective of this publication is the reduction of the probabilities of a fuel explosion within the boiler furnace by the provision of instrumentation, interlocks, and control valves applicable for the fuels being used and for its intended operation.

Scope

A boiler within the scope of this publication is a self-contained firetube or watertube boiler which is oil or gas fired, and is built in conformance with the ASME boiler construction code to furnish hot water or steam for heating, processing, or power. Further, boiler-furnaces considered within the scope of this publication are limited to those fired by a single oil or gas burner, or by a burner designed to use either fuel, having capacities greater than 12 horsepower (approximately 400,000 Btu per hour) and less than 300 horsepower (approximately 12,500,000 Btu per hour or 10,000 lb of steam per hour). For larger units, refer to NFPA 85, "Standard for Prevention of Furnace Explosions in Fuel Oil- and Natural Gas Fired Single Burner Boiler-Furnaces."

Only a single fuel is considered on main burner operation. Transfer to an alternate fuel is to be accomplished by shutting off first fuel and then following a normal light-off sequence including prepurge, before firing the alternate fuel. An interlock is to be provided to require this procedure prior to any change of fuel.

General

The importance of a well maintained industrial boiler or furnace installation is recognized by insurance companies. Where a plant does not have a staff trained in the complete maintenance of their boiler, a maintenance contract should be established with the boiler company or other organization which is knowledgeable regarding all systems associated with it.

Recommendations Applicable to Burner Management Systems

1. All operating, control, and supervisory devices to be approved and listed for their intended use by a nationally recognized fire-testing laboratory, such as Underwriters' Laboratories, Inc.

2. Electric equipment:

   a. Installation of electrical equipment to be made in accordance with applicable sections of the National Electrical Code.

   b. Safety-control and interlock circuits to be single-phase, ac, two-wire, one side grounded, operating at a nominal 120 volts, having all contacts in the ungrounded conductor and with suitable overcurrent protection.

   c. The safety-control circuits to be arranged so that a short circuit to ground or opening of a supervisory circuit or limit contact will result in safety shutdown. All such devices to be placed on scheduled cleaning program to ensure their reliable operation.

3. Low water cutoff to be provided and interlocked to cause a nonrecycling safety shutdown when a low water condition exists. This low water cutoff is to be in addition to any water level indicators used to maintain water level in the boiler.

4. Excess pressure cut-off switch to be provided for a steam boiler and interlocked to cause a nonrecycling safety shutdown in case of abnormally high steam pressures. This switch is in addition to operating pressure controls.

5. Excess temperature cut-off switch to be provided for a hot-water boiler and interlocked to cause a nonrecycling safety shutdown in case of abnormally high water temperatures. This switch is in addition to operating temperature controls.

6. Preignition purging to be provided for the combustion chamber, boiler passes, and breeching. The airflow to be supervised during the entire purge period. The purge shall provide at least four air changes of the above volumes accomplished at an airflow rate of not less than 60% of the airflow which occurs when the boiler is operating at maximum firing rate.

7. Fuels to be free from all residue and other foreign materials. Suitable and adequate cleaning equipment such as filters, strainers, drip legs, or water separators to be provided where necessary.

8. Fuel pressure supervision to be provided by pressure switches interlocked to accomplish a nonrecycling safety shutdown in the event of:

   a. High fuel gas pressure, or

   b. Low fuel gas pressure, or

   c. Low fuel oil pressure.

9. Forced and induced draft fans (including combustion air blowers) to be supervised to ensure safe minimum air flow through the combustion chamber and boiler passes. This is to be accomplished by the provision of both an airflow proving switch and a motor starter interlock arranged to provide a nonrecycling safety shutdown.

10. Suitable limit switches to be provided and interlocked to ensure that the fuel–air proportioning dampers and burner controls are in the low-fire start position before the burner can be fired.

11. Stack dampers should be avoided, but where necessary, they should be interlocked to be open fully during purge, light-off, and operating periods.

12. Spark-ignited interrupted flame igniter to be provided. It is to have stable flame over the entire operating range of the main burner. The proved igniter to be adequate and properly located to reliably light the main burner. Gas igniters having ratings above 120,000 Btus per hour to be equipped with two safety shut-off valves with vent line between them. Igniter burners using No. 2 fuel oil may be permitted for ignition of fuel oil-fired main burners. *Note:* Under certain circumstances intermittent igniters or supervised manual igniters may be permitted. Permission for these special conditions should be obtained through a regional office of Industrial Risk Insurers.

13. Nonrecycling combustion flame safeguard, suitable for the fuel being used,

to be provided to supervise the igniter and main burner flames. The combustion safeguard is to provide a self-check of its electronic components at least once each ignition cycle. Combustion safeguards providing continuous self-checking should be provided for important boilers or for boilers having prolonged flame-on operating periods.

14. Timed trial-for-ignition to be provided for both the igniter and main burner. Igniter burner trial-for-ignition period not to exceed 10 sec. The main burner trial-for-ignition period to be as short as practical but not to exceed 15 sec.

15. Automatic restart after an electrical power failure is permitted provided, upon restoration of electrical power, the control system is instantaneously reset to provide a normal automatic light-off sequence including preignition purging; otherwise, a nonrecycling safety shutdown is to occur.

16. Alarms indicating interruption to the safety-control circuit to be provided and transmitted to a suitable location where there is constant attendance.

17. When two or more boilers are connected to a common breeching or stack, each connection is to be provided with dampers. Supervision and interlocks to be provided for the dampers requiring them to be in the open position during the purge, light-off, and firing periods.

18. Safety shut-off valves to be provided for the main burner supply lines.

   A. For gas fuels

      1. A safety shut-off valve of the manual* reset type to be provided in the main gas line to the burner. Another safety shut-off valve to be provided downstream from the manual* reset valve and as near

---

*Motor-driven reset safety shut-off valves to be used for automatic boilers.

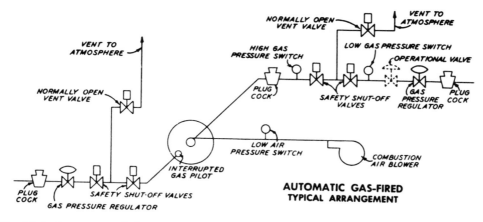

Fig. 11-1. Recommended gas piping. (Courtesy of Fireye Division, Allen-Bradley Co.)

the burner as practical to prevent excessive fuel flow into combustion chamber after shutdown. A normally open, fully ported, electrically operated valve to be provided in a vent line connected between two safety shut-off valves. The vent pipe shall be run to outside atmosphere.

The above combination achieves:

a. Additional security against fuel leakage into the combustion chamber when the burner is off. The vent valve relieves any leakage through the seat of the first valve and prevents the development of a pressure upon the seat of the second valve. Any leakage through the seat of the second valve without an acting pressure would be due to gravity flow and in some instances an induced draft within the boiler-furnace.

b. The independent valves and valve operators provide two opportunities to achieve a valve closure when a shutdown is indicated by the safety system.

2. Permanent and ready means for making easy, accurate, periodic tightness tests of the main burner gas safety shut-off valves to be provided in the piping. This includes appropriate bleed valves located downstream of each safety shut-off valve, and a lubricated, manually operated plug cock located downstream of both safety shut-off valves.

B. For oil fuels:

1. A safety shut-off valve or valves to be provided. Two safety shut-off valves should be provided when No. 2 oil is used or when:

a. Oil pressure acting upon the seat of the safety shut-off valves exceeds 125 psi.

b. Oil pump operates without main oil burner firing regardless of pressure.

c. Oil pump of a combination gas/ oil burner operates during gas burner operation.

2. Supervision of the atomizing medium, interlocked to close the oil safety shut-off valve, is required. Supervisory devices which are effective in detecting abnormal conditions in the atomizing systems to be used. (A large percentage of

fuel-oil fired boiler explosions have been traced to failures involving the atomizing systems.)

3. A circulating by-pass to the storage tank to be provided downstream of an oil preheater to prevent overheating of the oil during burner "down" periods.

4. Low- and excess-temperature switches to be provided for all preheated oil systems to ensure satisfactory fuel oil temperatures for proper burner operation. The supplier of burner equipment to indicate proper temperature limits.

C. Flame Safeguard Control Panels

Now that we have established the requirements for the equipment for lighting off and maintaining safe combustion with automatic controls, or in some cases, by carefully followed manual procedures, we shall investigate the methods by which this is done automatically.

Various firms closely related to the boiler trade over the years have generally followed the developments established by the insuring agencies and the code committees. The result has been numerous systems being made available to perform the functions required with the minimum of trouble and maximum of safety.

Figures 11-3, 11-4, and 11-5 give a typical program for a gas-fired burner. This procedure has been simplified to contain only the essential elements, but for purposes of illustrating the method, it is satisfactory.

The total program has been split into three sections to make it easier to follow. Please remember that each section of the system feeds into the next one, to take the process through from the "Call for steam" to either of two modes of shut down.

Starting with Fig. 11-3, with a cold system, the power must be available to the panel, before the "Start" button is activated, thus calling for steam. Before the burner will light, however, there are several steps that must be covered first, to ensure safe operation.

Now that the "Start" button has been activated, the interlocks must all be in "Go" position in order for the fan switch to close for the purging cycle to commence. This is shown in Fig. 11-3 and the first part of Fig. 11-4, as the "Fan Starts" sequence.

After the purging cycle is completed, the system enters the pilot ignition phase. If, after the spark for igniting the pilot has not produced a pilot flame, the entire system shuts down, and the operator is back to "square one," meaning that the whole cycle must be restarted from the beginning.

If the pilot flame is detected by the system,

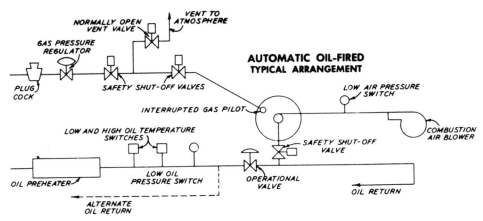

Fig. 11-2. Recommended oil piping with gas pilot. (Courtesy of Fireye Division, Allen-Bradley Co.)

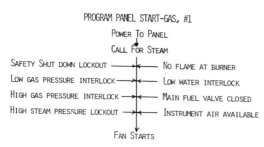

Fig. 11-3. Typical fan purging control sequence.

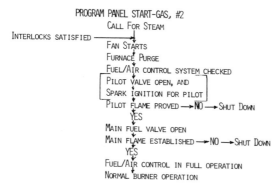

Fig. 11-4. Typical ignition phase control sequence.

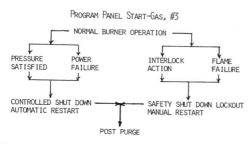

Fig. 11-5. Typical automatic shut-down phase control.

the main fuel valve closes slowly, and when the main flame is out, the pilot is extinguished, and the postpurge fan cycle is started.

If the pressure control on the steam main has throttled the main fuel valve to the burner as far as it will go, and the steam pressure control still calls for further fuel pressure reduction, then the main gas valve closes completely, and the system goes into the postpurge phase, as before.

In the last mode just described, the panel resets itself to permit automatic start-up, should the steam pressure control call for more pressure.

During normal operation, should the main flame fail, or any of the safety interlocks detect a defect in the circuit or the burner system, then the main gas valve closes completely, the system goes into the postpurge phase, and the entire system shuts down. It must then be started manually, but only after the trouble has been located and cleared.

The preceding description is subject to many variations and we caution the reader to consult the instructions for the particular control system in use. The variations will generally depend on the size of the system, the fuel used, the code and insurance carrier requirements, and the personal desires of the engineer who designed the steam plant and wrote the specifications.

The panels available will vary from just the barest system of timers, relays, switches, and flame detection, to one of the more complex systems with pilot lights, alarms, flowmeters, and power meters.

Figure 11-6 shows a typical programmed control panel as supplied by one manufacturer for this service. It is one of the simpler arrangements, some of which may include other controls for the boiler system in addition to the flame safeguard controls.

The new generation of boiler control and monitoring systems have been designed around the electronic module concept, and are more compact and able to perform more varied functions. Typical of this trend are the two systems described briefly on pages 335 and 336 of the Appendix.

then the main fuel valve opens on low flame position. If the main flame does not prove out, then again the operator is back to "square one." However, if the main flame is proven, then the fuel/air control and the entire system of controls come into normal operation, with the burner coming up to full flame.

Figure 11-5 shows the various methods by which the boiler may be shut down. First, if the operator desires to shut the boiler down manually, all the operator has to do is push the "Stop" button, the system will detect this as a power failure, and the panel will enter into the controlled shut-down phase. This means that

Fig. 11-6. Typical burner programming panel. (Courtesy of Fireye Division, Allen-Bradley Co.)

It should not be necessary to remind the reader that the programmed control system is a complicated arrangement of controls, and that only one skilled and trained in its maintenance should be allowed to service it.

Servicing of the burner programming panels has in some cases caused problems. One difficulty is deciding who is to be responsible for the panel's maintenance. There are only three logical choices; in-house boiler plant operating personnel; in-house electrical personnel; or outside service organizations, such as the panel supplier. The final decision will no doubt be based on number and complexity of the control panels involved, training and expertise of in-house personnel, and availability of reliable outside servicemen.

The problem entails not only the testing and repair of the equipment involved, but also the supply of all necessary parts and the keeping of all service records.

Whatever the decision, there should be an all-out effort properly to train or instruct whoever is to perform the maintenance functions in the technical aspects and in the plant policies regarding spare parts supply and inventorying. Also, the policy concerning the entire servicing operation should be in the form of a written memo, kept on file in the engineering department, and posted on the bulletin board in the appropriate place. This is to ensure a continuation in the policy when personnel turnover takes place.

D. Flame Detectors

From the previous discussions, it is obvious that one of the key elements in the entire combustion control system is the device for detecting the presence of a flame, both at the pilot burner and at the main burner. Here, again, there have been many improvements over the years in the approach to this problem, and we have several makes and styles available, all designed to do a specific job, and all fairly reliable.

Flame detectors for boilers operate on two major principles, which are

1. *Flame rods*, in which a rod inserted into the flame becomes an active part of an electrical circuit. The elevated temperature of the flame causes the rod to act as a rectifier when an alternating current is imposed across the circuit.

2. *Optical detectors*, such as infrared sensors, ultraviolet sensors, and visible light rectifying phototubes, each of which has distinctive characteristics, as will be described in the following paragraphs.

Figure 11-7 shows the operating ranges of the various types of flame detectors used on boiler flame safety systems. The bottom scale covers the light wavelengths involved, starting with the shorter ultra violet (UV) rays, through the visible rays, into the longer infrared rays.

The flame rod would appear to be a universal selection, since it covers the entire range of light waves. They operate on the ability of ionized gases in a gas flame to rectify an alternating current to produce a direct current, which operates the flame safety system. They

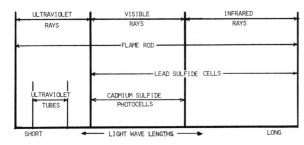

Fig. 11-7. Flame spectrum versus scanner response.

can be used only on gas flames, and must be very precisely placed to do the job properly. They can be used only on the smaller burners, and care must be taken to ensure the porcelain insulator is kept clean and dry. Too hot a flame will cause the flame rods to droop and shut down the system.

The cadmium sulfide photocells are not used too frequently, since they operate by sensing visible light rays, regardless of where the light rays come from or the intensity of the rays, which causes the photocells to be very fickle.

The lead sulfide cells operate by detecting the range of flickering frequencies from the visible through the infrared ranges. They must be placed to see the region of highest turbulence in the flame, which is usually the base of the flame, and that is true for all expected firing rates. Therefore, the primary combustion zone to be monitored by the scanner must be the only one within the field of view of the scanner.

Probably the most popular flame detector is the ultraviolet tube, which detects only the short rays emitted by all flames, and probably one of the better known ones, as well as one of the earliest to be placed on the market, is the "Purple Peeper," so called because of its distinctive color.

These detectors will pass a direct current through the cell as long as UV rays enter the cell. When UV rays stop entering the cell, the flow of current stops, and the burner shuts down. They are insensitive to infrared and visible rays, as well as reflected rays from hot refractories. They must be sighted into the root end, or base, of the flame, and must be shut out of the circuit when the spark igniter operates, because the igniter emits UV rays.

We shall borrow published descriptions of flame detector installations from one of the major suppliers of packaged flame programming and management systems, the Allen-Bradley Co., maker of the "Fireye" series of flame safeguards.

FLAME RODS

The flame rod assembly is a "spark plug" type unit. It consists of a $\frac{1}{2}$-14 pipe taper threaded

mounting base, a KANTHAL flame rod, a glazed porcelain insulating rod holder, and a spark plug connector for making an electrical connection. Standard length of the flame rod is 12 in., but it may be shortened to any desired length.

The flame rod assembly is located so as to monitor only the gas pilot flame; however, if desired, the electrode may be located so that it operates from the main gas flame or from both the pilot and the main gas flame. It is mounted on a $\frac{1}{2}$ in. NPT coupling screwed onto a $\frac{1}{2}$ in. NPT nipple welded on the boiler plate, the pipe length being $\frac{1}{2}$ in. from the edge of the refractory material around the burner throat.

The following precautions should be noted:

1. Avoid bending the flame rod if at all possible.

2. Keep flame rod as short as possible.

3. Keep flame rod at least $\frac{1}{2}$ in. away from refractory.

4. Flame rod should enter the pilot flame from the side so as to safely provide an adequate pilot flame under all draft conditions.

5. If used to supervise gas pilot on an oil burner, the rod should be far enough from the oil spray to prevent oil impingement and burning of the rod.

6. If the flame is nonluminous (air and gas mixed before burning), the flame electrode tip should extend at least $\frac{1}{2}$ in. into the flame, but not more than half way through.

7. If the flame is partly luminous, the flame electrode tip should extend only to the edge of the flame, or not more than $\frac{1}{2}$ in. into it. It is not necessary to maintain absolutely uninterrupted contact with the flame.

8. Carbon deposits on the flame rod indicate poor combustion and the fuel/air ratio should be adjusted.

9. It is preferable to angle the rod downward to minimize the effect of sagging and

prevent it from coming in to contact with any object. Because of the basic requirements of a rectification system, an adequate ground surface for the pilot flame must be provided. The grounding surface in actual contact with the flame must be at least four times greater than the area of the portion of the flame rod in contact with the flame.

Only proven types of ground adapters should be used to provide adequate grounding surface. High-temperature stainless steel should be used to minimize the effect of metal oxidation. This assembly may be welded directly over the pilot or main burner nozzle.

Flame rod installations are generally done on an individual basis, and it is essential to adjust the flame rod and ground area ratio to provide a minimum signal strength of 2 microamperes.

INFRARED DETECTORS

To give you an idea of the proper way to install optical detectors, we refer you to Figs. 11-8, 11-9, and 11-10, and the material in the following paragraphs.

The scanner for the surface mounting should always be located on a swivel mount. Conducted heat should be avoided by installing a heat nipple. Hot gases and lens contamination should be blocked by supplying 4 cfm of clean air, or more, through $\frac{3}{8}$ in. Wye fitting on the scanner flange or by use of an external Wye on

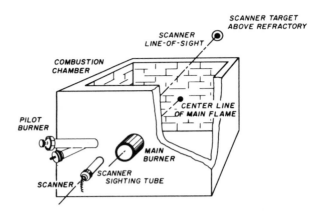

Fig. 11-9. Proper arrangement of burner and scanner, #2. (Courtesy of Fireye Division, Allen-Bradley Co.)

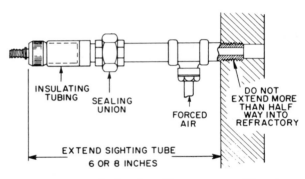

Fig. 11-10. Methods of cooling scanner. (Courtesy of Fireye Division, Allen-Bradley Co.)

the sight pipe. See Figure 11-10. Air should be supplied at 1.0 psi at the $\frac{3}{8}$ in. fitting or 5 in. wc above wind box pressure if the external 1 in. × 1 in. × 1 in. Wye is installed (plug $\frac{3}{8}$ in. opening). To restrict field of view and to reduce air flow and maintain air block, install a stainless-steel orifice disk within the swivel ball. Suitable disks are available and are secured within the swivel ball with spring rings.

The scanner should have unrestricted view of flame as far as possible, and interfering vanes or other hardware may need to be cut away or notched to afford clear field of view to base of flame at all firing levels.

For best results, the fiber optics extended scanner head should be welded into an air passage as close as practical to the fuel delivery nozzle. The head has been made extra heavy to accommodate field welding. If unable to attach it inside an airway of a fuel compartment,

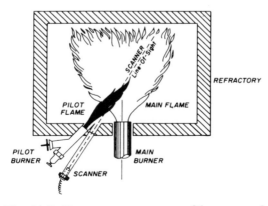

Fig. 11-8. Proper arrangement of burner and scanner, #1. (Courtesy of Fireye Division, Allen-Bradley Co.)

attach the head to the nearest air compartment. The idea is to approach the fuel burner center line, intercept the zone of ignition, and follow any motion of fuel and air compartments. Frequently these move up and down as much as 30°. Leave adequate slack in the flexible carrier tube to accommodate such motion, then pass the rear portion of the tube through the windbox wall at the convenient location and weld it firmly in place, sealing windbox air within.

THE ULTRAVIOLET DETECTOR

The detector is a sealed, gas-filled, ultraviolet-transmitting envelope containing two electrodes, which in use are connected to a source of AC voltage. When UV radiation of sufficient energy (at wavelengths shorter than those in sunlight at the earth's surface) falls upon the electrodes, the current flow starts and ends abruptly and is known as an "avalanche." A very intense source of UV radiation will produce several thousand avalanches or pulses per second. With less radiation there will be fewer pulses per second, varying in number and occurring randomly. Upon total disappearance of flame, the detector output ceases, except for very infrequent single pulses caused by cosmic rays, to which the circuitry does not respond.

The pulses from the detector tube are passed to a transistor-type pulse amplifier and are integrated in a capacitor. When the integrated voltage reaches a predetermined trigger level, a transistor switch energizes the flame responsive relay. If the integrated voltage falls below a predetermined level for a period of between 2 and 4 sec, the transistor switch turns off and the flame relay is deenergized.

For general instructions in installing the UV style of scanner, we refer you to Fig. 11-8 and 11-9, and the following "Fireye" UV scanner instructions.

1. If the scanner is to monitor both pilot and main flame, the sight pipe on which the scanner mounts must be aimed such that the scanner sights a point at the intersection of main and pilot flames. This usually places the sight tube in an area between the main and pilot burners. An acceptable location must ensure the following:

   a. Reliable pilot flame signal.

   b. Reliable main flame signal.

   c. A pilot flame too short or in the wrong position to ignite the main flame reliably shall not be detected.

   Since oil and gas flames radiate more UV from the base of the flame than from further out in the flame, this fact should be considered when sighting to satisfy the preceding three requirements.

2. If combustion air enters the furnace with a rotational movement or sufficient velocity to deflect pilot flame in the direction of rotation, sight the scanner 10–15° downstream of the pilot burner.

3. If in any operating position register vanes will interfere with the desired line-of-sight, the interfering vane(s) should be trimmed to ensure an unobstructed viewing path.

4. When a satisfactory sighting position has been confirmed by operating tests, the sight tube should be firmly welded in place. If the swivel mount is used, the ball position should be secured by tack welding the ball to the socket hub.

5. Two important operating requirements are that the scanner viewing window should be kept free of contaminants (oil, smoke, soot, dirt) and the scanner temperature must not exceed the maximum rating. Both requirements are ordinarily satisfied by continuous injection of purging air through the tapping provided. If further temperature reduction is needed, the sight tube may be extended with a length of either metal or nonmetallic (heat-insulating) pipe, not to exceed 1 ft total length.

6. In installations where individual UVP-4S systems are used to monitor main and pilot flames, the main flame scanner should be sighted so as not to detect pilot flame.

LEAD SULFIDE SCANNERS (FIRETRON CELL)

The system does not "detect" hot refractory. However, excessive steady radiation reduces flame signal. The same effect results from excessive scanner temperature. To avoid nuisance shutdowns, it is important to avoid sighting hot refractory and to keep scanner temperature low (never over 125°F). Otherwise, installation of the Firetron^c Cell is as described under the methods for the infrared and ultraviolet scanners.

Of the above three types described, the flame rod and the lead sulfide infrared cells are used to the greatest extent. The cadmium sulfide photocells are very rarely used, since they have difficulty distinguishing between the burner flame and other light sources.

Flame rods are used only on the smaller gas flames for the most part, while the lead sulfide infrared cell is the best all-around scanner to use. It does require more maintenance, and is more sensitive as to its location, but it can be used on any flame.

As you may have noted from the foregoing, the two main problems with the use of scanners are the temperature limitations and the accumulation of dirt on the tip. Both of these problems may be solved by purging the scanner with cooling air from the draft fan discharge or from the plant air system. A very small amount of low-pressure air is required, and this is usually specified by the supplier of the scanner. See Fig. 11-10 for details.

E. Steam Pressure Control

The major or primary control for the boiler is the steam pressure as it leaves the boiler nozzle, since this is the main purpose of having a boiler plant on the premises, to supply steam in a definite pressure range and amount for the plant process. All of the other controls covered in this chapter are supplementary to this main control of the steam pressure.

Starting with the pressure dictated by the

plant requirements, it is necessary to establish a minimum and a maximum pressure setting for the pressure control. Generally speaking, there are three rules to follow in this procedure:

1. Supply steam at the least possible pressure to satisfy the demands.

2. Set the pressure control to operate at the largest possible differential between the on and off settings.

3. Do not set the upper or shut-off point above the design working pressure of the boiler.

Small boilers are usually operated at the burner on-and-off modes, with no attempt at modulation while the burner is firing. This is the simplest method, but also the least efficient, as we shall see later in this book. It may be the only mode possible if the boiler is so small that modulating burners are not available for it.

The steam pressure control in this case is simply an on-and-off pressure switch, similar to that shown in Fig. 11-11, which utilizes a mercury switch as the activating element for the control circuit. The mercury switch element is shown in simplified form in Fig. 11-12. The pressure control shown is one of the more common makes on the market. There are oth-

Fig. 11-11. Mercoid pressure switch. (Courtesy of Mercoid Corp.)

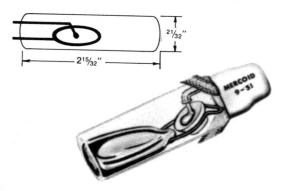

Fig. 11-12. Mercoid switch element, SPST. (Courtesy of Mercoid Corp.)

ers in use, but the principle of an on-and-off cycle is the same, and they all provide the same function. These switches actuate the main fuel supply valve through the electrical circuit for the programming panel and then to the valve operating motor or solenoid. The air supply for combustion may be on a separate circuit or may be of the natural draft type, depending on the

boiler size, governing codes, or insurance requirements.

As the boiler gets larger, the style of pressure control changes and becomes more complicated, mostly because it is more economical to provide more expensive controls, which will improve the efficiency of the boiler. The next step up the complexity ladder is to provide a burner control that will vary the rate of firing to suit the demands of the plant as near as possible. There are burner limitations that must be considered, and we shall cover them soon. Figure 11-13 shows a common method of combining steam pressure control with firing rate of the burner and with the combustion air requirements. This system has been in use for years, and is very reliable, simple, and easy to adjust and maintain. Properly maintained and properly set or calibrated, the controls can operate for months without requiring any tinkering or adjustments.

The system shown in Fig. 11-13 is used on a

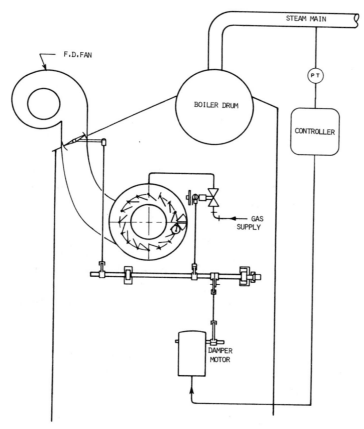

Fig. 11-13. Jackshaft style combustion control. (Courtesy of Bailey Controls Co.)

mechanical draft fan system, with the air controlled by a leaf damper in the air supply duct. The air may also be controlled by shutter dampers on the fan inlet, which is often done on larger units. In the case of natural draft systems, the air may be controlled by leaf dampers in the stack or breeching, supplementary to the weighted damper, or the leaf damper may replace the weighted damper and thus control the natural draft direct. In all cases, of course, governing codes must be followed.

The jackshaft contains cams that may be adjusted to vary the travel of the fuel and damper operating arms as the jackshaft rotates. These cams are calibrated when the burner is first put into operation, and are then checked for calibration at regular intervals to be sure that conditions have not altered the system requirements. Of course, as the fuel varies, and any change in the draft system is made, the cams must be recalibrated to ensure the best efficiency possible from the boiler–burner combination.

As the boiler–burner units increase in size, the basic jackshaft system may be replaced by electronic or electrical proportioning linkage in place of the mechanical linkage shown. The principle is the same, however, and there should be no difficulty for the boiler operator to switch from a system using the jackshaft arrangement shown to one using electrical or electronic means to proportion the air supply with the firing rate.

Getting back to the basic steam pressure control function, medium to large sized combinations usually follow the variable firing rate mode for the burner. In this method, an attempt is made to have the burner firing rate follow the steam demand load; as the steam demand changes, the firing rate and the air supply to the burner are varied accordingly by one of the methods described previously. However, there is a definite limitation for each burner in the minimum firing rate which it is capable of producing without becoming unstable or uneconomical in operation. This limitation is commonly referred to as the "turn-down ratio," and it varies with the type of burner and the fuel.

To explain the minimum firing rate limitation, we must look at the burner orifice through which the fuel flows into the furnace. The flow of fuel is proportional to the square of the pressure drop across the orifice, which means that to reduce the fuel flow to one-half of maximum design flow, it is necessary to reduce the fuel pressure ahead of the orifice to one-fourth of the maximum design pressure. To reduce the fuel flow to one-third of maximum design, the inlet fuel pressure must be reduced to one-ninth of the maximum design pressure, and to reduce the fuel flow to one-fourth of maximum design, the inlet fuel pressure must be reduced to approximately one-sixteenth of the full design pressure. These ratios are approximate only, but they are close enough to explain why it is seldom possible to reduce a burner output to less than one-third or one-fourth of the maximum design output. Remember, that the fuel inlet pressure for gas burners is seldom over 1 psig at full design flow. Reducing the gas pressure to one-sixteenth of that value will probably bring the flame front back to the burner tip itself, which will reduce the life of the burner and could at least soot up the burner and thus increase the maintenance problem. Also, and this is the main difficulty, reducing the flame too low may cause "popping," and possible low flame, or even an explosion.

In burners that we are discussing here, the normal turn-down ratio for gas will probably not be any more than 3 to 1 (minimum output one-third of maximum design). In the case of fuel oil, the turn-down ratio will be more like 2 to 1. However, it is possible to change the size of the burner tips in fuel oil burner guns, and thus alter the amount of fuel that may be handled by the gun. In fact, it is quite common to place two oil burner guns in one burner front, and thus obtain more flexibility in firing rates, as the burner guns may be cut in or out to follow the load changes, and thus obtain a higher turn-down ratio from the one burner front. In larger boilers with multiple burners flexibility is often obtained by cutting entire burners in and out to follow the load changes.

Let us now consider the matter of on-and off pressure settings for the steam pressure control. We have already stated the basic requirements or recommendations to be followed. Relating those requirements to an actual boiler, we make the following suggestions to be used as a guide only, for lack of any other recommendations, such as those made by the boiler supplier. The operator is warned that the burner and boiler combination may not tolerate them in all cases, as a great deal depends on the plant load demand pattern, and the style and design of the boiler, burner, and the controls.

Some suggested on–off pressure settings (in psig) are: 12–15, 25–30, 40–50, 85–100, 110–125, 135–150, 180–200, 275–300, 360–400.

F. Dual Fuel Firing Controls

In Chapter 7 we described a burner arranged for firing either gas or oil. This is quite common in today's boiler installations, as the result is more flexibility in obtaining the maximum efficiency for the least amount of fuel money. In those areas where the cost of fuel oil and the comparable cost of fuel gas may oscillate from week to week, it makes sense to be able to take advantage of whichever fuel is the cheapest. Also, there may be times when one fuel is in short supply, or there may be keen competition between the suppliers of both fuels. In this case, being able to burn either fuel may help the plant management in obtaining the best rates from both suppliers. Also, with the energy crunch now being experienced, along with the demands of the air-quality people, it just makes good common sense to be able to switch fuels as the political winds switch.

Later in this book we will go into the matter of deciding which fuel is the cheapest to burn for the heating value attainable.

Looking at the burner management systems for combination burners, we find that there are several arrangements possible, depending on the particular fuels, burner size, and steam demand requirements. The simplest one is arranged for firing only one fuel at a time, and the transfer from one fuel to the other is ac-

complished manually by shutting off one fuel, switching to the other. The manner in which this is done will most likely follow this sequence approximately:

1. Shut the burner down, in the prescribed manner, making sure the fuel valves are closed and are not leaking fuel into the furnace.

2. Adjust the control system according to the manufacturer's directions to handle the alternate fuel.

3. Follow the prescribed light-off procedure for bringing the boiler back on the line with the alternate fuel.

It goes without saying that the combustion control system must be capable of adjustment to handle the two fuels with equal ability and with the simplest technique in making the transition.

Some burner management systems are capable of being switched to alternate fuels without shutting down the burner, even though the controls can handle only one fuel at a time. It must be done with care, following the maker's directions, and taking care not to produce a fuel-rich mixture. To accomplish this, the following minimum requirements must be met.

1. The various interlocks and flame supervision and safety shutoff valves must be capable of handling both fuels.

2. The fuel selector switch must have positions for

   Gas firing, lock-out on oil firing.

   Oil firing, lock-out on gas firing.

   Gas–oil firing, providing all interlocks for both fuels are satisfied, including light-off position.

There are some burner arrangements in use that permit a mix of oil and gas to be fired simultaneously in any proportion. They are more complex, as the proper fuel/air ratio must be maintained in any combination available. To accomplish this, the fuels must be metered, the air must be metered and controlled, and the various interlocks and flame supervision and safety shutoff valves should remain in service for the fuel or fuels being fired. Chapter 15 contains detailed instructions for switching fuels without shutting down the boiler, for those systems properly designed for this mode of operation.

During dual-fuel firing, the interlocks for one fuel should not bypass the interlocks for the second fuel, otherwise, improper combustion will result.

G. Testing Safety Shut-Off Valves

Safety devices are worthless if they are not kept in good operating condition at all times, and your safety shut-off valves are no exception to this rule. Figure 11-14 illustrates the usually accepted method of performing periodic checks of a dual safety shut-off valve arrangement, and it also permits you to check other important items in the fuel gas train, as we shall explain.

Figure 11-14 contains a "dead man switch," which is used to operate the vent valve, and is very handy if the vent valve is not too accessible. It is not entirely necessary, however.

To test the tightness of the safety shut-off valves, follow the general procedure given here.

1. Close the safety shut-off valves.

2. Open main cock "C," and the vent valve.

3. Note the pressure gauge reading between the gas pressure regulator and the first safety shut-off valve. It should stabilize at what is known as the "lock-up" pressure, if the regulator is designed to shut off tight.

4. Open the gas cock "D."

5. Connect the rubber hose with the test valve to the nipple "A."

6. Close the vent valve, and place the lower end of the test hose in a cup of water.

7. Note any bubbles coming from the end of the hose, which indicates the first safety shut-off valve is leaking.

8. Move the hose to nipple "B," and cap off nipple "A."

9. Place the end of the hose in the cup of water, and close gas cock "D."

10. If bubbles appear in the cup coming from the end of the hose, then you know that the downstream safety shut-off valve is leaking.

11. The pressure gauge in that section of piping should read 0 psig, if it is accurate.

There are two things to keep in mind in this setup. First, the boiler draft fan should be on when this test is being performed, to remove any leaking gas that enters the furnace. Second, the vent line should lead to a safe location.

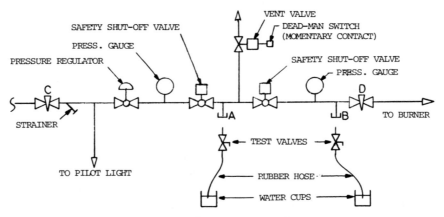

Fig. 11-14. Testing IRI gas supply system.

H. Boiler Alarm Systems

The extent to which boiler plant alarm systems are installed and utilized depends on the size of the plant, the philosophy followed by the plant designers and operators, and the insurance underwriter's requirements. We shall in this section give some basic ideas and suggestions covering this phase of the boiler plant's control scheme.

The following list contains the boiler conditions that we feel should be considered in the alarm system in some manner:

High steam pressure (or high water temperature for hot water boilers)

Safety shutdowns activated, consisting of:

Low water condition

High water condition

Loss of flame

Low fuel pressure

High fuel pressure

Loss of fuel oil atomizing medium

Loss of combustion air

In the medium to small plant, the high-steam-pressure condition does not normally activate a separate alarm, as when the safety valve blows, the high-pitched shriek is heard throughout the surrounding area. Thus, no separate alarm is required, unless the boiler is located some distance from the main plant and the operator is expected to roam far from the boiler. More on this later, however. Also, the combustion control panels often include local alarms operated by some or all of the preceding situations.

In the larger boiler plants, the alarms are often incorporated within an annunciator system, which consists of a panel with individual lights, each connected to one of the preceding condition sensors. When any one of the preceding dangerous conditions activates one of the alarms in the panel, then its window on the panel lights up, an alarm bell rings, either close by or remote from the boiler room. The more sophisticated panels indicate which condition caused the initial shutdown. They often have "acknowledge and cancel" switches, so that the alarm may be shut off while the operator is locating and correcting the malfunction. It is also an excellent idea to include in the annunciator alarm array a high-steam-pressure alarm, to give advance warning of an impending safety valve blow.

The operator may find it necessary to switch the annunciator panel off during the start-up phase of the combustion control system. This depends on the design and complexity of the system and local and insurance code requirements.

As an absolute minimum, we urge the installation of a low water alarm on all boilers, regardless of the code requirements for the area.

CHAPTER 12
BOILER WATER LEVEL CONTROL

A. Basic Requirements

From the beginning of the industrial revolution when steam boilers were first used, the problem of feeding water to the boiler at a steady rate to keep the boiler producing steam without burning it out or without causing an accident was always one of the primary considerations. Of course, when labor was cheap and steaming rates relatively slow for the size of the boiler, the feedwater valve could be throttled manually to give the approximate feed rate required, since there was always someone tending the boiler plant who could keep a steady eye on the water level. Furthermore, steam pressures were low enough so that obtaining sufficient feedwater pressure to supply the boiler was no particular problem. As pressures increased, the steam pump was designed to pressurize the water supply, and the injection principle, described in the next section, was invented to enable the operator to inject water into a boiler under pressure when the water pressure was at or below the steam pressure.

Level controls, thermostatically and float operated, came along with the development of higher pressures and faster steaming rates and the recognition of the dangers of low water levels. Today, we have some very sosphisticated control systems on the larger boilers. However, the controls on the size of boilers we are discussing in this book are still relatively simple, but basic to the industry.

Let us first describe what takes place inside a boiler drum as the water level changes due primarily to the change in steam flow. We covered this matter in a very rudimentary fashion in the first portion of this book, and we shall elaborate upon it here.

We shall assume that your boiler is steaming along at a reduced rate, the burner is modulated down to the low end of its output, and the steam pressure is nearing the upper limit of its setting, say, at 125 psig. The boiler is in equilibrium at this point, the drum water and the steam are both at 353°F, which is the saturation temperature of steam at 125 psig.

Assume now that someone opens a control valve wide open on a large process unit in the plant, and the steam flow rate is so great that the steam pressure drops in seconds to about 100 psig. From the steam tables in the Appendix, we see that the saturation temperature of steam at 100 psig is only 338°F, 15°F below the actual temperature of the water in the drum.

As the drum water has excess heat stored within it, approximately 15 BTU/lb, the water immediately starts to produce flash steam throughout the entire water volume in the boiler, by producing minute bubbles of steam from the bottom up to the surface of the water. This causes the water to swell up, and the water level in the drum rises, giving a false indication to the level control of the requirement for water fed into the boiler. In other words, as the steam flow increases, there is a need for an increased feedwater flow. However, the rising level due to the swelling may cause the feedwater level control to close down, especially if the level control system is too sensitive.

When the drum water has reduced its temperature to the saturation point of the new steam pressure, 338°F in this case, the water resumes its proper density, owing to the reduction in the rate of bubble release, and the water level drops to below its normal operating point, since by this time considerable water has left the boiler in the form of steam.

Again, if the level controls are too sensitive, the controller will open the feedwater valve too far, causing an upsurge in the water level, possibly over its upper safe limit. This may be followed by a fast closing of the valve, in an attempt to catch up to the change in water level. The net result can be a hunting or oscillating effect between the water level and the control. This is known as a surging water level, and the result may be carryover of water into the steam main. We shall cover this in more detail later in this book.

The reverse situation—a fast closing of the steam valve to the process—generally does not cause any difficulties, other than possibly a lifting of the safety valve occasionally.

The ordinary float type of control as found on smaller boilers usually has sufficient sluggishness or lag built into it to resist this surging action. For larger boilers, the level controls are altered in various ways to compensate for the surging effect. We shall describe one method later in this chapter.

Today's package boilers have been designed to be operated with the level controls furnished with the package. It is not recommended that the operator attempt to alter these controls or replace them with another style unless first obtaining the approval of the boiler supplier. The characteristics of the boiler and its steaming qualities must be properly matched to the operating characteristics of the water level control system. The boiler supplier has spent considerable amounts of time and money in determining the best combinations of accessories for the supplier's boiler. Before you blame the style of controls and accessories for im-

proper behavior, it is well for you to check everything else first.

All boilers should have three levels marked on their outer casing or drum ends. They should be clearly marked and labeled, indicating the highest level, the normal designed operating level, and the lowest level for safe operation.

The lowest level is at least 2 in. above the uppermost tube ends in the drum, to prevent the tube ends from being subjected to direct heat from the furnace or the hot gases passing through the tubes or around them. This is to prevent them from being burned out.

For a firetube boiler, the critical point is obviously the upper edge of the uppermost tubes, since the entire tube and tube joints must be surrounded on the outside with water to keep the temperatures down.

For water tube boilers, the critical point is the entrance to those tubes that feed water down from the drum. If any of the down-flowing tubes are uncovered, water cannot enter them, and there is no circulation in those tubes, with a high probability of burned out tubes.

Generally speaking, the higher the steaming rate for the amount of water stored in the boiler, the more critical is the water level control system.

There is one more safety device covering low water conditions in the boiler that should be mentioned here and that is the fusible plug. This plug is placed in firetube boilers as a last line of defense, in case all other safety devices have failed. It is designed to melt out and allow any remaining water in the boiler to spray down into the flame in the furnace, should it become hot enough from lack of water on its upper side. It is placed at a point below the low water setting, where there is hot furnace gases on the lower side, and normally water on the upper side. Ideally, it should melt out before the tubes are exposed, or just as they become exposed, so that there is sufficient water remaining to put out the flame. If this is not possible, then it will be placed where the steam will do the most good in extinguishing the flame. They are used in the old-style horizontal return tubular boilers, the Firebox-style and the locomotive-style firetube boilers. They are not practical in most watertube boilers, since the boiler drum does not come in contact with the hottest flue gas.

There is more discussion on fusible plugs in the Appendix.

B. Feeding Water to the Boiler

There are several methods of feeding water to a boiler while it is steaming under pressure, but they may be divided into two general methods. The most common one, of course, is by means of one or more pumps. Not nearly so well known, since it's not used as much as it has been in previous years, is the steam injector. We shall cover both in this section, but first we shall cover the usual code requirements.

Most codes state that all boilers over 500 ft^2 of heating surface shall have at least two ways of feeding water to the boiler.

Feedwater piping to the boiler shall have a check valve near the boiler, with a stop valve between the boiler and the check valve. When two or more boilers are being fed from a common source, there shall be a stop valve between the check valve and the source of water.

When globe valves are used as stop valves in the boiler feedwater line, the flow of water shall be from under the disk.

Feeding water to a single boiler is relatively simple, and there are several methods of doing it, depending on many factors. First, of course, the boiler supplier should be consulted for recommendations. If the supplier is reputable, the supplier will come into your plant and make suggestions after looking over your operations. If such service is not available for any reason, then we suggest you refer to one of the boiler feedwater pump suppliers, who are usually well versed in the requirements. There are many package-style boiler feedwater tank and pump units available, which would undoubtedly serve the purpose very well.

The check valve previously mentioned must be reliable, because if this check valve leaks, no end of troubles may result. Many boiler operators have been plagued to distraction as a result of the water level in the boiler dropping too rapidly, owing to the feedwater check valve not doing its job properly. The result is rapid cycling of the feedwater pump, trying to keep up with the demand created by an inefficient system.

Another problem often encountered in some of today's high steaming rate boilers is the rapid fluctuations, or surging, to be found at times when certain conditions occur simultaneously in the boiler system. This situation may be due to a combination of a high steaming rate, very little contained water volume in the boiler, a sensitive feedwater level controller operating with an "on" and "off" cycle, with possibly a feed pump that has too high a capacity. This causes the water level to rise and fall with a definite rhythm or cycle, sometimes sufficiently to cause the water feeder high and low level alarm switches to go on and off very rapidly. The feed pump may be turned on and off at the same time, with only a few seconds between cycles, which is very hard on the pump starting equipment.

The best cure we know of for this problem is to change the method of feeding to a continuous style, with the pump running continuously, and a modulating feed valve to maintain the proper level in the boiler. In switching to this system, remember to select the pump with care, since it may require a certain amount of water to be recirculated to the feed tank at all times.

Packaged boiler feedwater units, with hotwells, piping, and base, are available in standard size ranges and in either single or duplex pump arrangements. The duplex system is the most reliable, since each pump is capable of supplying the total demand of the boiler. This type of system may be provided with an alternator in the starting circuit, so that the pumps will alternate operation on each start-up cycle. If the duty pump does not handle the load for any reason, then the second pump will come on. In this case, each pump will get just half the service that a single pump would get. Also, with the alternator, the total life of the system is much longer before requiring repairs.

Along with the chosen feedwater pumping package, suitable boiler level controls must be provided, of course. The combination shown in Fig. 12-4 is a typical arrangement, and available in several styles and level settings to accomplish the task. We strongly urge the boiler plant operator to call in one of the suppliers of boiler controls, who will be familiar with the local codes, to assist when a change is contemplated.

We mentioned the steam injector at the beginning of this section. This was quite popular when steam pressures were moderate. Steam injectors are still available and still do an excellent job, if they are chosen and operated properly.

The steam injector feeds water into the boiler by using the steam from the boiler that is being fed. The device is attached at the side of the boiler drum, with a check valve between the injector and the boiler drum. There is a drain line from the injector, containing a check valve allowing drainage out, but preventing air from being drawn into the injector. The feedwater is piped into the side of the injector, as shown in Fig. 12-1.

When boiler feed water is to be injected into the boiler, first open the water supply valve, then open the steam valve. Next, throttle the water supply valve until the injector quits spilling through the overflow or is reduced to a minimum.

When this condition is reached, the injector will feed water into the boiler at a fairly good rate, and will do so until the steam or feedwater supplies are shut off.

The injector is a very useful device, and may serve as the second source of feeding water to the boiler. Steam injectors have been used very little recently, owing mainly to the difficulty of automating them. The operation as described is strictly manual, and all attempts to make them automatic so far have proved disappointing. The main problems have been the degree of reliability required under automatic control and the limited operating range of a multiple jet system.

C. On-and-Off Feedwater Pump Controls

Most of the smaller boilers have on-and-off feedwater pump controls. There are several makes and styles on the market, and we shall describe two of them, both of which are of the float operated design. There are other styles in use, such as the submerged electrode type, but the general principles of use are the same. The only difference among them is minor problems in maintenance. For instance, the electrode style may be subject to loss of effectiveness from grease or corrosion products coating the electrodes. Floats, on the other hand, may spring leaks from corrosion, or collapse from high surges of pressure from water hammer conditions. However, if this happens, there very likely will be other troubles encountered from the water hammer.

The on-and-off controller turns the boiler feedwater pump on and off to meet the level requirements of the boiler, regardless of what the water level in the hotwell or storage tank happens to be. Therefore, the tank should be sized such as to ensure always having sufficient water within it to keep the pump supplied and to prevent the boiler from becoming starved, which could be disastrous or, at least, inconvenient. At the beginning of this book, we explained the principle of surge capacities and their relationship to the steady or intermittent

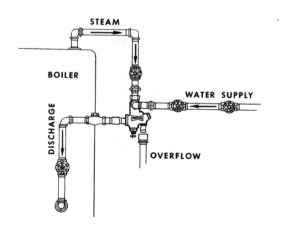

Fig. 12-1. Boiler feedwater injector. (Courtesy of Penberthy Houdaille.)

demands from a constantly hunting system, such as the boiler cycle. This is a prime example of that principle in operation. The hot-well must have sufficient water within it at all times, but at the same time it must never be so close to the overflow point that any condensate returns cannot be taken in without spilling over the vent or overflow. Within reason, it is better to have an oversized hotwell than one which is too small or borderline.

Figure 12-2 shows one of the more common float operated water level controllers found on package boilers, as supplied by the ITT McDonnell and Miller firm.

Mounted at the boiler water line, this controller starts and stops the pump as the boiler level itself dictates—not according to the rate at which condensate accumulates in the receiver. Consequently, it holds the boiler level between the close limits recommended by the boiler manufacturer to maintain maximum steaming efficiency.

In addition to its use as a pump control, the No. 150 also serves as a dependable low-water fuel cutoff for automatically fired boilers. Besides the pump control switch, it has another which can be wired to interrupt the circuit to the burner should an emergency of any sort permit the boiler water level to drop into the danger zone. This second switch can also be utilized to provide a low-water alarm if de-

The design of the No. 150 always keeps all operating parts up out of the water. Construction is completely packless, and uses a special monel bellows to seal operating parts from the float bowl. The float is of heavy gauge copper alloy, and is capable of withstanding pressures far beyond the rating of the No. 150. Switches are mercury type specially designed for high-temperature service. The body is of heavy close-grained cast iron.

Since operating parts of the No. 150 are not affected by heat, the junction box is fully enclosed to keep out dust and dirt. The junction box cover is removable to provide access to the terminal panel and switch inspection.

Figure 12-3 shows a more sophisticated version of this same control, being arranged for installation with a code-designed water column. This model has the same basic operating mechanism as the one previously described, with the same switch combinations. The only difference is the float chamber being drilled and tapped for a water column.

Figure 12-4 shows a typical use of both controllers on one package steam boiler. In this case, the two units, each with three switches, are combined in such a way as to utilize four switches, as shown in the Fig. 12-5 wiring diagram. Other combinations are available and possible, depending on the code requirements

Fig. 12-2. McDonnell and Miller No. 150 level control. (Courtesy of ITT Fluid Handling Division.)

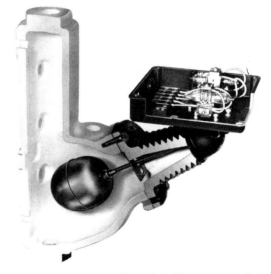

Fig. 12-3. McDonnell and Miller No. 157 level control. (Courtesy of ITT Fluid Handling Division.)

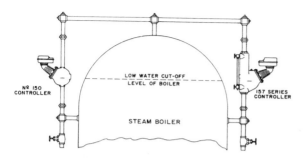

Fig. 12-4. Installation of Nos. 150 and 157 level controls. (Courtesy of ITT Fluid Handling Division.)

and the requirements of the boiler supplier, as well as the insuring agency.

Figure 12-6 shows another style of float operated controller for boiler feedwater service. Its operation is as follows: A permanent magnet (1) is attached to a pivoted mercury switch (2). As the float (3) rises with the water level, it raises the magnet attractor (4) into the field of the magnet. The magnet snaps against the nonmagnetic barrier tube (5), tilting the mercury switch. The barrier tube provides a static seal between the switch mechanism and the float. When the water level falls, such as with a low-water condition, the float draws the magnet attractor below the magnetic field. The magnet swings out and tilts the mercury switch to the reverse position, actuating the low-water alarm and operating the burner cutoff circuit.

The magnetic field is the only "linkage" between float movement and switch actuation.

Magnetrol multiswitch-type boiler controls are designed for the dual functions of boiler feed pump control and low-water cutoff and alarm. Uniquely, a third switch can be added at any time in the field for additional feed pump control or high-water alarm.

Dry contact switches are available for excessive vibration applications or for shipboard use. Pneumatic pilot switches are also available.

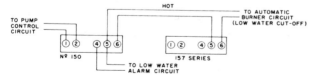

Fig. 12-5. Wiring diagrams, Nos. 150 and 157 controls. (Courtesy of ITT Fluid Handling Division.)

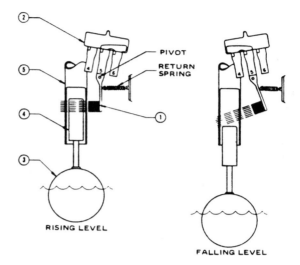

Fig. 12-6. Magnetrol boiler level controller. (Courtesy of Magnetrol International Inc.)

Standard Magnetrol boiler controls are available custom tailored to meet special levels or operating sequences for virtually any application specification. A typical example includes the use of five individual switch mechanisms to provide high-water alarm, individual control of two boiler feed pumps, low-water alarm, and low-water cutoff.

Water-column-type instruments are available with tandem floats that function independently to provide widely spaced high- and low-level switch operation such as high-water alarm and low-water alarm.

D. Modulating Level Controllers

Larger boilers often have modulating water level controls, and these are especially desirable on watertube boilers with a small volume of water contained in the drum combined with a high steaming rate—in short, a high turnover rate. The water level in this arrangement is usually very sensitive, and should the level extremes exceed the limits designed by the boiler supplier, the result may mean trouble. In this case, the modulating style of level control is called for.

One of the earliest designs of water level control for the slower steaming rate boilers is the thermostatic level control. One of the designs currently available is shown in Fig. 12-7, as supplied by Copes-Vulcan, Inc. Other styles and makes are available, of course, using this same basic principle of operation. They all work very well, as long as they are maintained properly, and the boiler is not expected to perform beyond its design steaming rate or rate of load change. If the load changes more rapidly than the control was designed to handle, then the water may drop or rise to a dangerous extreme. This control is a little slower than float or electrode-style controllers in following the water level changes.

The thermostatic feedwater regulator feeds continuously to a working boiler, stabilizing the drum water level within predetermined safe limits. It is fully automatic, simple, dependable, and rugged. Maintenance costs are negligible.

The actuating element is a thermostatic expansion tube in which the water level moves up or down as the boiler water level rises or falls. As water level drops, the steam temperature causes the tube to expand. It is fixed at the heel end, so only the head piece moves. A bell crank lever at the head multiplies the movement, providing necessary travel to operate the control valve. An adjustment nut at the heel piece end can be turned to provide the desired drum level while the boiler remains in operation.

The expansion tube is made of a special alloy

having a high coefficient of expansion. The alloy is selected according to specified operating conditions. The tube is in tension. The assembly, consisting of tube, head and heel pieces, and bell crank, is securely mounted on heavy channel supports, which protect the tube throughout its entire length and ensure a rigid installation.

Connection between the thermostat and valve is through a tension-relief strut. The tension relief protects the regulator from any undue strain when the valve is closed and the expansion tube is contracted.

The feedwater-regulating valve operates through a horizontal rotating lever shaft. It is weight loaded. In an emergency, the lever can be used to open or close the valve by hand.

Full travel is accomplished with only 30° rotation. A permanently lubricated stuffing box seals the liquids inside the valve. An outboard ball-bearing stuffing box is used with steam and water pressures greater than 325 psi. This stuffing box supports the outer end of the lever shaft and counteracts the higher end thrust present at high pressures.

The valves are available in ANSI Classes 250 through 600 lb standards. Cast gray iron bodies are provided for the 250 lb, cast steel for higher pressures. Cage guided trim is used, with V-ports machined in the cage and accurately selected for specified flow conditions. The cylinder-type piston is of 410 hardened stainless steel. High valve lift gives close control and reduces wear on fittings. It can be set to give full valve travel with any specified water level variation.

The maximum pressure drop across the valve is 50 psi. If the pressure drop varies no more than ±25%, the feedwater regulator will hold drum level within ±2 in., exclusive of shrink and swell on load changes.

So far, we have been describing what is known as the single-element style of level control, shown diagrammatically in Fig. 12-8. This simply means that the boiler feedwater requirement is being detected by only one element, or detector. This consists of the actual level of the water in the boiler drum. In the size range we are covering in this discussion,

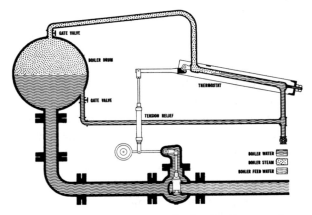

Fig. 12-7. Thermostatic boiler level control. (Courtesy of Copes-Vulcan Inc.)

this is satisfactory, and is probably the only one you will have to handle or maintain. Figure 12-8 shows the water level being controlled through a feedwater-regulating valve by a controller, which in turn receives its actuating signal from the level transmitter on the water column. This, for the smaller boilers described so far, reduces to a series of level switches on the water column operating either a feedwater pump or the feedwater-regulating valve direct, without requiring a controller. As the boilers get larger, and the pumps or valves get larger, the control has to be more complicated, as shown in Fig. 12-8. The principles remain the same, however.

It is expected that some readers will come in contact with the more sophisticated system shown in Fig. 12-9, so we show it here for your edification. Note that a flow transmitter has been added on the steam main, which registers the steam flow to the plant. This control is the primary element operating the controller and thus, the feedwater regulating valve. The level transmitter is a supplementary control element, modifying the basic signal to the controller. The system compensates for the affect on the water level caused by rapid surges in steam flow, as explained in the first section of this chapter.

There is a third level of boiler feedwater level control sophistication known as the three-element system, which is used on boilers in the much larger industrial and power utility fields. This third element in the control system monitors the flow of boiler feedwater, and prevents the feedwater-regulating valve from starving the boiler during the periods of rapid steam demand. This is the resultant action in both the

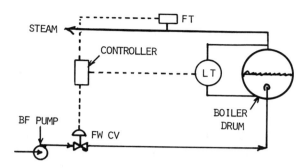

Fig. 12-9. Two-element boiler level control.

two-element and the three-element systems. In both of these systems, the injection of relatively cool feedwater into the foaming drum tends to quench the bubbling water and thus stabilizes the water level at the new steaming rate and reduced pressure and temperature caused by the increased steam flow.

Returning to the level of sophistication that we were discussing, the next one up the ladder is the single-element modulating control. One of the more common ones is shown in Fig. 12-10, as made by the Magnetrol Co. This style requires a source of instrument air for operation for the feedwater-regulating valve. The level control element merely senses the change in water level, then transmits this level change in

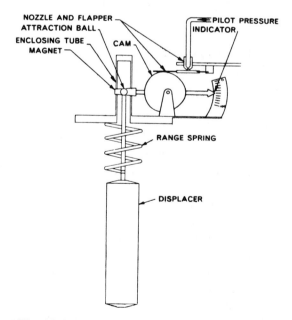

Fig. 12-10. Operation of Modulevel control. (Courtesy of Magnetrol International, Inc.)

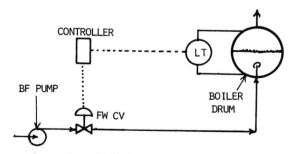

Fig. 12-8. One-element boiler level control.

the form of an air flow under very small pressures to operate the valve.

The description of its operation is as follows: Displacer response to boiler water level change positions the attraction ball within the pressure-tight enclosing tube. As the external magnet follows the attraction ball, the cam rotates, repositioning the pilot flapper, which, in turn, varies the pilot pressure signal. A pilot relay amplifies this signal to provide the necessary proportional output pressure range. Magnetrol's magnetic coupling design completely isolates the control mechanism from the level sensor and provides friction-free input motion to the controller.

Modulevel controllers largely ignore agitated or turbulent level conditions, maintaining output pressure and control valve stability under all operating conditions.

The Modulevel controller responds instantly to significant boiler water level changes and simultaneously produces an air pressure signal directly proportional to level change. The air signal positions a diaphragm-actuated control valve resulting in fast, smooth feedwater flow regulation. And, with Modulevel, feedwater flow rate is changed only by changes in boiler water level. The Modulevel controller is totally unaffected by steam temperature or pressure variation.

The Modulevel water column mounts in a normal position on the boiler, adding no awkward appearance or clutter. Every Modulevel controller is factory precalibrated, ready to operate.

Figure 12-11 shows a typical boiler water level control system utilizing the Modulevel controller. We shall go into the features of it very briefly, since most of them are quite evident by now, if the reader has been paying close attention to our discussion.

Item #1 is a pressure regulator to provide a constant supply of low-pressure dry air to the Modulevel control head.

Item #2 indicates the extreme difference between the normal operating water level in the drum and the absolute lowest level before trouble may be expected. The lower arrow points to the lowest visible point in the gauge

glass. If the water is below that point, the burner should be off, and the operator should be attempting to find the problem.

The upper arrow in Item #2 points to the try-cock, which is, by Code, located at the normal operating water level in the drum. The water column will be described fully in the next chapter.

Item #3 consists of float switches, each of which may have more than one switch set at different levels. These are emergency switches, generally. The upper setting may be set to alarm when activated, since it is set above the level of the top try-cock in the water column, and represents the point at which carryover may be expected from the drum. The lower switch is set below the lowest try-cock level, and is an alarm to signal trouble before the water sinks out of sight in the gauge glass.

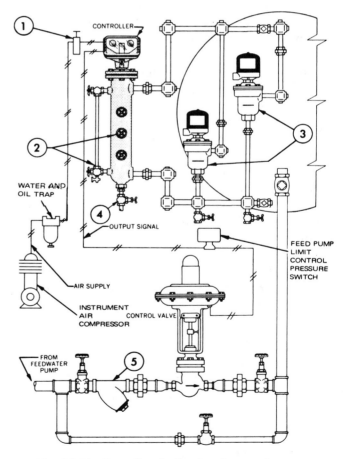

Fig. 12-11. Complete boiler level control system with Modulevel controller. (Courtesy of Magnetrol International, Inc.)

Item #4 is an approved drain cock, with the discharge piped to a drain, visible from the operator when he opens the drain valve, so that the operator may be sure the line is clear and is draining.

Item #5 is a strainer, with a fine mesh screen to keep any particles of dirt from entering the control valve. If this strainer is not kept clean, there is a good possibility of the control valve failing to close, thus causing a high water level in the boiler drum.

The air pressure leaving the Modutrol control head is varied indirectly with the change in water level in the drum. As the water level drops, the air pressure leaving the control head rises, opening the control valve to feed more water to the drum. As the water level rises, the air pressure fed to the control valve drops, causing the valve to close. The control valve (usually referred to as the Feedwater-Regulating Valve) is generally not expected to shut off tight, so that there may be a low steam demand situation where the control head is attempting to shut off the feedwater supply, the water level is rising, and it is thus out of control on the low-flow side. There is a feed pump limit control pressure switch in the air loading line to the control valve. This switch shuts off the boiler feedwater pump when the air loading pressure drops to a preset point, and will thus prevent flooding the boiler.

The system just described is only typical, and not necessarily the one on your boiler. There are many variations of it, depending on the local governing codes, the insuring agency requirements, and other considerations.

E. Feeding Multiple Boilers

So far in this chapter, we have covered only the methods of feeding water to single boilers. There are many installations where several boilers are supplied with feedwater from one hotwell and pumping system. The individual boiler arrangements are the same as just covered, but the central water supply must be sized and altered in the controls and piping to accommodate more than one boiler's demand.

We shall now cover the basic problem of the high-pressure steam boiler, going into the basic theory and requirements first. Then, we shall give you the benefit of one control supplier's experience in feeding multiple low-pressure steam boilers from one hotwell and feedwater pump system. Generally, there is very little difference between the high-pressure and the low-pressure system. By studying these two systems, you should get a very excellent idea of some of the problems and methods of avoiding future troubles.

A typical feedwater supply system to a multiple boiler hook-up is shown in Fig. 12-12. This illustrates two boilers requiring a fairly constant source of water, which calls for a constantly running pump to keep the supply main up to pressure at all times. The pressure in the water supply main should be at least 25 psig above the boiler pressure, which permits the feedwater-regulating valves to feed water into the boilers properly. Each boiler on the line has a level controller that regulates the water level through a feedwater-regulating valve. There should be a check valve at each such valve and on the discharge from the feedwater pump. Also, according to most codes, there should be a shut-off valve between the boiler check valves and the pump.

The 25 psi differential between the boiler operating pressure and the boiler feed pump discharge pressure is an approximate average for the small boiler installation. This differential should be chosen to overcome the pressure drop required for the feedwater-regulating valve when feeding into the boiler, and the frictional resistance from the piping and fittings from the pump to the furthest boiler. Usually, in most boiler plants, the feedwater-regulator pressure drop is the major factor.

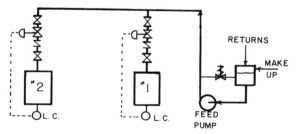

Fig. 12-12. Feeding multiple boilers.

Here again, we urge the boiler plant operator to check with suppliers when making a change in equipment, or when there is any question concerning adequacy of the equipment.

On the discharge side of the pump there must be a means of preventing the pump from running against a shut-off condition. This is usually done by installing a pressure-relief by-pass valve as shown in Fig. 12-12, which opens at a pressure low enough to prevent the pump from overheating when no water is required for either boiler. This valve should be selected with the assistance of the pump supplier.

On boilers having "on" and "off" style level control switches, these switches are usually connected to electric-motor-operated feedwater valves. When either of the boiler level controllers call for feedwater, the electric supply valve is opened, and the boiler feedwater pump is turned on from the same electrical circuit. As long as any of the electric feedwater supply valves are open, the pump will be running to supply the demand. When all valves are shut off, the pump is off, also.

We have described a two-boiler system, but the same general discussion will fit any number of boilers installed on a common system.

There are several refinements of this basic feedwater system, but we need not go into them all here. You should understand the various problems involved, so we shall list the steps required for supplying the water to the boiler, as follows:

1. When the feedwater-regulating valve opens to supply water to the boiler, there should be sufficient water pressure ahead of the valve to ensure a proper flow into the boiler.

2. The boiler feedwater pump should have proper protection against a closed discharge line, otherwise the pump may be damaged.

3. The feedwater pump must have sufficient water pressure and flow at the inlet to the pump to prevent the pump from being starved, which may damage the pump.

4. The controls should be arranged so that only the boilers requiring water will get it when they need it, and at no other time.

5. The feedwater check valves shall be a reliable type, to prevent water from one boiler from backing up into another boiler on the same supply system, or back into the feedwater pump when it is shut down.

6. The pump or pumps must be sized to supply all of the boilers on the system with the rated flow in pounds per hour should all boilers demand water at the same time. The total pump capacity in gallons per minute is calculated by dividing the steaming rate in pounds per hour by 500 for each boiler, adding up the gpm demands for all boilers on the system, and multiplying by 2 or 3 to permit pump cycling.

7. All boilers on one feedwater pumping system should be of the same operating pressure for best operation.

8. The boiler feedwater tank should be sized to hold from 5 to 15 min of storage, based on the total actual steaming rate determined in item 6.

And now, by means of Fig. 12-13, we shall let ITT McDonnell and Miller Co. explain how they recommend that multiple boilers be supplied with water from one common hotwell and feed pump. As you go through this description, keep in mind the principles involved and the reason for using each piece of equipment installed, and then you will be in a position to apply them to your multiple boiler installation.

It is sometimes assumed that when two or more boilers are installed, it is a simple matter to arrange the steam piping and boiler feed piping in such a way as to produce equal water levels in all boilers. Although this objective is realized in a few installations, it is more generally true that other steps must be taken to obtain reasonable water level regulation.

Even though steam and water equalizing lines between boilers provide free flow, it is possible to have widely different water levels owing to small pressure differentials. These pressure

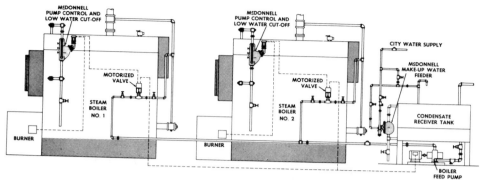

Fig. 12-13. Typical multiple boiler feedwater system. (Courtesy of ITT Fluid Handling Division.)

variations are due to differences in steaming rates caused by the inability to adjust the firing devices to produce an equal output at any given moment in two or more boilers.

Although the diagrams of the water level regulation of steam boilers illustrated in this book are for single-boiler installations, the same basic methods can also be used for multiple-boiler installations using a common or single-boiler feed pump. It is only necessary to install a regulating valve between the boiler feed pump and each individual boiler of a multiple-boiler installation.

A typical multiple-boiler installation is illustrated in Fig. 12-13.

This illustration shows two or more low-pressure steam boilers installed with a common boiler feedwater tank and electrical boiler feed pump. An ITT McDonnel and Miller No. 157 Pump Controller and Low Water Cut-off is installed at the normal boiler water line of each boiler, and a motor-operated valve is installed in the feed line to each boiler.

When any boiler requires water, the controller opens the motorized valve in the feed line. When the motorized valve reaches the full open position, it closes an end switch, which in turn starts the boiler feed pump. When the normal boiler water level is restored, the controller closes the motorized valve and the end switch of the valve stops the boiler feed pump.

With the boiler water level controlled in this manner, it becomes obvious that it makes little difference as to the size of boilers installed, or the relative positions of the normal water lines, as with this arrangement the water level in each boiler is individually controlled.

Should an emergency condition arise, such as a fuse failure in the boiler feed pump circuit, the electrical low-water cut-off switch in the controller will stop the burner on a further drop in the water level.

For addititonal material on boiler feedwater systems we refer you to page 305 in the Appendix.

CHAPTER 13
BOILER TRIM

A. Code Requirements

When we refer to "boiler trim," we mean the minimum accessories that are required to monitor the boiler properly and to operate it. Mainly these consist of the steam pressure gauge, the level glass, the water column, the test cocks, the safety valves, and the blowdown and drain valves. The test cocks are for the purpose of testing for boiler water level when the gauge glass is dirty, broken, or plugged or for some reason is suspected of being out of action.

First, we shall see what most codes have to say about the boiler trim.

Each boiler shall have a steam pressure gauge, which should be calibrated at least once a year. The dial of the pressure gauge shall be graduated to approximately double the pressure at which the safety valve is set, but in no case to less than $1\frac{1}{2}$ times the safety valve setting.

A valve or a cock shall be placed in the gauge line adjacent to the gauge. An additional gauge or cock may be placed in the gauge line near the boiler, provided it is locked or sealed in the open position.

For a steam boiler the gauge or connections shall contain a siphon or equivalent device, which will develop and maintain a water seal that will prevent the steam from entering the gauge.

Each boiler shall be provided with a valved connection at least $\frac{1}{4}$-in.-pipe size for the exclusive purpose of attaching a test gauge when the boiler is in service, so that the accuracy of the boiler pressure gauge can be tested.

The boiler water column pipe shall be a minimum of 1-in.-pipe size connecting the column to the boiler.

Each boiler shall have at least one water gauge glass for observing the water level in the boiler. Boilers operated over 400 psig shall be provided with two water gauge glasses, which may be connected to a single water column or connected directly to the boiler drum.

The lowest visible part of the water gauge glass shall be at least 2 in. above the lowest permissible water level. If the boiler is of the horizontal firetube type, the water gauge glasses shall be installed so that when water is at the lowest reading in the water gauge glass, there shall be at least 3 in. of water over the highest point of the tubes, flues, or crown sheet.

Each boiler shall have three or more test cocks of not less than $\frac{1}{2}$-in.-pipe size. The test cocks shall be located within the visible length of the water gauge glass, except when the boiler has two water gauge glasses located on the same horizontal plane indicating the same water level. They shall be operable from the boiler room floor.

The pipes connecting the water column with the boiler shall have no apparatus connected to them that will permit the escape of an appreciable amount of water or steam from the water column. You may connect such trim as water column drains, steam pressure gauge, feedwater regulator, or damper regulator.

No doubt you have seen some water columns that look like a pipefitter's nightmare. This tendency is prevalent on some small package boilers. However, there is a minimum size limit (1 in.) to which some of the required water column equipment may be supplied, which at times makes for rather ungainly looking water columns. Of course, the safety of the boiler is more important than appearance, and as long as such ungainly looking water columns are properly piped and applied, they should be left alone, except for regular maintenance.

There are many gadgets on the market designed to help the boiler operator, but they should be used with caution. The fittings covered in this chapter are all that are really needed for good boiler plant workers to monitor the boilers under their care. It is not good to place too much reliance on automatic gadgets when there is so much at stake.

We strongly suggest that you obtain your own copy of your State's boiler code, and spend some time going through it. Much of it probably pertains to the design of boilers; however, there are many parts that will be of help and guidance to you as an operator or maintenance man.

B. Steam Pressure Gauges

One of the most important instruments on the boiler is the steam pressure gauge, for it registers the actual pressure of the steam, above the atmospheric pressure surrounding the gauge. For pressures in the ranges that we are discussing here, the ordinary type of gauge is known as the bourdon tube gauge.

Figure 13-1 shows the basic internal works of the bourdon tube gauge. The bourdon tube is a flattened tube of oval cross section as shown, and the entire assembly is bent in the form of a semicircle. One end, which is anchored, or fixed, is open to the pressure from the boiler. The opposite end is free to move, which it does with application of pressure to the inside of the

tube. The free, or moving end, is attached to a linkage, which in turn rotates the needle or pointer on the face of the dial containing the pressure scale.

The bourdon tubes are available in various metals, the three most common, which are used on low to medium steam pressures, being brass, stainless steel, or bronze. The stainless-steel tubes are usually used for all pressures above 300 psig.

The usual location for the pressure gauge is on the top of the water column, since this is a convenient place to attach it, is close to the steam space in the boiler, and is in a location suitable for ready reference.

The bourdon tube gauge may be placed on a gauge board at a remote location, but when this is done, two pressure gauges are called for: one at the boiler on the water column and an additional one on the gauge board. If the gauge is placed on the remote board, then it is recommended that a gauge cock be placed at both ends of the connecting tubing.

Most of our readers are familiar with the siphon arrangement, or the "pigtail" as it is commonly called, ahead of the gauge. This is for the purpose of collecting condensate and forming an isolation chamber between the gauge and the steam drum. This ensures that only lower temperature water will reach the bourdon tube, thus prolonging the life of the gauge.

Very little else must be said about the pressure gauge, except that it should be large

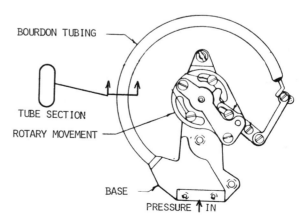

Fig. 13-1. Bourdon pressure gauge principle.

enough so that the operator need not squint or strain to read it. Also, the numbers and the needle should be large enough and of such pattern that they can be picked out at a glance from the operating floor. Some earlier design gauges were known for their figures and the engraving on the face, along with fancy pointers, which were beautiful to see, but not always too easy to read.

Some steam pressure gauges being furnished today are equipped with an auxiliary scale inside the pressure scale, giving the saturated steam temperature for the equivalent pressure. This makes it very handy for the plant that is mainly concerned with the steam temperature required for a particular process. However, it can lead the operator into a false sense of security, which we shall cover later in this book.

The California State Code states that the steam pressure gauge shall be calibrated once a year, which is common with most state codes. There are firms equipped to perform this service, who will also certify their work. If there are sufficient other pressure gauges in the plant to make it worth while, the purchase of a simple test and calibration kit may be a good thing to consider.

C. The Water Column

We have already shown one version of a boiler water column, in Chapter 12, Fig. 12-3, with an attached water level controller. The standard water column consists of an enlarged water chamber with tappings on top and on the bottom for vertical installation, provisions for a water level gauge glass, and three tappings on the side for test cocks. These three test cocks are of the manual style, and are the only approved guarantee of water level indication. The gauge glass is not considered 100% reliable, since it is subject to plugging and dirt and may be unreadable, and the gauge valves may jamb closed.

The position and use of the test cocks have already been explained in Chapter 12, and the only thing we need add here is the possible use

of chain operators for them when the water column is too high to be reached conveniently, such as when installed on some of the larger watertube boilers where the drum is frequently mounted higher than about 8 ft from the floor to the drum centerline, which places the water column too high to be reached from the floor.

There are three general types of water level gauge glasses for use on boilers.

The first one, and the most common type, is the ordinary tubular glass direct-reading style made of Pyrex glass, and used only up to about 350 psig steam service. The main point we wish to make concerning this one is the susceptibility to breakage from thermal shock and from outside forces, such as being struck a blow. It is necessary to protect the gauge glass by vertical rods, plastic encasing panels, or a cover of wire mesh reinforced glass. See Fig. 13-2.

The next type of gauge glass you will encounter is the flat glass transparent type. This one has a level-indicating tube made up of two strips of flat, thick Pyrex glass, clamped between steel slabs, designed to permit light rays to pass through from front to back. This is shown in Fig. 13-3, along with a section of the construction. Appropriate gaskets must be used between the glass and the steel slabs, and suit-

Fig. 13-2. Low-pressure gauge glass. (Courtesy of Penberthy Houdaille.)

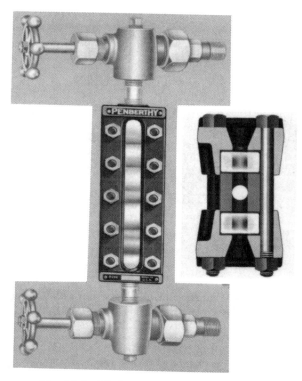

Fig. 13-3. High-pressure gauge glass. (Courtesy of Penberthy Houdaille.)

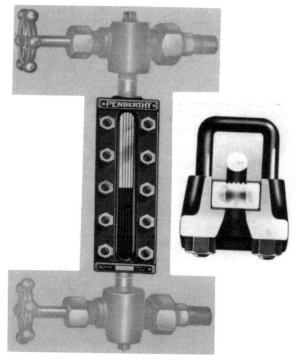

Fig. 13-4. High-pressure refractive gauge glass. (Courtesy of Penberthy Houdaille.)

able end pieces top and bottom must be provided. Also, this type often requires an auxiliary light source behind it to increase visibility of the water level inside the glass.

The third type of gauge glass is known as the refractive style, and it is shown in Fig. 13-4, with a cross section of its construction. The glass is a flat slab of Pyrex glass, similar to the transparent type, but only one is required in the front portion of the glass. This strip of flat glass has its rear side, the one exposed to the boiler water, inscribed with vertical grooves. Behind this glass strip is a hollow vertical column for the water, which has its water side coated black. Light entering the glass from the front is reflected to the viewer by those vertical grooves above the water, and that portion then appears clear.

Light entering the glass in that portion containing water is transmitted directly to the black surface, and that portion appears black. There is then a very clear difference between the wet and the dry section of the glass column.

Figure 13-5 shows a typical valve designed

especially for use on gauge glasses. Note that it has a ball check built into it. This ball check is designed to fall away from its seat by gravity during normal conditions, when no water or steam is flowing through the valve. Should the glass get broken or should a major leak develop in the packing and mounting of the glass, the flow of water or steam from the boiler out to the leak will cause the ball to be forced against its seat, closing off the flow.

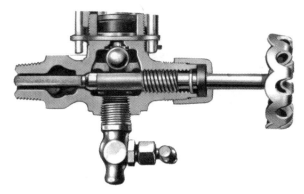

Fig. 13-5. Self-sealing gauge cock. (Courtesy of Penberthy Houdaille.)

The valve shown here is mounted on the bottom of the glass. A somewhat similar style valve may be installed at the top, in which case the test or drain cock may be removed and the steam pressure gauge installed.

There are several designs of valves on the market for this particular service, and they all have their particular place and use.

Figure 13-6 shows the installation of a typical water column on a firebox style of boiler. The installation on a package firetube boiler is similar.

Figure 13-7 shows a similar installation on a watertube boiler.

Both of these illustrations are simplified considerably, as often there are other accessories attached to the water column, which is permissible under very rigid conditions. The water column is the usual point of attachment for such auxiliaries as high- and low-water alarms and cut-outs. Each boiler manufacturer has a unique method of applying these accessories, and as long as they follow the applicable codes and do the job they were designed to perform, it is best not to tamper with their location, without first checking with the supplier.

Note that in Figures 13-6 and 7, there is a valved blowdown line at the bottom of the water

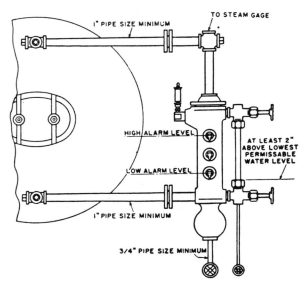

Fig. 13-7. Water column on a water tube boiler. (Courtesy of Clark-Reliance Corp.)

column. When performing the daily chores around the boiler, we strongly urge you to blow the water column down first, then the gauge glass. This will remove the larger particles of dirt first, so they will not be blown into the gauge glass when it is blown down.

D. Safety Valves

Probably the most important valve on the boiler is the safety valve. It can mean the difference between life or death, or at least, the difference between an operator keeping a job and losing it! It is so important that all boiler codes devote considerable space and attention to it. We give you brief points of the general requirements of the safety valves here.

Each boiler shall have at least one safety valve, and if the boiler has more than 500 ft^2 of heating surface, it shall have two or more safety valves.

Electric boilers are treated the same as fired boilers, except that each boiler over 500 kW power input shall have two or more safety valves.

One or more safety valves on the boiler shall be set at or below the maximum allowable

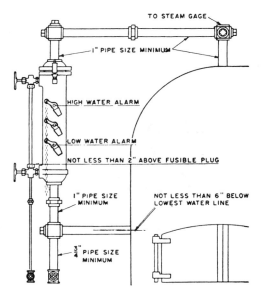

Fig. 13-6. Water column on a firebox boiler. (Courtesy of Clark-Reliance Corp.)

working pressure. If more than one safety valve is installed, the highest pressure setting shall not exceed the maximum allowable working pressure by more than 3%.

The safety valves shall be direct spring-loaded, pop-type, ASME standard construction and approved. The popping, or opening, pressure shall not be changed more than 10% above or below the popping pressure marked on the valve at the factory for settings up to 250 psig. For settings over 250 psig, the safety valve setting shall not be changed more than 5% above or below the factory setting.

The capacity of the safety valve or valves shall be such that they will discharge all the steam that can be generated by the boiler without allowing the pressure to rise more than 6% above the highest pressure at which any valve is set, and in no case to more than 6% above the maximum allowable working pressure.

The blow back, or blowdown as it is sometimes called, which is the difference between the opening and closing pressures of the valve, shall be set and sealed at the factory. It is usually about 4% of the set pressure or popping pressure.

Every superheater that is an integral part of the boiler shall have one or more safety valves near the steam outlet on the superheater, and these valves are to be set to open before the safety valves on the boiler drum.

The safety valve escape pipe must be supported independently from the valve and must be well drained and freeze-proof.

The safety valve capacity may be checked by making what is known as an accumulation test, providing the boiler does not have a superheater or a reheater. An accumulation test is performed by shutting off all normal steam outlets on the boiler, forcing the fires to the maximum capacity, and noting the pressures at which all the safety valves on the boiler open and close. Under these conditions, the maximum pressure reached by the boiler shall not be more than 6% above the highest setting of any safety valve, and in no case more than 6% above the maximum allowable working pressure of the boiler.

As you can surmise by the preceding requirements, weight-loaded safety valves are not permitted.

Perhaps the best way to illustrate the preceding rules is to give an example of a package Scotch Marine Boiler suitable for use at 125 psig operation. The boiler is designed for 150 psig maximum allowable working pressure, and it has a rating of 125 boiler horsepower, which means its expected output would be:

$$\text{Output} = 125 \text{ bhp} \times 32.3$$

$$= 4{,}050 \text{ lb/hr at } 125 \text{ psig}$$

$$(185°F \text{ feedwater temperature})$$

The total square feet of heating surface would be, assuming that the boiler has been rated using 5 ft^2 of heating surface per boiler horsepower:

$$\text{Square Feet} = 125 \text{ bhp} \times 5 \text{ ft}^2 = 625 \text{ ft}^2$$

As there are more than 500 ft^2 of heating surface in the boiler, at least two safety valves will be required.

With a maximum allowable working pressure of 150 psig, the highest setting of either safety valve would then be

$$\text{Maximum setting} = 150 + 3\%$$

$$= 150 + (150 \times 0.03)$$

$$= 155 \text{ psig setting}$$

It is improbable that an accumulation test would be run in the plant after the boiler is installed, owing to its small size and also to the acceptance being enjoyed by this style of boiler. However, just in case one were to be run, let us assume that the boiler is capable of an output of 4050 lb/hr at a pressure of 6% above the maximum allowable working pressure, then the total relieving capacity of both safety valves must be 4050 lb/hr at a maximum pressure of 150 + 6% = 159 psig.

Each safety valve, then, would be chosen to pass a minimum of 2025 lb/hr at an inlet pressure of 159 psig. The opening, or popping, pressure on the safety valves would be between 125 psig, which is the working pressure, and 155 psig, which is the maximum setting permissible by the Code. It is advisable to stagger the

settings on multiple-safety-valve systems, so
that they do not all pop off at once. The lowest
one should be set about 10% above the high-
pressure-alarm setting, with the remaining ones
set in steps above that.

We give here three more paragraphs from the
Code for your guidance, as found in the Cleaver-
Brooks Training Center Manual:

> No valve of any description shall be placed
> between the required safety valve or
> valves and the boiler, nor on the dis-
> charge pipe between the safety valve and
> the atmosphere. When a discharge pipe
> is used, the cross-sectional area shall be
> not less than the full area of the valve
> outlet or of the total of the area of the
> valve outlets discharging thereinto and
> shall be as short and straight as possible
> and so arranged as to avoid undue
> stresses on the valves.

> All safety valve or safety relief valve dis-
> charges shall be so located or piped as to
> be carried clear from running boards or
> platforms. Ample provision for gravity
> drain shall be made in the discharge

> pipe at or near each safety valve or
> safety relief valve and where water of
> condensation may collect. Each valve
> shall have an open gravity drain through
> the casing below the level of the valve
> seat. For iron-and-steel-bodied valves ex-
> ceeding 2 inch size, the drain hole shall
> be tapped not less than $\frac{3}{8}$ inch pipe size.

> When a boiler is fitted with two or more
> safety valves or safety relief valves on
> one connection, this connection to the
> boiler shall have a cross-sectional area
> not less than the combined area of inlet
> connections of all the safety valves with
> which it connects and shall also meet the
> requirements of paragraph PG-71.3.

Figure 13-8 shows a typical ASME-approved
safety valve for high-pressure boilers, which is
what we have covered so far in this section. The
ASME Code refers to boilers in this pressure
range as Power Boilers, as opposed to Heating
Boilers, which are 15 psig steam or under, and
hot water boilers under 160 psig and 250°F.

Figure 13-9 shows the manual lifting device
for this safety valve, which is necessary for pe-

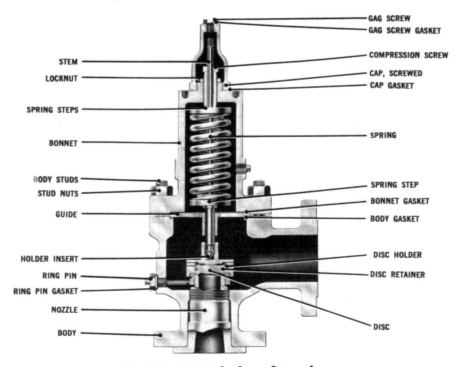

Fig. 13-8. ASME boiler safety valve.

riodic lifting of the disk to be sure that the valve inlet and seat are clear of any buildup of scale, dirt, or other obstruction. This lifting of the disk by the manual lever does not take the place of an annual pressure test performed by closing the steam outlet from the boiler and firing it until the safety valve lifts, and thus noting its popping pressure and the reseating pressure.

All repair or resetting of safety valves must be done by shops and facilities that are properly equipped and carry the authorization stamp of the National Board of Boiler and Pressure Vessel Inspectors.

Figure 13-10 illustrates the approved method of running the discharge piping from the high-pressure safety valve to a safe point away from the plant personnel. There are several minor variations of this method, but they all must meet the code requirements. The important points to keep in mind are:

1. Condensate from a simmering or leaking valve must not be allowed to accumulate above the valve seat or in the discharge line. A suitable drain system must be provided for it.

2. The discharge piping must be freezeproof.

3. The discharge piping must direct the steam to a safe point, and must not damage the surrounding structures.

4. When multiple valves are used, they may be installed on either individual drum nozzles or a short header, with as little piping as possible between the safety valves and the drum, and no intervening valves.

5. On higher pressures, care must be exercised in the installation of the safety valves to resist the reactive force exerted upon the inlet piping from the high-velocity discharge hitting the elbow at the base of the discharge pipe. This force acts in the opposite direction from the valve discharge, and places a high stress upon the valve inlet piping and the drum nozzle.

6. On high-pressure service, there may be a

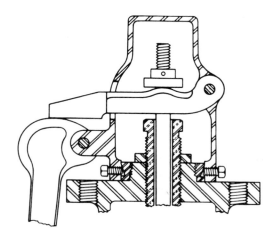

Fig. 13-9. Manual testing device. (Courtesy of Cleaver-Brooks Co.)

noise problem to be dealt with. Discharge mufflers are available for this purpose.

7. Expansion of the discharge piping must also be considered when the safety valve discharges. This is the purpose of the slip joint shown in Fig. 13-10, which will ensure there being no thrust or stress put on the valve discharge from the thermal expansion effects.

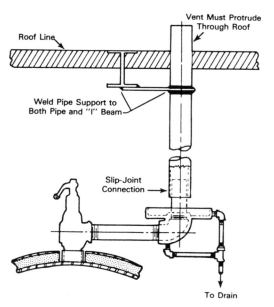

Fig. 13-10. High-pressure steam safety valve piping. (Courtesy of Cleaver-Brooks Co.)

On low-pressure steam heating boilers and hot water boilers, the safety valve discharge piping is much simpler, as the energy being released is much lower. However, there is the danger from scalding, which must be dealt with.

Figure 13-11 illustrates the much simpler piping arrangement required for low-pressure boilers and hot water boilers.

Safety valves for low-pressure steam are the same general design as for high pressure, except that bronze may be used in their construction.

Relief valves for hot water boilers are rated in capacity by the thermal input to the boiler, in BTUs per hour. This is primarily due to the large divergence in operating pressures and temperatures available for hot water boilers. The BTU rating is a very useful and much more uniform method than attempting to rate them by the various combinations of pressure and temperature and their relationship to the expected emergency discharges.

The relief valves must be of sufficient capacity to discharge all of the hot water built up in the boiler when the burner is firing at maximum capacity, and the water lines to and from the boiler are shut off.

E. When Safety Valves Are Neglected

All codes and manufacturer's instructions for boiler operation stipulate that the safety valves

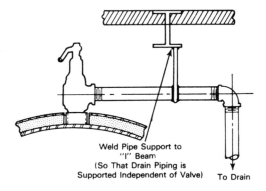

Weld Pipe Support to
"I" Beam
(So That Drain Piping is
Supported Independent of Valve)   To Drain

Fig. 13-11. Low-pressure steam safety valve piping. (Courtesy of Cleaver-Brooks Co.)

must be kept in safe and operable condition, and must be tested periodically, usually once a week at least. This is a reasonable requirement, since the safety valve is the one last safeguard against a possibly disastrous explosion.

However, we have to recognize that conditions in some plants are not as good as they should be, and that there are safety valves installed that are in dubious condition, and have not been tested for weeks or months. This usually comes about from past neglect and sloppy attitudes on the past operators. And no one since has mustered the nerve to correct the deplorable situation, but has merely carried on the dangerous policies of the former operators.

How does the responsible operator function under these conditions, being ordered to continue the mistakes of the past, without losing job or life?

The true answer to that question depends on the operator's position in the plant hierarchy, the operator's knowledge of the plant political situation, and the value the operator places on his or her job and life. We can only offer a few suggestions.

First, the operator should be well versed in what the local governing codes actually require and what the plant's insurance underwriter expects of the boiler plant. With this knowledge to back up his or her position, the operator should go on written record with the plant management, stating that an unsafe practice in violation of the code is being followed, and that he or she would like to see the situation rectified in the interest of all concerned. The statement should also list what steps must be taken to correct the problem. This approach is applicable to any condition in the boiler plant that does not meet the code requirements.

In the case under discussion here, the remedy is to replace the faulty safety valve at the earliest shutdown possible, send the faulty one out to be repaired, and place it into stores as a spare.

Another approach is for the operator to place in the log sheet at the same time each week, a statement such as, "the safety valves on boiler number _____ have not been tested this week, owing to plant policy prohibiting it." This will continually bring it to the attention of those higher up who check the log sheets, and will protect the operator should an accident occur.

There is one more step that we offer, with the strict understanding that it is not recognized by any code or state safety orders that we are aware of, as a valid substitute for a properly maintained and tested set of safety valves. It will, however, serve as some protection for the operator and may even save lives.

In addition to the alarms mentioned at the end of Chapter 11, a reliable pressure switch may be installed on the steam header or next to the pressure gauge connection. This pressure switch should be set at just over the upper pressure setting on the combustion controls, and connected into a very loud, distinctive sound-ing alarm bell. It may be located in the boiler room, but also at various locations throughout the plant where the proper officials are sure to hear it. It should be tested at least once a week, at the same time each week, and preferably while the Safety Committee is in session!

We have given these suggestions, and we leave it to you, the responsible operator, to decide to what degree the campaign for a safe boiler plant should be carried.

Be aware that recent court actions often strip away the cover given individuals in large firms, so that more frequently the blame and the penalties are placed on the people responsible for giving bad instructions and to those who follow them without question. So you may have more at stake than just your job and your life.

CHAPTER 14
THE BOILER ATTENDANT

A. Qualifications of the Boiler Attendant

It goes without repeating, of course, that the best equipment may fail miserably in its expectations if not properly operated and maintained. And yet, there are many areas in the country that have no legal requirements on the statute books governing the training and competency of the people who operate the boiler plants in the private industrial plants in the area. If the boiler plant happens to be in one of the state or federal institutions, then there usually are some minimum requirements that the operators have to meet in order to be employed.

Fortunately, the picture is not as bad as it appears, for all industrial plants carry some form of accident, fire, and liability insurance. Before they may be insured, the insuring agency usually exercises considerable authority over the basic minimum requirements of the plant equipment, operation, and layout. These insuring agencies have helped immensely in bringing a considerable degree of order and safety out of what could otherwise become a chaotic situation. The National Board of Boiler and Pressure Vessel Inspectors have played an important part in reducing the hazards in industry. There are other agencies that have helped in the battle, also, and almost without exception, they all take their lead from the ASME Boiler Code.

In this chapter, we shall devote some space to the discussion of the basic minimum abilities in operating a typical industrial boiler plant.

The first requirement for an effective and efficient boiler plant operator is the proper attitude toward his or her job and the equipment. This is true of any worker in any position, of course, but we must remind all prospective boiler plant operators that they are dealing with a highly concentrated source of energy. This requires a healthy respect for its power and the consequences should it get out of control. The boiler plant is no place for sloppy workmanship and a devil-may-care approach to the job.

The California State Boiler Code has a few statements of a logical nature that set down the qualifications of the boiler operator. The standards that they have stipulated as a guide for an employer to determine the competency of a boiler plant operator are as follows:

1. He shall be able to explain the functions and operations of all controls on the boiler or boilers.

2. He shall be able to light off the boiler or boilers in a safe manner.

3. He shall know all possible methods of feeding water to the boiler or boilers.

4. He shall know how to blow down the boiler or boilers in a safe manner.

5. He shall know what would happen if the

water was permitted to drop below the lowest permissible operating level.

6. He shall know what would happen if the water in the boiler were carried too high.

7. He shall know how to shut down the boiler or boilers.

If you read those requirements over very carefully, you will no doubt agree with us when we state that if a boiler plant operator knew only the things he or she needed to know to meet those requirements, he or she would not be much of a boiler operator. Obviously, there is much more to the job than the seven points listed. By the time you get through reading this book, you will see what we mean.

Many of today's industrial steam plants are not large enough to be able to afford specialization in the duties of their employees to the extent that some of our giants in the industries are able to do. The medium to small steam plant requires a broader knowledge and experience from its boiler plant employees. There are many more of this latter category in existence than there are of the former. We feel that this is probably an advantage to our industrial complex, since the engineer with the broad background of experience is often better material for advancement to higher positions in the plant than is the specialist.

Unfortunately, there appears to be at this time no uniform national body or code set up as a guide to the training or experience requirements for the operators of boilers or boiler plants. There are many state, county, and city codes set up throughout the country, but few of them are what may be considered complete coverage of the problem. There are still too many areas where the training and selection of the boiler plant operators is left up to the discretion of the employer. And, in too many cases, the employer appears to be more interested in getting help as cheaply as possible, without regard to the safety of either property or lives.

The trend is changing for the better in this respect, however. There is a gradual drift toward a tightening up of requirements for the employment of fully qualified boiler plant operators. All we can say is that this trend is long overdue. Too often, we have seen evidences of a bad fire or explosion due to the employer or plant manager attempting to save the owners about $6.00 an hour on boiler plant employees—which is the approximate difference between a plant guard and a qualified watch engineer.

There is one more requirement for the well-rounded boiler plant operator that we feel should be added to those already listed, and that is a true understanding of the part the boiler plant and the operator play in the overall plant system. The production of steam is not an end in itself, but is a very essential component of those total elements required to assist the firm in turning the raw materials into finished marketable products. This simply means that teamwork is required, a genuine feeling of cooperation, and a realization that the Production Dept., or its equivalent, is rightfully the one to call the shots in the plant. Regardless of the firm's end product, this is usually the case.

We do not mean to imply that the Production Dept. has the right to make unreasonable demands upon the boiler plant personnel, as few firms will permit that situation to exist. It is the responsibility of the boiler plant manager to at all times make the remainder of the plant fully aware of the limitations built into the steam-producing department and its personnel. In this respect, common sense and prudence should govern the boiler plant manager's actions, within the existing framework of the physical and political makeup of the firm.

B. Boiler Supervision

Now that we have set down the minimum requirements for the training and competency of the personnel who will be operating the boiler plant system, the next step is to determine just how much supervision a boiler should receive. This is not always easy to stipulate, since each steam plant system is unique and, therefore, must receive individual attention in devising the operating procedure manual or schedule.

Again, all we can do in this discussion is cover the essentials necessary for a typical average steam plant system.

In this matter, the California State Boiler Code has delved as deep into the minimum requirements as they did for the qualifications of the operator. The following summaries adapted from their Code will show you just what is expected of today's boiler operator as a minimum.

All boilers that come under this code shall be under the direct supervision of a responsible person who shall see that the boiler is operated safely by a competent attendant and that the boiler and its appurtenances receive the proper maintenance.

While the boiler is in operation, it shall not be left unattended for a period longer than it will take the water level to drop from the normal operating level to the lowest permissible water level in the water gauge glass or on the indicating device or recorders when the feedwater is shut off and the boiler is forced to its maximum capacity.

This requirement is modified if the following conditions exist:

1. The boiler is equipped with an audible alarm that will operate when the water reaches the highest and the lowest permissible operating level. If the boiler has no fixed steam or water line, the alarm must ring when the highest permissible operating temperature is reached.

2. The audible alarm shall be loud enough that it can be plainly heard at the most remote point from the boiler that the attendant is required to work.

3. The boiler is equipped with a low-water safety device that will shut off the fuel to the burner or burners when the water reaches the lowest permissible operating level. For boilers having no fixed steam or water line, the shut-off shall occur when the highest permissible operating temperature is reached. This device shall require manual resetting unless the pilot is equipped with a full safety pilot control.

4. The attendant shall personally check the operation of the boiler, the necessary auxiliaries, and the water level in the boiler at least once every hour, and oftener if necessary for the safe operation of the boiler plant. The operation of the automatic controls shall be checked at the beginning of each shift.

5. All float chambers of automatic controls shall be dismantled for inspection at the annual boiler inspection, or whenever the boiler inspector deems it necessary.

That is all this code contains covering the supervision of the boiler. As we stated previously, the average medium to small steam plant requires that the boiler plant operator be capable of much more than this minimum.

We wonder just how many boilers in today's industrial plants are being operated with the minimum required supervision. Is yours?

C. Training the Boiler Attendant

Now that we have a rudimentary idea of what is required in the way of operating the average industrial boiler system, just how do we go about producing our "perfect" attendant to perform this function?

Where does the aspiring boiler plant operator go for education in the managing of those larger boiler plants, which dot the industrial areas of our larger cities and counties?

Fortunately, owing to the rapid growth of our industrial complex, there have been various schools established in those same centers where one finds the need for the operators. At least, this is true to a large extent in the older, well-established industrial areas. In the centers of industries that have just recently—in the last few years—sprung up, there may be a problem, since the educational institutions in our society are not always able to keep pace with industry's requirements. The educational bureaucracy moves much slower than our industrial entrepreneurs.

The local union organizations are often able to organize suitable schools to meet local demands, and they have in many areas proven to be a very reliable source of trained boiler plant operators. Often, they work in collaboration with one of the local educational institutions, such as trade schools or high schools. These combinations are probably the best source, and the first one to approach should one wish to further his or her career along this line.

In some areas, there are what is known as "Junior Colleges," or some such similar school systems, which have courses of instruction for this level of industrial worker. They may be partially or wholly funded by either the state or federal government, as an incentive to produce jobs for the local economy. Often, they will be a part of the adult education program or a retraining program.

The state universities seldom reach this level of education for our industrial workers, usually concerning themselves with the higher level of professional worker, such as engineers, builders, architects, or industrial engineers.

For those aspiring boiler plant operators who live beyond the reach of any of the preceding organized instructional programs, there remains only one or two alternatives.

The first route to check into is the availability of a correspondence course or similar method of attaining a working knowledge of the boiler plant, such as one of the self-help courses now on the market. The next possibility is to obtain the necessary textbook, hire on as an apprentice or helper at a local boiler plant, and start a course of self-education on your own initiative. This may seem to be the hardest, but in the end, the result should be very rewarding.

The most effective way to learn the subject of boiler plant operation, we feel, is a combination of formal classroom work, accompanied by a few hours each day on the actual boiler plant site, either as a helper, an apprentice, or simply as an oiler, grease monkey, or whatever the "Boss" wants to call you. Many of our best operators (sometimes referred to as "stationary engineer") learned on the job, it is true. However, if they were to be questioned about it, they would all have to admit that they also spent a lot of time reading the technical literature and anything they could obtain on the subject of boiler plant operation, and thus they taught themselves. No doubt, it took them years of work and studying for them to get their experience and reputation.

As we stated elsewhere in this book, this textbook was written to give the young aspiring boiler operator a first step up the ladder to competency in this field. It is our fond hope that many who read this book will be spurred on to bigger and better levels of education and accomplishment.

One other suggestion we wish to make here, and one which should help the boiler plant apprentice on the rise up the ladder of success, is continually to be on the watch for literature that will upgrade his or her knowledge, competency, and worth to an employer. Never stop learning. Subscribe to at least one of the major trade magazines on the market, and read it carefully for ideas from other operators who have solved problems with which they have been faced and which may help you in some future difficulty. There are no "trade secrets" in this field, fortunately, and many operators freely exchange ideas through the trade journals and through engineering and operating organizations. One of the major ones, and one which has enjoyed an excellent reputation for years, is the National Association of Power Engineers, whose headquarters is at 176 W. Adams Street, Chicago, IL 60603. They have been in existence for over 100 years, and have chapters in most of the cities of any appreciable size in the country. They produce various educational pamphlets, which are of a very high quality.

In the past, the field of steam boiler plant technology has progressed steadily, but at a slower pace than other facets of our technology. The plant operator, with a little initiative, has been able, in most cases, to keep abreast of developments. This picture may change, now, however, as the energy crunch, and the environmental restrictions have been responsible for a whole new field of mechanical and electrical additions to the boiler plant.

The fields that show promise of being the next areas requiring mastering by the plant operator are pollution control, plant instrumentation, and computers. Many plants are very much involved in these fields, so let the boiler plant operator entering the market today be aware of the future demands on his or her abilities.

D. Training for Emergencies

This is another phase of boiler plant operation that is often overlooked until it is too late. When a plant institutes a safety program, the boiler plant should be one of the first areas included. Many industries today have reached the point, fortunately, where they hold regular safety meetings, and they make rules that are being enforced. The steam plant is one of the prime sources of fires in industrial plants, so it makes sense to show some concern for the engineering personnel.

Even if your plant does not have a safety program, this is no reason why you cannot carry on your own personal program. As you make your rounds over the area under your supervision, study each piece of equipment, and keep asking yourself what can go wrong with it. Once you have determined what trouble to expect from your equipment, it is then a simple matter to plan a program of action to take should that trouble develop. Learn to anticipate trouble, learn what each piece of equipment is expected to do, and learn how it does its job. Know the function of each piece of accessory on it. Develop the ability to visualize what is taking place inside a piece of equipment, behind the steel walls, as indicated by only the outward appearances visible to you.

As an example of what we mean, we shall list here a few accidents or conditions that can occur at any time; some of them could even happen to you. We will not tell you what action to take to correct the difficulty nor will we tell you how urgent the problem is. These things you must decide for yourself.

1. You are about 200 ft from the boiler under your care, and you hear the safety valve popping.

2. A worker is carrying a length of pipe past your steaming boiler, and the pipe demolishes the level gauge glass on the boiler.

3. You have opened the inside blowdown valve on the boiler, and are closing the outside blow valve after blowing down the boiler, and the valve stem breaks, while the valve is still open.

4. An electrician accidentally trips out the circuit breaker to the control panel on your steaming boiler.

5. You suddenly note that the steam pressure gauge indicates a steam pressure 10 psi above the safety valve popping pressure, but the safety valve has not yet lifted.

6. The V-belt drive on your draft fan breaks while the boiler is under full firing rate.

7. You are firing the boiler with oil, and a slug of water comes through the line and kills the fire, but oil flow resumes almost immediately.

8. You enter the boiler room and smell a gas leak.

9. You are tightening up on a leaking flange bolt on the steam line, and the bolt breaks, and you cannot shut the system down.

10. You enter the boiler room and note that the water is out of sight in the bottom of the level gauge glass, but no alarms are ringing, and the boiler is still operating in an otherwise normal manner.

To review what is expected of you as a responsible boiler plant operator, in case of an emergency in your plant, let us list our suggested program:

1. Know what can possibly go wrong with each item of equipment in your plant.

2. Know how to detect trouble in each piece of equipment.

3. Have a well-rehearsed plan of action in mind when trouble does develop, and be sure others who may be involved are also aware of the plan and what is expected of them.

4. Know what spare parts must be kept on hand to correct the trouble, and know where they are kept.

5. Be aware of your participation in any other emergency that may take place in other areas of the plant.

6. Be sure your plans are known by management, and approved by them.

7. In addition to all of the preceding responsibilities, the boiler operator should be aware of his or her part in case of area-wide catastrophies, such as block fires, earthquakes, windstorms, floods, etc. Be sure you have a plan of action to safeguard your plant when any of the above takes place and that the plan is tied into any that the plant has formulated.

8. As you should be aware by now, the key word is *responsibility*.

CHAPTER 15
GETTING THE BOILER ON THE LINE

A. Prestart Check: Water and Steam Systems

Let us assume that it is early Monday morning. You have just come on duty to start up the boiler in the plant and get steam up in time for the first shift operation. The boiler was shut down Friday night, and there have been a few overtime workers in the plant over the week end, who did not require steam for their work.

We shall list the primary items that you should check, and the proper procedure for getting the boiler on the line and steam up to pressure. This will be only a general guide, since there may be some items which must be modified or omitted, or some new ones inserted, depending on your particular situation.

1. Take a tour of the boiler room and check to be sure no unauthorized people have been in the place and caused damage, such as a fork truck knocking into the equipment or piping. Is everything as you left it on Friday night?

2. Check the water level in the gauge glass to be sure it is slightly below the midpoint of the glass, and be sure all gauge glass valves are open.

3. Test the true water level by opening the three test cocks, starting at the top one and working down.

4. Open all water supply valves in the feed-water pump systems.

5. If the water level tests low, open the manual feed valve to bring the water level up to slightly below the midpoint of the gauge glass.

6. Turn on the main switch to the boiler feedwater pump.

7. Check the make-up water supply system into the hotwell. Check it all the way back to the source, where it enters the boiler room. Be sure it is ready to feed water into the hotwell automatically.

8. The boiler vent valve, superheater drain, and any steam line vents should all be open to bleed the air out as steam is raised.

9. The main steam stop valve bypass should be open into the distribution main. When the main is up to pressure and temperature, the bypass valve should be closed, and the main steam stop valve opened.

10. Open all steam header drain valves or by-passes around steam traps.

11. Be sure the water treatment system is in full operating condition.

12. If a steam atomizing system for heavy oil firing is installed, be sure the drain for

it is open, to discharge any accumulation of water.

Some of the above items may sound infantile, but just ask yourself: on Monday morning, are you absolutely sure you left everything in proper order when you left Friday night? Any one of the above valves left closed when it should be open, or vice versa, could cause you much trouble when you attempt to get steam up and on the line.

We shall take each item in turn, where applicable, and discuss it.

1. Many boiler rooms are wide open to the remainder of the plant, and cannot be secured during off hours. Some boiler plant equipment may even be in an area that is considered an open thoroughfare by the plant personnel. In these cases, the operator must be on guard to be sure that no one has tampered with the boiler or its accessories while the operator was absent. Warning! Watch your tools and supplies in this case, also.

2. Self-explanatory.

3. The test cocks are the only reliable method of checking the level of the water in the boiler. We suggest you start at the top and work down, leaving the top one open until the water level has been verified. This is to relieve any vacuum that may be present in the boiler, which would also alert you to the fact that the vent valves are not open.

4. Also check the water supply pressure ahead of the boiler feedwater pump and ahead of the water treatment plant or hotwell.

5. Caution: do not overfill. Experience will teach you just how much water should be in the gauge glass before you light-off. Too much water will cause priming and carryover when the burner is on, since the water will swell due to expansion, and in some boilers this may carry the water level too high for proper operation.

6. Self-explanatory.

7. Float-operated level control valves have a habit of getting stuck when left down on their seat for any length of time. Also, the float may corrode and leak, causing it to fail to close when it should. A leaking float will overfeed, and may send water out the overflow or out the vent.

8. As steam pressure reaches about 5–10 psig, the vents should be closed, starting at the boiler, and proceeding out to the ends of the main. Close them only when steam issues from them under a fair jet, with only steam or vapor visible, and no air. Superheater drains should be left open until steam is being sent into the plant system.

9. The steam distribution main should be warmed up slowly to prevent violent waterhammer. If there is no valved by-pass around the main steam stop valve, then the valve will have to be cracked open for warmup, then opened all the way when the main is at pressure and temperature for plant operation. This is hard on the valve seat and will increase maintenance.

10. As the start-up condensate is discharged, and steam begins to issue from the drains, they should be closed. From then on, the steam traps should handle the condensate that forms from normal operation.

11. Self-explanatory.

12. This is important, since any water remaining in the steam atomizing line could cause a flame-out of the burner, with the possibility of a disastrous explosion when the fuel oil lights from the hot refractory.

We shall now give you the start-up procedure published by the Hartford Steam Boiler Inspection Company. These procedures are, again, purely general in nature, and it is left up to the individual operator to use both of the procedures given in this section for devising his or her own system to match the equipment.

Preparing Boiler for Service—Before filling a boiler for service, make certain that it is free from scale, oil, tools, and other foreign matter and that there is no one in the boiler. Manhole and handhole gaskets should be new or in good condition. Use a graphite paste or other suitable paste to prevent sticking. Special gaskets should be installed in accordance with the manufacturer's instructions.

Filling the Boiler—Do not refill a boiler while it is hot. Treated warm water should be used, if available, and the boiler should be vented to permit the air to escape, while filling. Fill to the normal working level, or higher where directed by the manufacturer.

Starting the Boiler—Start the boiler early enough so that steam pressure may be raised without forcing the boiler. Before starting a fire, check all parts of the fuel-burning system as far as possible. Make sure all dampers and other draft equipment are in working order and in proper position. Do not use highly flammable oils or gasoline in starting fires where solid fuel is used. On gas-, oil-, or pulverized-coal-fired boilers, all ignition equipment and electrical interlocks, if used, should be checked to determine that they are in proper operating condition. DO NOT TURN ON THE FUEL SUPPLY AT ANY TIME UNTIL THE FURNACE HAS BEEN VENTED THOROUGHLY. If a lighting torch is used, make sure it is in the proper position. If the fire goes out, the fuel supply should be shut off immediately. No attempt should be made to restart the fire until the furnace has been ventilated thoroughly and the cause of the flame failure is found and corrected.

Cutting-In the Boiler—In placing a boiler on the line with others that are already in service, keep its stop valve closed until the pressure in the boiler and the steam main are approximately equal. See that the steam piping is thoroughly drained before the valve is opened. Open the valves slightly; then, if there is no unusual jar or disturbance, complete the opening slowly. Close the valve at once if there is the slightest evidence of an unusual jar or disturbance in the boiler or piping.

Water Level—THE MOST IMPORTANT RULE IN THE SAFE OPERATION OF BOILERS IS TO KEEP WATER IN THE BOILER AT PROPER LEVEL. NEVER DEPEND ENTIRELY ON AUTOMATIC ALARMS, FEEDWATER REGULATORS, OR WATER LEVEL CONTROLS. When going on duty, determine the level of water in the boiler. The gauge glass, gauge cocks, and connecting lines should be blown several times daily to make sure that all connections are clear and in proper working order. The gauge glass must be kept clean because it is of extreme importance that the water level be accurately indicated at all times. IF THERE IS ANY QUESTION AS TO THE ACCURACY OF THE WATER LEVEL INDICATED, AND THE TRUE LEVEL CANNOT BE DETERMINED IMMEDIATELY, THE BOILER SHOULD BE REMOVED FROM SERVICE AND ALL WATER LEVEL INDICATING ATTACHMENTS SHOULD BE CHECKED.

B. Prestart Check: Fuel Systems

Here, again this list is purely a guide, being all-inclusive to cover all fuel oil and fuel gas supply systems to the boiler. As a result, all of the following points may not be applicable to your situation, but you should pick and choose the ones that suit your systems.

1. Check the furnace for obvious signs of fuel leaks due to failure of the fuel valves to shut off tight.

2. Be sure the gas supply pressure is up to the required amount for proper feed into the fuel train of the boiler.

3. Check the fuel oil temperatures. Light oil may not be critical, but heavy fuel oils will have to be brought up to the proper pumping temperatures before the final heating system can receive the fuel oil.

4. Be sure all light-off fuel valves are in final position for the light-off sequence. The vent valves between shut-off valves should be closed.

5. Be sure all control panel lights are in "GO"

position. If not, check the faulty condition and clear it off before proceeding further.

Our discussion of the preceding points follows, with comments and explanations covering them:

1. If there is an obvious fuel leak, the draft fan should be started immediately to clear out the furnace. After a few minutes of operation, the leaking valve should be located and repaired. First, of course, the gas line shut-off valve should be closed. And puddles of fuel oil accumulating in the furnace must be cleaned up, with the draft fan running to clear away the fumes. The leaking fuel valve must be fixed before proceeding with light-off.

2. Self-explanatory.

3. If the atomizing fuel oil heater is steam fed, and the boiler you are lighting off is the only one, or the first one in the battery, then it will be necessary to light off on gas or light oil, until steam pressure is sufficient for the heater to raise the heavy fuel oil up to atomizing temperature. The fuel oil pump inlet suction in the fuel tank should have an auxiliary electric heater, if conditions require it, to enable the heavy fuel system to get on the line without steam.

4. Once the fuel supply valves are lined up for light-off, it is a good idea to give one final check to be sure none of them are leaking fuel into the furnace. Minor leaks generally are no problem, and may be expected from control valves that are not usually tight shut-off valves.

5. Self-explanatory.

C. Prestart Check: Air Systems

What we mean to cover here is the combustion air system and the instrument air system, as well as plant air used for atomizing the fuel oil.

1. Boiler room intake air dampers should be open and clear of all obstructions. If heating of the intake air is required, the heating coils should be in full operation, if steam is available.

2. Be sure both the primary and secondary air dampers are set and clear in readiness for automatic operation.

3. Check the damper control system to be sure it is ready to operate.

4. Check the flue dampers for readiness to operate.

5. Be sure the draft fans are clear, run free, and ready to operate.

6. Check the instrument air supply lines. Be sure they are clear of collected water and are up to pressure.

7. If air is being used for atomizing the fuel oil, clear it of all condensed water, and check its pressure. It is well to leave the drain valve open a small amount to drain off any water that may have accumulated in the air line. This is important for oil firing, since water in the atomizing line could cause a flame-out.

We shall not make any additional comments on the preceding points, since they are all self-explanatory.

D. Manual Light-Off on Fuel Gas

It is rather difficult to write a precise and clear-cut set of rules and procedures for manually lighting off a boiler, without knowing the complete system of burners, fuels, controls, and draft system involved. We will give you a general procedure, suitable for adapting to your particular set of conditions and equipment.

At this point we issue a general warning, regarding both manual and automatic light-off of a boiler. When the process of ignition of either the pilot or the main burner is taking place, NEVER stand in front of the burner. It may under some conditions be projected out of the holder with some force, which could badly injure anyone standing in front. The burner front is one of the weakest parts of the fireside of the boiler, and any of the appurtenances in the burner front could become a dangerous missile should a furnace explosion take place.

Equally as important, be sure that no one else is standing in front of the boiler when lighting-off is taking place.

First, a few words on the natural draft system start-up.

The natural draft system should be considered similar to an induced draft system, for the natural draft stack simply replaces what is normally the function of the induced draft fan. The only difference is that the natural draft stack induces a negative draft in the boiler by means of the difference in weight between two columns of atmospheric air, one of which is ambient temperature, and the other column which is lighter due to its increased temperature. Instead of starting the draft by pushing a switch to an electric motor, in the case of the natural draft stack it is necessary to heat up the air in the stack. This is only true in the older installations, and those on oil firing of larger size. In the small units, and especially those on fuel gas, the draft requirements may be so low as to require a draft-limiting device in the base of the stack, to cut down the draft. One of these devices is known as a draft diverter, common on most small gas-fired appliances, described in Chapter 4, Section C.

However, in those installations involving a tall natural draft stack, it may be necessary to heat the air in the stack to start it drawing. This may be done by means of a gas burner in the base of the stack, or simply by means of a small wood or kerosene fire, prior to lighting off the boiler burner.

The general procedure for manually lighting off a boiler on natural gas follows.

1. Start the draft fan and purge the furnace for the prescribed length of time, usually a minimum of four complete air changes at a fan flow rate of not less than 50% of its normal design rate. This may vary by local codes or the insuring agency's requirements.

2. Switch on the electric igniter, and while watching through the peep hole in the burner front, open the gas valve to the pilot burner.

3. Watch the pilot long enough to ensure a stable pilot flame.

4. Turn off the electric igniter.

5. Set controls for low fire start.

6. While watching the burner through the peep hole, turn on the main gas valve slowly, and watch for a stable flame.

7. When a stable main flame has been established, set the controls for full automatic firing.

8. Turn off the pilot fuel line valve.

If steps 3 and 7 do not prove out within a few seconds of opening the fuel gas valve, then shut the valve and turn off the igniter. The entire procedure from step 1 must be repeated, until a stable main burner flame is established.

E. Manual Light-Off on Light Oil

This procedure is very similar to that previously given for lighting off on fuel gas. We are assuming that the pilot flame is on gas and that the light oil is sufficiently volatile to be lit off from the gas pilot flame. This may require a fairly large pilot flame.

Steps 1–4 are the same procedures given for fuel gas above.

5. Set controls for low fire, and open up all fuel supply valves to the fuel pump, and up to the last valve entering the burner.

6. Start the fuel pump, and if there is a re-

circulating oil line back to the fuel tank, be sure the fuel oil is recirculating.

7. While watching the pilot flame through the observation port, open the last fuel oil supply valve into the burner gun, slowly.

8. Watch for a stable flame at the burner tip.

9. Set controls for full automatic operation, once the flame has been stabilized.

10. Turn off the pilot fuel line valve.

If the pilot flame or the main flame does not get established and stablized within a few seconds of turning on the fuel line, shut down completely and repeat the procedure from the beginning, until the main burner is producing the required flame.

F. Manual Light-Off on Gas, Switching to Heavy Oil

In this case, the usual arrangement is to use a combination gas and heavy oil burner, similar to that shown in Fig. 7–10. The flame is established, and the boiler put on line with the use of fuel gas, following the general procedure given in Section D. We shall take this procedure as a start, and continue it for the case of switching the burner to heavy fuel oil. Furthermore, we shall assume that the fuel oil system is the usual one for this type of service. There is a suction heater in the fuel storage tank to bring the oil up to pumping temperature, a final fuel oil heater using steam to heat the oil up to the proper temperature for atomizing, and steam is used for atomizing at the burner tip. The oil circulating system has arrangements for a continual recirculation of hot oil to the storage tank. The procedure, then, continues as follows:

1. Once the flame is established on fuel gas, allow the boiler to come up to pressure,

closing all the steam vents and drains as the pressure rises.

2. Turn steam into the suction heater in the storage tank.

3. When the oil in the tank is hot enough for pumping, open all the oil line valves through the pump, into the final heater, and through the recirculation system back to the storage tank.

4. Turn on the fuel oil pump, and check the fuel oil system for proper pressures and temperatures.

5. Reduce the burner firing rate to minimum flame on gas.

6. Open the condensate drain on the atomizing steam line into the burner, and clear all water from the line. Leave the valve open slightly, if there is no steam trap on the bottom point of the line. If there is a steam trap installed, close the drain valve completely, and let the steam trap take over the job of getting rid of the condensate in the line.

7. Watching the flame through the observation port, turn on the atomizing steam slowly, part way only, making sure that the flame is not affected.

8. Turn on the flow of fuel oil into the burner, slowly, watching the effect upon the flame.

9. Slowly increase the flow of fuel oil and atomizing steam while simultaneously cutting out the fuel gas.

10. When the flame is stable on heavy oil, place the burner combustion control system back into full operation.

11. Secure all fuel gas supply valves.

Of course, if this procedure results in a flameout at any time, then it is necessary for you to shut everything down, and start at the beginning of the light-off procedure on fuel gas.

Some larger systems go through an intermediary step on light oil, lighting off on gas, switching to light oil, then to heavy oil. In this case, the procedure is simply a combination of the three given for all of the fuels.

When using air for atomizing the oil, the procedure is the same. There are also systems where atomizing may be done by either air or steam, and the switch may be accomplished during full firing.

G. Dual Fuel Firing Switch-Over

In Chapters 6 and 11 we covered briefly the subject of combination burners, which are capable of firing either gas or oil, one at a time, or both simultaneously. We gave you a brief idea of what is involved in making the transition from one fuel to another, depending on the style of burner and the combustion control system. We also emphasized, as we have done many times in this manual, that the manufacturer's directions for your particular system must be followed carefully, and we issue the same qualification in giving you the following procedures for switching fuels while the boiler is in operation and on the line, for those boiler–burner units properly equipped to permit it.

First, we shall cover the procedure of switching from gas to oil, assuming that the boiler has been on gas.

1. Place the pilot light in service, following method in Section D, items 1 through 4, and observe it for several minutes to ensure that it is stable.

2. Be sure the manual oil valve at the oil gun is definitely closed, by backing off on it slightly and then closing it.

3. Place the fuel oil system in readiness sufficiently to satisfy all of the system's interlocks.

4. Place the oil atomizing system in operation, up to the last valve to the burner gun.

5. Drain all condensate out of the atomizing system, and check the condensate removal system for proper operation.

6. Switch the combustion control system over to "Manual" operation.

7. Following the burner and control supplier's instructions, place the oil control valve in the normal position for lighting off.

8. The fuel transfer switch should now be placed in the "oil–gas" position.

If all is going according to plan, the oil safety interlocks should now be satisfied, and the oil safety shutoff valve will open, placing full oil pressure upstream of the manual oil valve at the burner gun.

9. Slowly close the manual gas shutoff valve, watching the pressure carefully until it starts to drop, at which point the gas flow rate is being controlled by the manual valve instead of the normal gas flow control valve.

10. This is the critical point in the operation, and it requires very careful attention on the part of the operator. With one hand open the manual oil valve and simultaneously with the other hand slowly close the manual gas valve, permitting the oil to be ignited by the gas flame. Increase the oil flow and decrease the gas flow steadily, watching the results through the observation port to maintain a constant flame in the furnace, until the oil valve is fully open and the gas valve has been closed. During this operation, the fuel/air ratio must be maintained at all times, either by watching the flame or by whatever method has been provided in the equipment. The air flow rate should be held constant during the switch-over process described above.

11. Switch the fuel transfer system into the "oil" position, which will cause the gas safety shutoff valve to close.

12. It is now time to place the combustion control system and burner firing rate in normal operation, as required by the equipment furnished.

WARNING!

1. Do not simply walk away from the boiler at the end of the preceding operation. Continue to observe the operation for some time to ensure that nothing goes wrong with the sytem.

2. Do not stand directly in front of the burner while making the change in fuels described. It may take a little ingenuity on your part to accomplish this, but your life may very well be at stake.

Now the procedure for changing from oil to gas will be given, with the same provisions, warnings, and qualifications understood to be applicable.

1. Place the pilot light in service, as covered previously, and observe it for several minutes to ensure stable firing.

2. Check the manual gas valve at the burner to be sure it is definitely closed.

3. Place the gas fuel system in readiness sufficiently to satisfy all of the system's interlocks.

4. Switch the combustion control system over to the "Manual" position.

5. Following the burner and control supplier's instructions, place the gas control valve in the normal position for lighting off.

6. The fuel transfer switch should now be placed in the "gas–oil" position.

If all is going according to plan, the gas safety interlocks should now be satisfied, and the gas safety shutoff valve will open, placing full gas pressure upstream of the manual gas valve at the burner.

7. Slowly close the manual oil valve, watching the pressure carefully until it starts to drop, at which point the oil flow rate is being controlled by the manual valve instead of the normal oil flow control valve.

8. This is the critical point in the opera-

tion, requiring very careful attention on the part of the operator. With one hand slowly open the manual gas valve and simultaneously with the other hand slowly close the manual oil valve, permitting the gas to be ignited by the oil flame. Increase the gas flow and decrease the oil flow steadily, watching the results through the observation port to maintain a constant flame in the furnace, until the gas valve is fully open and the oil valve has been closed. During this operation the fuel/air ratio must be maintained at all times, either by watching the flame or by whatever method has been provided in the equipment. The air flow rate should be held constant during the switch-over process described above.

9. Switch the fuel transfer system into the "gas" position, which will cause the oil safety shutoff valve to close.

10. It is now time to place the combustion control system and burner firing rate in normal operation, as required by the equipment furnished.

11. Close down the steam atomizing system, opening the condensate drains.

NOTE! The same warnings apply to this operation as was described previously for switching from gas to oil firing.

H. Start-Up on Automatic Controls

The procedures we have given you so far are based on manual start-up, following the general recommendations of the various insurance agencies. You may never have to follow this procedure yourself, since many of today's package boilers come equipped with all of the operations for start-up incorporated into one control package, completely programmed and timed to handle the entire process of starting up. In this case, all that is necessary for you to do, after performing all of the prestart checks given earlier in this chapter is to push the "Start" button, and stand by in case of trouble. If the control panel should kick out for any reason, you should be on hand to start the cycle over again and to correct any deficiencies that may be causing the panel to recycle.

You should know the complete cycle timing and procedure programmed into the control system. In case it is necessary for you to switch to manual start-up, there will thus be a minimum of change in the procedure. One of the most important things you should know are the times allowed in the programmed procedure for proving the pilot and main flames. If you must start the boiler on manual procedure, pending repair of a faulty control panel, then it is important that you follow those same times when attempting to light off both the pilots and the main flame.

It is a good idea to have the complete program typed on a card and posted in a handy place, such as on the control panel front or inside the cover. Follow it exactly when on manual start-up. To deviate from it appreciably is inviting disaster.

Of course, the programmed control panel does not release you from the responsibility of performing all the other things that have to be done to get the boiler on the line. You still must manipulate the vents and drains, and monitor the system for at least 1 hr after start-up, as described in the next section.

Of course, the actual time of lighting off should be entered into the log sheet, with any trouble encountered, along with the initials of the person responsible for the light-off. Also, if fuel is switched, the time and the switch-over fuel should be entered.

I. First Hour after Light-Off

So now the boiler is on the line, steam pressure is rising, and it's time to grab a cup of coffee in the cafeteria?

Wrong! Stick around for a while, your morning chores are not over.

You are sending steam out into the plant, and you are certain that everything appears well in your boiler room. But how about the end users of that steam? Do you know what is happening to it out at the other end of the steam main? How do you know that there isn't an emergency taking place in the plant's process equipment? Just because you have performed your duty well in getting the boiler on the line, does not mean to say that everyone in the remainder of the plant have performed their work with the same diligence. Remember, it is the first hour after light-off when most problems appear in the plant processes. The equipment has been shut down for several hours, air has leaked into the steam spaces, corrosion has been taking place, and, to top it all off, the valve and equipment manipulation at start-up in the morning is not always done properly. The net result is that then is the time when trouble is very apt to appear.

Although you have done your best to get the plant operating again for another profitable day, you had better be extra alert for at least an hour after light-off.

Some of the things you should be watching or doing during that first hour are given here, some of which should be done several times during the first hour:

1. Watch the water level very closely, and check it with the test cocks.

2. Check the boiler feedwater system completely for proper operation.

3. Check the burner flame visually through the peephole.

4. Watch the stack for presence of smoke.

5. Watch the steam pressure fluctuation in correlation with what is normal and what is happening in the plant.

6. Be sure all air systems are operating properly and normally.

7. Check the water treatment system, and replenish the chemicals, if necessary.

8. At least once a week, if instruments are available (and they should be), check for proper flue gas composition and boiler operating efficiency.

Now that everything has been checked and is operating properly, you can go to the cafeteria, draw a cup of coffee, and sit down and relax for a few minutes. Then it's back to the boiler room to commence the daily routine.

C H A P T E R 1 6
REGULAR OPERATING PROCEDURES

A. General Requirements

The boiler is now on the line, steaming merrily, and you are sitting comfortably in the cafeteria with a cup of coffee, well pleased with the world.

What's next? You may ask yourself.

We shall now discuss the ordinary, everyday operating procedures that have proven to be the best practice for a long and trouble free life of the boiler plant. This discussion will be in general terms, first, followed by more specific directions as published by firms in the regular business of supplying, insuring, or testing of boilers. These instructions will be quoted directly from their published literature, and will consist either of their entire material on the subject, or a few pertinent selections. There are additional instructions in the Appendix.

It would be nice if we could devise one all-encompassing set of rules of procedure for you to follow in your specific installation, but that would be very difficult to do. There are so many different styles and makes of boilers, burners, and control systems on the market, and so many different operating codes and insuring agencies, that we are forced to take the approach that follows.

By studying each approach given in the following sections of this chapter, you will pick up a number of very useful hints or ideas, which are no doubt applicable to your installation. So we recommend you study each one carefully,

and keep asking yourself if that particular point or idea can be of use to you, and help keep you out of trouble.

Thus, by combining these ideas with the instructions that you have received from the boiler manufacturer, the end result should be a very valid and safe set of procedures. In addition to which, they will be a set of rules that you devised, and thus you will be familiar with them. There should then be no reason why you should not be able to embark upon years of successful boiler plant operation, with a minimum of difficulties to beset you.

This material is to supplement any instructions that you have on hand from the boiler supplier, and should not be construed as replacing those instructions.

We shall first give an over-all rundown of the procedure, as follows:

1. Water column and water gauge glass blowdown—at least once each day.

2. Inspect all joints for tightness and leaks—once a day.

3. Check high pressure and instrument air systems for lubrication and moisture removal—once a day.

4. Check stack temperature—several times a day.

5. Check for visible stack emissions—several times a day.

6. Regulate continuous blowdown rate in accordance with instructions from water-treatment personnel—at least once a day, as required.

7. Bottom or mud blowdown, intermittent—once a shift, or as directed by water-treatment personnel.

8. Clean all fuel strainers—once a week.

9. Test all controls, safety devices, and safety relief valves—once a week.

10. Clean oil burner guns and torches—once a month.

11. Check burner tips, igniter, and diffusers—each day.

12. Fill out the log sheet accurately and in detail—as directed.

B. The Cleaver-Brooks Approach

This first set of instructions will be from the Cleaver-Brooks manual entitled "Cleaver-Brooks Training Center Manual." It covers both steam boilers and hot-water boilers. The actual burner and combustion systems are similar, the main difference being in the controls. Steam boilers are controlled by steam pressure, while hot-water boilers are controlled by the temperature of the circulating water.

BOILER ROOM CARE

A summary of boiler room maintenance should consist of the following:

A. General Maintenance

1. Know your Equipment! This information is available from manuals and field service engineers.

2. Maintain Complete Records! Each individual component should be listed on an index card as to model, serial number, and date of installation. Get part number information from manufacturer or field engineer.

3. Keep Written Operating Procedures Up dated! The ideal time to originate this procedure is when equipment is placed in use. A detailed start-up procedure is essential to standardizing the boiler room routine. Posted procedures consistently remind you of things to be done and help the inexperienced man to assume his proper role.

4. Good Housekeeping, A Must! Housekeeping is the one unalterable truth as to the quality of boiler room maintenance as well as of top management of the company.

5. Keep Electrical Equipment Clean! The most common cause of nuisance electrical control problems is failure to maintain this equipment properly.

6. Keep Fresh Air Supply Adequate! Proper combustion and burner operation require adequate air. Filters must be kept clean, and in severe winter areas it may be necessary to heat the room to an acceptable ambient temperature.

7. Keep Accurate Field Records! Do not depend on the supplier solely. A system of recording fuel consumption can keep you informed of any unusual fuel demands enabling you to spot a problem or waste before it gets out of hand.

B. Daily Maintenance

1. Check Water Level Controls! An unstable water level can indicate several problems such as excessive solids or treatment, contamination from oil, overload or mechanical malfunction.

2. Check Pressure or Temperature Controls! Excessive steam pressure or water temperature drop will alert you to excessive loading on the boiler.

3. Check Burner Operation and Its Controls! Maintaining top efficiency is the simple

and basic reason for having operating personnel.

4. Check All Motor Operation! By developing a routine on this any change in operation or bearing temperature will usually be caught in time to avoid a failure.

5. Check Auxiliary Equipment! There is a vast difference between "is it running" and "is it running properly." Take nothing for granted in this area as the auxiliaries can shut your operation down as well as a boiler failure.

C. Weekly Maintenance

1. Check Operation of Water Level Controls! By stopping the boiler feed pump and allow controls to stop burner under normal conditions.

2. Blow Down Water Level Controls! To purge float bowl of possible sediment accumulation. Operating conditions will dictate frequency of this check.

3. Check Safety or Relief Valve Operation by manually operating (opening) with not less than 80% of working pressure on the valve.

4. Check and Clean Burner Assembly by using the proper tools. The parts are precision made and must be handled carefully.

5. Check and Clean Pilot Assembly paying particular attention to the spark gap and to the insulator condition.

6. Check Flame Scanner Assembly for tightness and cleanliness.

7. Check Lubricating oil Level on atomizing air compressor assembly. Add oil or change according to manufacturer's recommendations.

8. Check the Gas Analysis with previous readings to determine if burner operation has been satisfactory. $$$ can fly out the stack if this is neglected.

9. Check Boiler Operating Characteristics by manually sequencing the unit. Close off fuel supply and check flame scanner reaction timing and flame failure timing. Restart and observe light-off characteristics.

D. Monthly Maintenance

1. Review Boiler Blowdown so that a waste of treated water is not occurring. Check water treatment and testing procedures.

2. Review Burner Operation! Take the flue gas analysis over entire firing range comparing these and the stack temperature readings with previous month.

3. Check All Combustion Air Supply inlets to boiler room and burner.

4. Clean Filters on atomizing air compressor. Check lubrication levels.

5. Check Fuel System to make certain that strainers, vacuum gauges, pressure gauges, and pumps are properly cared for.

6. Check All Belt Drives for possible failure. Tighten V-belt sheaves and make certain that belts operate with proper tension.

7. Check Lubrication Requirements of all bearing supported equipment. Do not over-lubricate electric motors.

8. Check Packing Glands on all pumps and metering devices. Proper tension on packing glands will extend life of equipment.

HOT-WATER-BOILER OPERATION

Peculiar as it may seem, some hot water systems are installed with, what appears to be, very little regard for the proper operation of the boiler. The following points should be considered carefully during the layout of a hot-water system:

Boiler Water Temperature—Boilers constructed in accordance with Section IV, Low Pressure Heating Boilers, of the ASME Boiler and Pressure Vessel Code can be operated with water temperatures up to 250°F.

Boilers for operation over 250°F must be

constructed in accordance with Section I, Power Boilers, of the ASME Boiler and Pressure Vessel Code.

Minimum Boiler Water Temperature—The minimum recommended boiler water temperature is 170°F. When water temperatures lower than 170°F are used, the combustion gases are reduced in temperatures to a point where the water vapor condenses. The net result is that corrosion occurs in the boiler and breeching.

This condensation problem is more severe on a unit that operates intermittently and that is greatly oversized for the actual load. This is not a matter that can be controlled by boiler design, since an efficient boiler extracts all the possible heat from the combustion gases. However, this problem can be minimized by maintaining boiler water temperatures above 170°F.

Another reason for maintaining boiler water temperature above 170°F is to provide a sufficient temperature "head" when No. 6 fuel oil is to be heated to the proper atomizing temperature by the boiler water in a safety-type oil preheater. (The electric preheater on the boiler must provide additional heat to the oil if boiler water temperature is not maintained above 200°F.)

Note: If the water temperature going to the system must be lower than 170°F, the boiler water temperature should be a minimum of 170°F (200°F if used to preheat No. 6 oil) and mixing valves should be used.

Rapid Replacement of Boiler Water—The system layout and controls should be arranged to prevent the possibility of pumping large quantities of cold water into a hot boiler, thus causing shock, or thermal stresses. A formula or "magic number" cannot be given, but it should be borne in mind that 200°F or 240°F water in a boiler cannot be completely replaced with 80°F water in a few minutes without causing thermal stress. This applies to periods of "normal operation" as well as during initial start-up.

This problem can be avoided in some systems by having the circulating pump interlocked with the burner so that the burner cannot operate unless the circulating pump is running.

When individual zone circulating pumps are used, it is recommended that they be kept running—even though the heat users do not require hot water. The relief device or by-pass valve will thus allow continuous circulation through the boiler and can help prevent rapid replacement of boiler water with "cold" zone water.

Continuous Flow through the Boiler—The system should be piped and the controls so arranged that there will be water circulation through the boiler under all operating conditions. The operation of three way valves and system controls should be checked to make sure that the boiler will not be by-passed. Constant circulation through the boiler eliminates the possibility of stratification within the unit and results in more even water temperatures to the system.

A rule of thumb of 0.5–1 gpm per boiler horsepower can be used to determine the minimum continuous flow rate through the boiler under all operating conditions.

Multiple Boiler Installations—When multiple boilers of equal or unequal size are used, care must be taken to ensure adequate or proportional flow through the boilers. This can best be accomplished by use of balancing cocks and gauges in the supply line from each boiler. If balancing cocks or orifice plates are used, a significant pressure drop (e.g., 3–5 psi) must be taken across the balancing device to accomplish this purpose.

If care is not taken to ensure adequate or proportional flow through the boilers, this can result in wide variations in firing rates between the boilers.

In extreme cases, one boiler may be in the "high fire" position, and the other boiler or boilers may be loafing. The net result would be that the common header water temperature to the system would not be up to the desired point. This is an important consideration in multiple boiler installations.

Pressure Drop through Boiler—There will be a pressure drop of less than 3 ft head (1 psi = 2.31 ft hd) through all standardly equipped Cleaver-Brooks boilers operating in any system which has more than a 10°F temperature drop.

C. Hartford Steam Boiler Inspection and Insurance Co. Approach

LOW WATER

In case of low water, stop the supply of air and fuel immediately. In the case of hand fired boilers, cover the fuel bed with ashes, fine coal, or earth. Close the ashpit doors and leave the fire doors open. DO NOT CHANGE THE FEEDWATER SUPPLY. DO NOT OPEN THE SAFETY VALVES OR TAMPER WITH THEM IN ANY WAY. After the fire is banked or out, close the feedwater valve. After the boiler is cool, determine the cause of low water and correct it. Carefully check the boiler for the effects of possible overheating before placing it in service again.

SAFETY VALVES

Each safety valve should be made to operate by steam pressure with sufficient frequency to make certain that it opens at the allowable pressure. The plant log should be signed by the operator to indicate the date and operating pressure of each test. If the pressure shown on the steam gauge exceeds the pressure that the safety valve is supposed to blow, or if there is any other evidence of inaccuracy, no attempt should be made to readjust the safety valve until the correctness of the pressure gauge has been determined.

BLOWING DOWN

If a continuous blow down is not provided, and particularly if the feedwater is of poor quality, the boiler should be blown down at frequent intervals to prevent serious scale accumulations. If it is necessary to blow down the waterwall headers, it should be done only when the boiler is under very light load, or no load, or only in accordance with specific instructions from the boiler manufacturer. When blowing down a boiler where valve and cock are used in the blow-off line, first open the cock and then open the valve SLOWLY. After blowing down one gauge of water, the valve should be closed slowly before the cock is closed, to avoid water hammer. DO NOT LEAVE THE BOILER DURING THE BLOWING DOWN OPERATION.

AUTOMATIC CONTROLS AND INSTRUMENTS

Automatic control devices should be kept in good operating condition at all times. A regular schedule for testing, adjustment, and repair of the controls should be adopted and rigidly followed. LOW WATER FUEL SUPPLY CUT-OFFS AND WATER LEVEL CONTROLS SHOULD BE TESTED AT LEAST TWICE DAILY IN ACCORDANCE WITH THE MANUFACTURER'S INSTRUCTIONS AND OVERHAULED AT LEAST ONCE EACH YEAR. All indicating and recording devices and instruments, such as pressure or draft gauges, steam or feedwater flow meters, thermometers, and combustion meters should be checked frequently for accuracy and to determine that they are in good working order.

C H A P T E R 1 7
BOILER SHUT-DOWN PROCEDURES

A. General Comments

The procedures given in this chapter are subject to the same comments that were made concerning light-off procedures in Chapter 15. We will give you the general chronological order of things to be performed, and will assume that the procedure is being performed manually. If the boiler is controlled by one of the programmed control panels on the market, the basic fuel and air control will all be timed and controlled by the panel, once the "Stop" button has been activated. However, as you will no doubt note upon reading this chapter, there is much more to shutting down a boiler than merely shutting off the fuel supply and purging the furnace. True, on many of the very small boilers, this may be all that is necessary, but this will apply only to a very small percentage of those boilers in service today.

The instructions that came with your boiler may not be as complete as the ones we give here, or they may be even more complete. Whichever is the case, we leave it up to you to work out your own procedure, taking into full account the applicable codes or insurance requirements. If any item given here is debatable, it is best to choose on the side of increased safety. Always keep in mind that most accidents occur when the boiler is being lit off or shut down, at times when you, the operator, are, or should be, in the boiler room. Therefore, your life and safety are at stake.

When a boiler is shut down, the time should be entered into the log sheet, along with all pertinent data concerning water level, boiler water chemical concentration, valves closed, etc., followed with the initials of the responsible person performing the work or in charge of the work.

B. Short Term Shut Down

There may be times when it is necessary or merely desirable to shut down the boiler for an hour or so. We do not recommend that this be done any more than necessary, since the result is an increase of fuel consumption when relighting. This will be explained later in this book. It is usually better to keep the boiler on the line at all times during the working shift, even if no steam is being used for short periods. The boiler during that time will merely recycle on the burner controls, and the boiler will be up to full steam pressure, ready for any demands to be made upon it once the plant process gets back into operation.

What could happen to require a short-term shut down? There are many things, such as emergency repairs to be made on the boiler, its piping, or service accessories such as the feed-water supply system.

CAUTION! Be extra careful in attempting to make repairs involving opening or tightening joints under steam or boiler pressure. Whenever possible, remove the pressure from the line or system under repair before touching a wrench or tool to it.

The following is the simple procedure to follow:

1. Turn off the manual fuel supply valves, including the pilot valve.

2. Open the vent or drain valve between the two manual fuel supply valves, if one is installed.

3. Keep the draft fan running for a minimum of 5 min, or for the length of time prescribed for your boiler.

4. After the draft fan shuts down, open the air and flue dampers to allow a free flow of air through the furnace and up the stack.

Under the preceding procedure we are assuming that it is not necessary to reduce the steam pressure in the boiler and steam supply system, but that it is desired to maintain some pressure on the system, to minimize the time it takes to get back on the line again with full pressure.

Under this procedure, it will not be possible to perform any work between the boiler and the first valve next to the boiler on any of these systems: the steam lines, feedwater lines, or the blow-down system. If it is necessary to perform service work on any of those valves next to the boiler, then it will require shutting down for a cold restart, which will be covered in the next section.

In all of the shut-down procedures given in this chapter we will assume that the fuel control valve is not of the tight shut-off design and that it will be necessary to shut off the manual fuel supply valves when the boiler is shut down.

Under some operating procedures written in the past, and still being followed in some areas, it was common practice to button up the boiler draft system when "banking" the boiler. The idea was to retain as much of the heat within the boiler as possible, since allowing a free flow of air through the boiler cools it down, and could cause thermal stresses in some boilers.

However, in recent years this practice has been ruled out in many areas. The reason for the change is the danger of fuel leaks into the furnace from leaking fuel valves, with the rapid accumulation of fuel in the furnace, and the possibility of an explosion should the fan not purge the furnace sufficiently prior to light-off.

C. Shut Down for Cold Restart

This is probably the most common method of shut down you will encounter since it represents the usual daytime only operation, or the five-day week operation. There are some plants that under near-ideal circumstances are able to keep the steam boiler and the distribution system hot enough overnight for a quick hot start in the morning, but this would be only in the larger plants. The plant size we are discussing here is such as to make it impractical, if not impossible, to keep steam up all night without having the boiler in full operation.

Generally speaking, the plants that find themselves able to "bank" the boiler overnight and still have steam in the lines at the start of the morning shift are those in which the steam demand is very large, with the process requiring that steam very close to the boiler. This reduces the steam line length, the total hot surface exposed to the cooling night air, and relatively large stored saturated water volume in the boiler.

We shall give here the manual procedure for shutting down the boiler plant for a cold restart, several hours to several days hence.

1. Reduce burner to minimum firing rate.

2. Turn off fuel supply to both main burner and the pilot.

3. Keep the draft fan running for at least 5 min, or through the prescribed length of time for your particular boiler.

4. If there are two manual fuel shut-off valves in series, after closing both of

them, open the vent or drain valve between them.

5. After the fan has been shut down, open the air and flue dampers to allow free circulation of air through the furnace and up the stack.

6. When the steam pressure has dropped to about 5 psig, open the boiler vents and the vents and drains in the steam-distribution system.

7. Bring water level up to the normal operating level.

8. When the boiler water has cooled to below 200°F, check the water for hydrazine concentration, cooling the sample to make the test.

9. Bring the hydrazine concentration up to between 10 and 20 ppm.

10. Close all water and steam valves to the boiler.

11. Shut off all electrical services to the boiler, the controls, and the feedwater pump.

This procedure is subject to judicious modifications, depending on your particular installation.

We have a few explanatory remarks to make on the above, as follows:

2. By fuel supply, we mean the automatic fuel control valve as well as the one or two manual shut-off valves.

10. This will depend on the method used to raise steam again and get back on the line. Many plants do not shut off the steam supply line when shutting down for a restart in a few hours. Some operators prefer to leave the steam supply valve closed to the distribution system until there is a good head of steam available to warm up the lines and clear them of condensate and air. On the larger distribution systems, it may be best to warm up the various distribution branches one at a time, depending on the plant sched-

ule for the various branch supplies. If the distribution system is relatively short and simple, it may be just as well not to close off the steam line when shutting down at night. Leave the entire steam system from the boiler out to the end of the mains open for a quick start-up in the morning. The thing to keep in mind is that all of the air and warm-up condensate must be drained out as fast as possible with the least amount of water hammer and other troubles when steam is turned into the system.

11. The extent to which the electrical services are shut off when closing down for a cold restart depends on the degree of security the plant requires. This could go as far as shutting off and locking the electrical service at the main breaker. In the average plant, however, it may only mean turning off the program control panel master switch on the panel front.

D. Wet Shut Down for Extended Periods

There may be times when it is desired to shut down the boiler for an extended period, in which no internal work is going to be performed on it. In this case, the procedure is only slightly different from that given in Section C. By extended period, we mean one that will last for more than one week.

In the procedure given in Section C, it should be followed for an extended shut down, with the following alterations to items 7 and 9:

7. Fill the boiler completely with water, until it runs out the open vent at the top of the boiler.

9. Raise the hydrazine concentration to between 100 and 200 ppm.

Add to the list of items, the following:

12. Check the water level and the hydrazine concentration weekly. The hydrazine level

should never fall below 100 ppm while it is in cold shut-down condition.

E. Shut Down for Internal Maintenance

This is probably one of the most important shut-down situations you will encounter, since it is one of those tasks that must be performed regularly in the boiler plant. The boiler may not require maintenance, but it must at least be opened for inspection.

The procedure to be followed here is similar to the first six items in the procedure given in Section C, except in the case of the boiler installed as one in a battery on a common header and breeching. This will require that the fuel valves, the blowdown valves, and the steam valves must all be either locked closed or tagged with a warning that valves are not to be opened without first contacting the plant engineer. In addition, the flue gas connection to the common breeching to the stack must be closed off in the same manner, with a blank inserted between flanges, or some such method of positive blocking. Do not rely on the dampers to effectively isolate the boiler from others on the breeching system.

Starting with item 7, the following alterations in Section C procedures are to be made:

7. When the boiler water has cooled to a temperature that is legal to be discharged to the sewer, open the drain on the boiler and drain all of the water out of the boiler drums and all of the associated piping and tubing. It may be necessary to dilute the boiler water with cold water to reduce its temperature.

8. Open all of the water-side inspection openings, manholes, etc.

9. Wash out the water side thoroughly.

10. Dry out the water side as fast as possible to prevent formation of too much rust on

the steel surfaces. Use fans, heaters, or any similar device available for this purpose.

11. Repeat steps 8, 9, and 10 on the hot gas side, being very careful not to get the soot-laden wash water in the eyes.

F. Dry Layup for Extended Periods

As an alternative to the wet layup just described, there may be times when it is desirable to take a boiler out of service and place it in a dry layup condition. This may be in expectation of a coming inspection, servicing that has been delayed, or reduction in plant demands.

The procedure is the same as steps 1 through 11 covered in Section E, with the following additional steps:

12. Fill several trays with dessicant dryer, and place them throughout the gas and water sides of the boiler.

13. Seal the boiler completely, excluding all possibility of outside air from entering the boiler.

14. Inspect the trays once a month, and replace the dryer as necessary.

There are three common dessicant dryers in use for this service: silica gel, activated alumina, and quicklime. The first two are the most expensive, but the safest to use. Quicklime is generally cheaper, but must be handled with care. Goggles and protective clothing should be worn. In case of direct contact of quicklime with the skin, a thorough and immediate shower is imperative.

There is an alternative to the use of dessicant dryers, and one which is often used on the larger industrial and power boilers, and that is blanketing the boiler water surface with an inert gas. Nitrogen is the one most often used, since it is the cheapest and most readily available.

However, this method requires that the boiler connections on the steam side be shut off tightly to prevent leakage. The nitrogen is kept at about 5 psig, and the water level in the boiler should be at normal steaming level, or close to it. The nitrogen pressure should be checked once a week, which requires the addition of a low-pressure gauge to the boiler drum. The hydrazine concentration must be maintained at the same levels as for a wet layup, since the wetted portions of the boiler are still in danger of corrosion.

C H A P T E R 1 8
RECORDS, REPORTS, AND DOCUMENTATION

A. Extent and Purpose

No matter how small your boiler plant may be, it is very important that you keep complete records of the activities and the equipment that make up your livelihood. No one's memory is good enough to be absolutely sure of remembering when definite maintenance procedures and repairs were performed. Also, in today's mobile work force, it is only a matter of forethought and common courtesy that you should keep sufficient written records to permit any successor to continue your work in the plant, with the least amount of upset to the remainder of the organization. How would you like to take over a going concern, complete with boiler plant, and not know the age and history of the equipment, and the last time it had been shut down and cleaned?

Of course, if yours is a branch of a larger plant complex, then you may be under strict company rules to produce complete prescribed reports and documentation on the activities of the boiler plant operation. The firm may have already supplied you with their standard forms for that purpose. If not, it is a simple matter for you to make up your own forms to be filled out regularly and kept in a convenient location for your retrieval, as well as being available for anyone who is authorized to look at them.

Even if only for your own convenience, it is

highly recommended that you keep the manufacturer's data, daily log sheets, maintenance records, and other such data as we shall list in this chapter.

There is a more practical reason for keeping complete records of your daily activities and experiences in operating the boiler plant. In case of trouble, such as an accident or an explosion, for your own protection, it is well to have complete records of the plant operation for the insurance firm that will probably be investigating the cause of the trouble. If there are no records for them to refer to, the first impression will be that the plant was improperly operated. You can be sure that the manufacturer has covered himself with all kinds of data and testing records on the equipment, so it is wise for you to do the same.

B. Manufacturer's Data and Literature

If your boiler is designed to either Section I or Section IV of the ASME Code, then it was inspected and tested at the factory to ASME requirements. The factory will then have available certificates attesting to that fact, containing all pertinent data. The only boilers usually exempt from this requirement are miniature boilers, and small coil-type flash generators used for steam-cleaning purposes.

These factory certificates are a part of the package of data that the manufacturer normally sends the customer before shipping the boiler or that are included in the shipment. This package of data should be reviewed by someone responsible for the integrity of the boiler plant, and then it should be kept in the company files in a safe and accessible location. The engineering department or the plant facilities department are the usual repository for this material.

Other data and literature available during the design, specification, and purchase of the boiler plant equipment, and which should be retained in the same file, are the following:

Manufacturer's Data Report Forms, a signed certificate attesting to the in-shop inspection of the quality and integrity of the pressure vessel and remainder of the boiler.

General arrangement drawing, with all pertinent dimensions of the complete package.

Bill of materials, including all accessories furnished by other than the boiler manufacturer. Contains complete model numbers and specifications of the accessories.

Installation instructions for the complete package.

Start-up instructions, including cleaning, testing in place (when used), and boil-out prior to start-up.

Operating instructions for the boiler and the complete package.

Maintenance instructions for the boiler and its accessories, including literature from the makers of the accessories.

Spare parts list, with prices, of the major items in the package.

Copies of all factory test data.

Descriptive literature of the boiler and its accessories.

Also, of course, similar data for all of the other boiler plant equipment purchased for operating the boiler plant should be included.

C. Contract Data

If the boiler in your plant was furnished in the usual manner, it was probably bid by several suppliers to an inquiry issued by your firm's purchasing department, based on specifications written by the plant engineering department or an outside engineering firm. The successful bidder then was issued a purchase order by your purchasing department. The installation was probably handled in a similar manner, by an outside contractor experienced in this class of construction.

It is an excellent idea to retain files of all of the major milestones in the above procedure, in case of questions or problems appearing later, such as method of installation, testing and cleaning procedures, etc. The minimum material to keep in your files for the project should consist of the following:

Specifications, with all addenda.

Successful bidder's proposal.

Purchase order to the contractor, with all supplements.

Copies of all attendant correspondence.

Copies of all start-up, cleaning, and testing procedures used.

Final inspection report.

Final letter of acceptance.

Copies of all guarantees on the equipment.

Any other material that is produced during the installation and start-up phases of the work, which you may feel important, should be added to the above files for future reference.

D. Daily Operating Logs

One of the most important activities of the boiler plant operator is the regular, conscientious entry into the operating log of all pertinent data and activities connected with the operation of the plant. This cannot be stressed too much, since it serves as the basic reference point for checking into the condition of the equipment should a problem arise. It also tells you how the boiler plant is reacting to activities in the process or steam using end of the plant.

The regularity or timing of the entries will depend on the complexity of the operation and the plant. Obviously, small hot-water heating boilers will not require entries to be made as often as a large industrial boiler plant installation. However, the importance of making those entries may be just as great, since many disastrous explosions and fires have occurred with small as well as large installations. One of the advantages of filling out a regular operating log is the ability to quickly spot any irregularities in the boiler operating results. It is a simple, and natural, matter to compare your present entry with those made previously in the same column. By glancing down the column each time an entry is made, trends in operating results may be spotted quickly, and trouble may sometimes be anticipated in advance of a dire emergency. As we shall point out later in this book, this is one method for spotting an increase of scale and soot build-up on the heat-transfer surfaces of the boiler. This is another advantage of keeping an accurate log sheet.

However, rather than attempt to sell you on the importance of keeping an accurate log of your boiler room activities, we are going to reprint the essential parts of the story, as published by the Hartford Steam Boiler Inspection and Insurance Company, in their Engineering Bulletin No. 70. Also included are reprints of the boiler log sheets to which they refer. These log sheets are available from their local office, or from their head office, located at One State Street, Hartford, CT 06102. Contact your Inspector serving your plant for cost and details.

A LOG IS THE ANSWER

The best method of determining that adequate attention is being given to the boiler and its control equipment is to provide a boiler log on which to record sufficient information to indicate that the boiler is receiving necessary care and maintenance.

For a log to be effective, it must provide a continuous record of boiler operation and maintenance. It should be required that a qualified person check the boiler at regular intervals and that the protective and operating devices be tested at sufficiently frequent intervals to determine that they constantly will be in good operating condition. Logs are effective only if properly used and accurately completed.

Two basic requirements must be met in order that a log be of any significance or usefulness. First, and foremost, the various tests specified must be performed on the equipment or apparatus as stipulated. Second, the results of observations of such tests must be entered with regularity in the log.

These two relatively simple steps are the very essence of a successful program of operating and maintenance logs. In order for any such program to have merit—as well as justification for its existence—it must meet both requisites. Moreover, these basic requirements must be practiced conscientiously and consistently.

The Hartford Steam Boiler Inspection and Insurance Co. previously requested suggestions for the development of boiler logs from its field inspectors located throughout the United States; men qualified to report on such practices and procedures. As a result of the replies received from these inspectors— representing a total of many thousands of man-years of experience in boiler inspection, accident prevention, and accident investigation—the company's engineering department has developed several types of boiler logs for varying operating conditions, which if properly used, should be effective in preventing boiler accidents and shutdowns.

LOW-PRESSURE STEAM AND HOT-WATER HEATING-BOILER LOGS

The log shown in Fig. 18-1 is recommended for low-pressure-steam heating boilers. It requires simple weekly checks, with one log sheet designed for a full year's operation.

The custodian, operator, owner, or other responsible person should record the tests performed each week, preferably on the same day of the week, thereby establishing a set schedule.

The necessity of actually making the tests and checks as outlined in the log, and not just filling in the spaces, should be discussed with the operator by the owner or management. The operator must realize the importance of the various tests and checks. The proper blocks should not be checked until the tests and observations have actually been made.

Remember: do not enter an item into the data section of the log until you have observed it or performed it. (Author's note.)

The safety valve should be tried in accordance with the suggested procedure at least weekly to determine that it is in operating condition. The valve should be tested when there is pressure on the boiler so that any accumulated dirt in the valve will not fall onto the seat during the test, and possibly damage the valve.

It is important that the water column and gauge glass be flushed out on a weekly basis in order that the actual water level in the boiler will be accurately indicated. Should there be any indication of obstruction in the connections, such as the slow return of water to the glass, the lines should be cleared.

Proper testing of the low-water fuel cut-out is probably the most important single test to be performed. The float chamber should be blown thoroughly during the tests; however, to determine if the cut-out is in proper condition, the water should be lowered by slowly opening the drain valve while the burner is in operation, thus simulating the gradual lowering of water in the boiler. If the cut-out does not function to stop the operation of the burner on this test, the float bowl should be flushed rapidly several times, and then the slow-opening test performed once again. If the cut-out does not operate properly on this second test, it should be dismantled and thoroughly overhauled at once by a reliable service organization.

The condensate pump and the entire return system should be checked at least once a week to determine that there is no leakage at pipe connections; that the pump is in satisfactory operating condition without appreciable leakage at the packing glands; that traps are functioning properly; etc. If the system is equipped with a condensate return tank, the float control and other parts of the return system should be carefully gone over.

It is also important that the operation of the burner be checked carefully each week to determine whether the burner starts properly and burns with a clean flame. A service man should be called immediately to service the burner, if it is not functioning properly.

The log shown in Fig. 18-2 is recommended for Low-Pressure Hot-Water Heating Boilers. The log provides for regular testing of the relief valve and the low-water fuel cut-out, when the boiler is so equipped. The log also provides for checking of the circulating pump, when the boiler is so equipped, and the operation of the burner. It is important that these items be checked weekly when the boiler is in service, with the space being initialed or checked by the operator, to indicate that the test was actually performed. The name of the

THE HARTFORD STEAM BOILER INSPECTION and INSURANCE COMPANY • HARTFORD • HARTFORD • CONNECTICUT 06102

LOW-PRESSURE Steam Heating Boiler Log

MANUFACTURER

BOILER NUMBER

YEAR

LOG THIS INFORMATION WEEKLY

MONTH	SAFETY VALVE TESTED WEEK					WATER COLUMN GAGE GLASS DRAINED WEEK					LOW WATER FUEL CUT-OFF TESTED WEEK					PUMP AND RETURN SYSTEM CHECKED WEEK					BURNER OPERATION CHECKED WEEK				
	1	2	3	4	5	1	2	3	4	5	1	2	3	4	5	1	2	3	4	5	1	2	3	4	5
JANUARY																									
FEBRUARY																									
MARCH																									
APRIL																									
MAY																									
JUNE																									
JULY																									
AUGUST																									
SEPTEMBER																									
OCTOBER																									
NOVEMBER																									
DECEMBER																									

TEST INSTRUCTIONS

SAFETY VALVE — pull try-lever to full open position with pressure on boiler. Release lever to allow valve to snap closed. **Caution**—All discharges must be piped to a safe place.

WATER COLUMN OR GAGE GLASS — open drain quickly to void small quantity of water. Water level should return quickly when valve is closed. **Caution**—All discharges must be piped to a safe place.

LOW WATER FUEL CUT—OFF — drain float chamber while boiler is running. This should interrupt the circuit and stop burner. **Caution**—All discharges must be piped to a safe place.

PUMP AND RETURN SYSTEM — check pump for proper operation, leaky packing. Examine traps, check valves, makeup float valves, and condition of condensate tank.

BURNER OPERATION — if burner starts with a puff or operates roughly, call your service man at once!

FOR IMMEDIATE REFERENCE, ENTER NAME, ADDRESS AND PHONE OF YOUR SERVICE MAN

SERVICE DATES

SERVICING

LOW WATER FUEL CUT-OFF—The low water fuel cut-off should be dismantled for a complete overhaul by a competent serviceman at least annually. The internal and external mechanism, including linkage contacts, mercury bulbs, floats, and wiring should be carefully checked for defects. See manufacturer's instructions. *Record service dates above.*

STOKER, OIL OR GAS BURNER AND CONTROLS—The stoker, oil or gas burner and all operating and protective controls should be thoroughly checked at least once every three months by a competent service organization. See manufacturer's instructions. *Record service dates above.*

UNUSUAL CONDITIONS SHOULD BE REPORTED TO THE COMPANY'S BRANCH OFFICE OR INSPECTOR AT ONCE.

4136 5/81 (ENG)

Fig. 18.1.

regular serviceman should be entered for immediate reference in the space provided, and the dates the burner and other equipment is serviced should be indicated on the log.

HIGH-PRESSURE POWER BOILER LOG—TWICE DAILY READINGS

The suggested boiler log illustrated in Fig. 18-3 is designed for use with high-pressure power boilers located in relatively small plants that do not have constant attendance or do not have a thorough checking system employing a boiler log at the present time.

This log is a simplified version of that shown in Fig. 18-4 and is designed to cover a full week of operation. While the items are to be recorded only twice-daily, the boiler should be checked at more frequent intervals while in operation.

A high-pressure power boiler should be checked at least hourly, observing the water level, steam pressure, feed pump pressure, feedwater temperature, condensate temperature, and flue gas temperature, with readings being recorded on the log at the same hour, twice-daily, to provide a consistent record of the boiler operation. Under some conditions it may be desirable to have the operator check the boiler on a more frequent schedule, such as every 20 or 30 min.

As high-pressure power boilers normally operate under a heavier load than low-pressure heating boilers, cycle more frequently, and are thus more subject to development of low water, it is extremely important that the low-water fuel cut-offs and water level controls be checked at more frequent intervals. This log provides for checking these protective devices twice-daily, preferably at the start of operation in the morning, and at noontime. At the same time, the gauge

glass and water column should be blown, and those tests recorded.

Also, within each 4 hr period, the feed pump, condensate return system, burner operation, and fuel supply should be checked and the general conditions recorded.

High-pressure power boilers usually require some feedwater treatment and therefore a space is provided for checking when the treatment is added. Also, the time that the boiler is blown down should be recorded daily. Blank spaces are provided on this chart for other items that the owner or operator may desire to check on either a 4-hr or daily basis.

The instructions on this form should be noted carefully. Any repairs or changes should be entered under the column "Remarks" so that a complete running-record of the operation of the boiler, including any repairs, will be available at all times.

MONTHLY CHECK SHEET

In addition to the daily checks covered by the log sheet, it is also desirable that certain monthly checks be made. These should include at least the items included in both high-pressure logs.

HIGH-PRESSURE POWER BOILER LOG—HOURLY READINGS:

The boiler room log illustrated in Fig. 18-4 is designed to cover one full day's operation.

This log is definitely the more desirable type for use with most high-pressure power boilers and should receive careful consideration by both management and operators.

In the first section of the log are those items that should be checked and recorded hourly, including checking of

THE HARTFORD STEAM BOILER INSPECTION and INSURANCE COMPANY • HARTFORD • CONNECTICUT 06102

LOW-PRESSURE Hot Water Heating Boiler Log

MANUFACTURER   BOILER NUMBER   YEAR

LOG THIS INFORMATION WEEKLY

MONTH	RELIEF VALVE TESTED WEEK					LOW WATER FUEL CUT-OFF TESTED (when provided) WEEK					CIRCULATING PUMP SYSTEM CHECKED (when provided) WEEK					BURNER OPERATION CHECKED WEEK				
	1	2	3	4	5	1	2	3	4	5	1	2	3	4	5	1	2	3	4	5
JANUARY																				
FEBRUARY																				
MARCH																				
APRIL																				
MAY																				
JUNE																				
JULY																				
AUGUST																				
SEPTEMBER																				
OCTOBER																				
NOVEMBER																				
DECEMBER																				

TEST INSTRUCTIONS

RELIEF VALVE—pull try-lever to full open position with pressure on boiler. Release lever to allow valve to snap closed. **Caution**—All discharges must be piped to a safe place.

LOW WATER FUEL CUT-OFF—check in accordance with manufacturer's instructions. **Caution**—All discharges must be piped to a safe place.

CIRCULATING PUMP SYSTEM—check pump for proper operation, leaky packing. Examine check valves and condition of other parts of system.

BURNER OPERATION—if burner starts with a puff or operates roughly, call your service man at once!

FOR IMMEDIATE REFERENCE, ENTER NAME, ADDRESS, AND PHONE NUMBER OF YOUR SERVICE MAN

SERVICE DATES

SERVICING

LOW WATER FUEL CUT-OFF—The low water fuel cut-off, if provided, should be dismantled for a complete overhaul by a competent serviceman at least annually. The internal and external mechanism, including linkage contacts, mercury bulbs, floats, and wiring should be carefully checked for defects. See manufacturer's instructions.

STOKER, OIL OR GAS BURNER AND CONTROLS—The oil or gas burner and all operating and protective controls should be thoroughly checked at least once every three months by a competent service organization. See manufacturer's instructions. *Record service dates above.*

UNUSUAL CONDITIONS SHOULD BE REPORTED TO THE COMPANY'S BRANCH OFFICE OR INSPECTOR AT ONCE.

4135 REV. 5/81 (ENG)

Fig. 18-2.

THE HARTFORD STEAM BOILER INSPECTION and INSURANCE COMPANY • HARTFORD • CONNECTICUT 06102

HIGH PRESSURE Power Boiler Log

| DAY AND TIME | WATER | | | | | PRESSURE | | TEMPERATURE | | | BURNER/ STOKER OPERA- TIONS | FUEL SUPPLY | WATER TREAT- MENT | BOILER BLOW DOWN | FEED WATER MAKE-UP | OPERATOR ON DUTY | REMARKS |
| | Level | Low Water Cut-off | Level Control | Column | Gage Glass | Feed Pump | Conden- sate Tank | Steam | Feed Pump | Feed Water | Condensate | Flue Gas | | | | | | | |

(Rows: MONDAY A.M./P.M., TUESDAY A.M./P.M., WEDNESDAY A.M./P.M., THURSDAY A.M./P.M., FRIDAY A.M./P.M., SATURDAY A.M./P.M., SUNDAY A.M./P.M.)

MANUFACTURER

BOILER NUMBER   LOCATION   WEEK BEGINNING

CHECK OR TEST AND RECORD TWICE DAILY

INSTRUCTIONS

Continued safe operation of a boiler depends on regular maintenance and testing of the boiler and its operating and protective controls. The tests and checks outlined above are designed to determine whether or not the boiler and controls are in good operating condition.

Should any check or test indicate that the device being tested or observed is not in good operating condition it should be repaired immediately. Record repairs or changes under "remarks" so that a complete record will be available for review at any time.

TESTING

WATER COLUMN AND GAGE GLASS—open drain valve quickly and flush water from glass and column. When drain is closed water level should recover promptly.

LOW WATER FUEL CUT-OFF AND WATER LEVEL CONTROL—drain float chamber when firing equipment is operating. Proper operation of the control should shut off the firing equipment and start feed pump. If controls are of probe or other type that require lowering of water level in boiler to test. DO NOT lower level to point below bottom of gage glass.

1467 4/81 (ENG)

Fig. 18-3.

THE HARTFORD STEAM BOILER INSPECTION AND INSURANCE COMPANY • HARTFORD, CONNECTICUT 06102

HIGH PRESSURE Power Boiler Log

MANUFACTURER   BOILER NUMBER   DATE

24 HOURLY READINGS—CHECK AND RECORD HOURLY

TIME OF DAY	WATER LEVEL	PRESSURE		TEMPERATURE			WATER			FEED PUMP / CONDENSATE TANK / BURNER OPERATION / FUEL SUPPLY				WATER TREAT-MENT	BOILER BLOW-DOWN	FEED-WATER MAKE-UP	OPERA-TOR ON DUTY	REMARKS
		Steam	Feed Pump	Feed Water	Conden-sate	Flue Gas	Low Cut-Off	Level Control	Column	Gage Glass								

TEST EVERY 8 HOURS (WATER: Low Cut-Off, Level Control, Column, Gage Glass)

CHECK EVERY 4 HOURS (FEED PUMP, CONDENSATE TANK, BURNER OPERATION, FUEL SUPPLY)

CHECK EACH SHIFT (WATER TREATMENT, BOILER BLOW-DOWN, FEED-WATER MAKE-UP)

Times: 12:00, 1:00, 2:00, 3:00, 4:00, 5:00, 6:00, 7:00, 8:00, 9:00, 10:00, 11:00, 12:00, 1:00, 2:00, 3:00, 4:00, 5:00, 6:00, 7:00, 8:00, 9:00, 10:00, 11:00

INST.

Continued safe operation of a boiler depends on regular maintenance and testing of the boiler and its operating protective controls. The tests and checks outlined above are designed to determine whether or not the boiler and controls are in good operating condition.

REV. LOG

Should any check or test indicate that the device being tested or observed is not in good operating condition it should be repaired immediately. Record repairs or changes under "remarks" so that a complete record will be available for review at any time.

The **log reviewer** should enter his Name, Title, and Date log was reviewed in space provided below. The log reviewer should **not** be the same individual who fills this log in on an hourly basis.

LOG TO TEST

WATER COLUMN AND GAGE GLASS—open drain valve quickly and flush water from glass and column. When drain is closed water level should recover promptly.

LOW WATER FUEL CUT-OFF AND WATER LEVEL CONTROL—drain float chamber when firing equipment is operating. Proper operation of the control should shut off the firing equipment and start feed pump. If

controls are of probe or other type that require lowering of water level in boiler to test. **DO NOT** lower level to point below bottom of gage glass.

LOG REVIEWED BY (Name & Title)

DATE REVIEWED

2055 6/81 (END)

Fig. 18-4.

the water level, steam pressure, feed pump pressure, feedwater temperature, return condensate temperature, and flue gas temperature. By a recording of these items hourly, the boiler will receive a reasonably thorough check at least at that frequency. Under some conditions it may be desirable to have the operator check the boiler on a more frequent schedule, such as every 20 or 30 minutes, recording the checked items at least hourly.

The log also provides for the testing of the operating and protective controls, including the low-water cut-off, water level control, water column, and gauge glass at 4-hr intervals. This will more nearly ensure the continuing proper operation of these important devices.

As indicated in the discussion of the chart in Fig. 18-3, the instructions on the form should be noted carefully, and repairs or changes should be entered under "Remarks" for further reference. The practice of making monthly checks as outlined in the "Monthly Check Sheet" is recommended here also.

Additional Advantages

Once the log sheet has been completed, whether it is a weekly or daily log, it should be filed for future reference. A review of the logs will provide much useful information. For example, it will indicate when to schedule cleaning of the fire side and internal surfaces of the boiler, thus preventing emergency shutdowns. Trends noted on changing conditions in the logs in many cases will warn of impending failures of boiler appurtenances or control devices. For instance, an increase in flue gas temperature over a period of time, without any increase in load, might indicate that the boiler heating surfaces are being fouled by soot or possibly internal deposits. Being forewarned by a

study of the chart, the operator can often determine just what the impending trouble may be and correct it before a failure occurs.

A review of the log sheets will also help determine the amount of overhaul or repair that is desirable or necessary during the annual shutdown period.

Proper completion of the log will ensure that the automatic control devices, such as the low-water fuel cut-off and water level control are kept in operating condition. As long as these protective devices are in operating condition, the chance of a low-water condition developing in the boiler is remote. The low-water fuel cut-off is a device that is not intended to operate regularly, since it is a "last-ditch" defense. However, should some other piece of equipment, such as a feed pump, fail or should some other condition develop to allow a low-water condition in the boiler, then the cut-out must operate to save the equipment.

Should an accident occur, complete log records may prove valuable in helping the operator and the inspector determine just what happened. A study of the records frequently points all too clearly to some testing or checking procedure that has been neglected. Once the cause of a failure has been established, it is usually possible to outline a program that will prevent a recurrence.

Complete records on a log will aid the boiler inspector in his efforts to prevent accidents or failures to equipment in service. Many times, an inspector coming in from the outside will note from a review of the log sheets a slowly developing condition that would not be so obvious to the boiler operator. By making suggestions for a change or by correcting the condition that is developing, a serious accident may be prevented.

CONCLUSION

Obviously, the reasons for establishing an operating and maintenance log program are both numerous and well founded.

The costs of maintaining any of the suggested boiler log programs are small, and in most instances negligible. By comparison to the probable alternative—an accident—neither owner nor operator can afford not to adopt such a program. For whatever reason such a program is undertaken, the results of a properly executed plan will basically remain the same—more efficient operation and the prevention of costly shutdowns.

BOILER LOGS AVAILABLE

In the interest of accident prevention and safe operating and maintenance practices, The Hartford Steam Boiler Inspection and Insurance Company offers without cost or obligation, the boiler logs discussed in this section. Primarily designed for insureds of the company, the offer is extended to all interested parties.

If your plant has need for additional logs, please contact any Hartford Steam Boiler branch office. Your Hartford Steam Boiler inspector will be pleased to discuss the matter with you and your plant engineer if you so desire. No one general log can exactly meet the complete needs of all plants; therefore, it is suggested that you consult with your Hartford Steam Boiler inspector on his next visit to your plant.

E. Maintenance Records

Of equal importance to the daily log sheets are the plant equipment maintenance records. On this form should be recorded all of the time and money spent to maintain the piece of equipment.

There are several good reasons why this record should be kept, and we give here some of the most obvious ones:

1. It will help pinpoint any constant source of difficulty that should be corrected.

2. With recurring problems, it will serve as a reference source for any new personnel working on the equipment.

3. If the time spent and the parts used, with their cost, are all recorded, it will enable the management to determine the cost-effectiveness of the equipment, and help in any decisions concerning its retention or replacement.

4. It may at times help straighten out any discrepancies in spare parts inventories.

5. Properly organized and kept, the maintenance record forms will help in alerting you to the approximate times when further maintenance will be required.

It is not necessary that you invest a lot of money to set up a system of record keeping, as long as the system you adopt is complete enough for your purposes, and is properly maintained. Even a simple notebook with pages devoted to the equipment will be satisfactory, as long as the relevant data and records are properly entered. Also, of course, the location for the records must be such as to ensure safe-keeping and ready accessibility.

We will list here a few of the more obvious items of maintenance that should be included in the entries in the maintenance logs:

1. Name and serial number of the equipment.

2. Date installed.

3. List of accessories, and their make and model numbers.

4. Electrical requirements.

5. Inspection dates, and name of inspector.

6. Dates preventive maintenance were performed, and name of personnel involved.

7. Spare parts used.

8. Time spent in performing the work.

9. Time equipment went out of service and time placed back in service.

F. Spare Parts Inventories

Closely allied to the various records and documentation we have covered so far in this chapter is the spare parts inventory. Any plant using mechanical equipment, boilers, or other operating equipment that tends to wear out and require servicing, will have to keep on hand spare parts to service that equipment. As the spare parts are part of the plant operating supplies, records should be kept of their cost, the source of supply, and their consumption, as well as the complete description for purposes of reordering.

There are many inventory systems for spare parts available, and some even are computerized. That is very well for the larger plants, or those which are a branch of a national network, but for the small to medium-sized plant, usually a simple form, kept in a file drawer, or a card file, is all that is normally required. It does not matter how complex or simple the system is, as long as it is adequate for your purposes, and is compatible with the records and inventory system used in the remainder of the plant.

Some of the items of information which should be entered on the form are:

1. Make and model number of the equipment the part is for.

2. Part identification; number and name.

3. Source of supply.

4. Cost.

5. Approximate delivery time required.

6. Number of parts required for each item of equipment.

7. Number to keep on hand for emergencies.

8. Where the part is stored.

9. Ordering dates; history and order numbers, chronologically.

10. Dates parts are withdrawn and placed into storage.

That completes the chapter on record keeping in the boiler plant. We hope this will give you sufficient information to enable you to organize your plant operations in an efficient manner and thus help to keep you out of undue hardships and chaotic situations. Remember, these are only guides. You are free to improvise as necessary.

C H A P T E R 1 9

BOILER PLANT SAFETY

A. Boiler Testing and Inspection

In this chapter and in Chapters 20 and 21 we shall cover subjects that are interrelated to a great extent: Safety, Boiler Troubles, and Inspection. Consequently, there will be some mingling of those topics among chapters. Therefore, we suggest that these three chapters be read as one continual project, knowing that there will be some repetition, which is unavoidable.

If a boiler is to be operated for years in a safe manner, it must be regularly inspected to determine its condition. This inspection must be performed by qualified inspectors, and they may be from one of several agencies supplying this service. Most of the larger cities have their own building and safety departments which regularly see to it that the boilers within its borders are in a fit condition to continue to operate. Also, some of the counties and states perform this function.

But by far the highest majority of the inspections are carried out by the insurance agencies, as we have mentioned previously. Regardless of who actually inspects the boilers, all power steam boilers should be inspected once a year. It is strongly suggested, of course, that the boiler plant operator become acquainted with the regulations covering the operator's particular area and situation, so that there may be no embarrassment later when the

plant manager is told that the plant is not complying with the law.

In addition to regular inspections, all boilers must be given a hydrostatic test if any repairs are made on the pressure portions of the boiler, and also whenever the inspector determines that a test should be made for any reason. This will be the case, for instance, if the boiler is an old one, and the inspector has doubts about its ability to continue carrying its original rated pressure. In fact, most boiler codes require the boiler plant supervisor or chief engineer to obtain permission from the local inspector before making any alterations on the pressure parts of the equipment that is covered by the codes.

The hydrostatic test will be made at one and one half times the maximum allowable working pressure of the boiler. The test pressure shall be controlled so that at no times during the test will the test pressure exceed the required amount by more than 6%. The test shall be made using water at ambient temperature, but in no case less than 70°F, and the test pressure is usually produced and held constant by means of a special test pump.

During the hydrostatic test of the boiler, the safety valves must be either "gagged" with a special clamp made for the purpose, or, if possible, the safety valves should be removed and the connection blanked off. The latter method is preferable, and in no case should the setting of the safety valve be altered during the running of the test. There is too much danger of damaging the valve stem when applying the gag or by screwing down on the compression spring that determines the setting or popping pressure of the valve.

The boiler plant operator will receive an official notice as to when the inspection of the boiler has been scheduled, and the notice will usually be at least 14 days in advance of the scheduled date. The plant operator has the opportunity to notify the inspection department if the date set is not convenient, and give reasons for asking for a rescheduling. If a rescheduling is allowed, it may not be for more than 30 days from the original date.

To prepare the boiler for inspection, the water must be drawn off and the boiler thoroughly washed. All manhole and handhole covers and wash-out plugs in the boiler, feedlines, and water column connections that are necessary for adequate inspection shall be removed. The furnace and the combustion chambers shall be thoroughly cooled and cleaned. Enough of the brickwork, refractory, or insulating material shall be removed to permit the inspector to determine the condition of the boiler, furnace, or other parts of the boiler. The inspector must be able to obtain sufficient data to enable a proper evaluation of the condition of the boiler to be made.

When opening up the boiler in preparation for inspection, we suggest you have the water treatment chemist or serviceman handling your plant present to obtain samples of any sludge, scale, or soot deposit from the boiler for his firm to analyze. This may help identify trouble in time for corrective action to be taken.

The steam gauge shall be removed for testing at this time, also.

Should the inspector wish to give a boiler a hydrostatic test, and if the boiler is one of a battery on one header, then precautions must be taken to ensure isolation from the other boilers, which may be under pressure at the time the test is being made. The connections from the boiler being tested must be blanked off, unless the connections are provided with double stop valves with a free drain between the valves.

WARNING! While any hydrotesting is in progress, take all necessary precautions to keep the area clear of all unnecessary personnel.

A qualified boiler inspector from experience is able to tell at a glance how the boiler has been treated during its life. The inspector has seen all conditions possible during his career, so nothing surprises him. He knows just what to look for in your style of boiler, and he knows where trouble may be expected to show itself.

It goes without saying, of course, that a good boiler inspector cannot be fooled or hoodwinked, so it is useless to attempt it. We suggest that you avoid falling into the trap, for instance, of displaying a bright, shiny boiler inside and expecting praise from the inspector. A bright, shiny surface can indicate that chemical action is eating away the metal, which may call for a hydrostatic test to determine how safe the boiler really is. A boiler may be too clean to satisfy some inspectors.

Remember, the boiler inspector is looking after your safety, as well as the safety of others, so it makes good sense to cooperate with him. Be fair in your dealings with him, and you will find it pays dividends in the long run, as inspectors are only human. More than one inspector has been known to call the boiler operator aside and give him some friendly tips that will help him improve his efficiency. They are an excellent source of help and information on how to get the best from your equipment. Don't hesitate to ask his advice on anything concerning the operation of the boiler. He does not have to give you his suggestions, but few of them will refuse if they have received your fullest cooperation during the inspection. We have known several boiler inspectors over the years, and every one of them were well qualified, and were always courteous and helpful.

B. Codes

There are several authorities that have issued codes for the installation and operation of boiler plants. Some of these codes may seem frivolous and unnecessary at times. This is far from the case, however. The plant operator should remember that the existing boiler codes came about in the same manner as most of our religious and national holidays. They are the result of considerable hardship and bloodshed in the past, and we should remember the fact with respect. And each plant operator and maintenance personnel should work toward the elimination of any further unnecessary bloodshed—including his own.

Most cities and counties will have codes or ordinances with which the steam plant operator and maintenance personnel is expected to be familiar. We cannot attempt to cover them all here, since they are too numerous to mention. However, we can give you an idea of what they will cover.

Local city codes usually are concerned with the effect of the steam plant upon the neighbors and the city's facilities, and in any phase in which the city may conceivably be held liable in case of any accident or misoperation.

County codes and rules are usually primarily concerned with factors covering a broader field than would concern the city. This includes things such as air and water pollution, and also factors normally under the jurisdiction of the city, if the plant is in the county territory and not inside the city.

The state is usually interested in safety, or any portion of the operational factors in the steam plant that would have an effect upon such state laws as workmen's compensation, sickness, accident, and health insurance, or the state labor code.

Many states have adopted all or parts of the ASME Boiler Code for their own use. Good examples are the states of California and Massachusetts. Generally speaking, the more industrialized and populous a state has become, the stricter its boiler code tends to become.

The federal government normally is not concerned with codes or rules governing the steam plant in the average industrial plant, unless the plant is part of the federal government's institutional system or unless the plant is doing critical work under contract to the federal government. In this case, it is not unusual for the government to dictate some of the working conditions, hours, and safety codes for the plant doing such contract work.

Any insurance agency that carries the insurance for the industrial plant having a steam system on its premises will have a very definite interest in how the steam plant is installed and operated. In fact, it is often the insuring agents who set the primary standards for the plant. In doing so, of course, they usually depend upon some of the existing codes, with a few of their own ideas thrown in. It pays to listen very carefully to them when planning any changes or enlargement of the plant, since not to do so may be dangerous and expensive on insurance premiums. For insurance agencies will put the matter on a practical economical basis. You either follow their suggestions, or else they will refuse to carry you, or if they do, it will be at higher insurance rates. So we suggest you cooperate with them to the fullest extent.

We wish to warn the boiler plant operator that the existing boiler codes in force in his area may not be the same tomorrow, or next year. The governing agencies are regularly updating the local codes to meet changes in the available equipment, processes, enforcing personnel, and experience with the existing codes. It is the boiler plant operator's responsibility to keep informed on current trends in his area, through his contact with other operators, the suppliers calling on him, the local inspectors, and the trade journals.

The leading authority in this country when it comes to installation and operation, as well as design, of boiler systems, is the American Society of Mechanical Engineers' Boiler Code. The sections that you are most interested in here will be Section I, covering high-pressure boilers, and Section IV, for low-pressure steam boilers.

A fired pressure vessel, or a fired steam boiler, is any pressure vessel in which steam is generated by the application of heat resulting from the combustion of fuel.

Some history concerning the founding of the ASME Boiler Code will help to establish its reliability.

Probably one of the larger contributing factors to the establishment of the ASME Boiler Code was the catastrophic boiler explosion at Brockton, MA on March 20, 1905—58 were killed and 117 injured causing the Commonwealth of Massachusetts to pass the legal code of rules for the construction of steam boilers in 1907.

The ASME Boiler and Pressure Vessel Code Committee was set up in 1911 by the American Society of Mechanical Engineers to formulate standard rules for the construction of steam boilers and other pressure vessels.

The purpose of the Code is to establish rules of safety governing the design, the fabrication, and the inspection during the construction of boilers and unfired pressure vessels. Another objective of the rules is to afford reasonably certain protection of life and property and to provide a margin for deterioration in service so as to give a reasonably long safe period of usefulness for boilers.

The ASME Boiler Code has gained recognition during the past 60 or more years as being the most reliable code to follow, owing mostly to its impartial nature and its proven reliability. It has at times been proven in error; but when this happens, the portion of the code in doubt is immediately investigated, and any necessary changes are made and published. As a result, today it is without a doubt the most trustworthy set of rules covering your steam plant. We sincerely hope it will continue to enjoy such an eviable reputation, and that it will not go the way of some of our other associations or agencies. We are referring to some of those institutes or associations formed by manufacturers of any number of specific classes of equipment, in an attempt to gain standardization in specifying and manufacturing much of the equipment used in industry today. Most of these organizations have done an excellent job and have performed a worthwhile service. Unfortunately, we have noted a trend toward relaxing some of the definite statements that they in years past have issued concerning design standards. In too many cases, the relaxation or easing up on the formerly positive at-

titudes has gone so far as to be almost useless. Too often today we see the phrase ". . . shall be adequately designed to perform the work for which it is intended," or similar words, which really have no practical value. Of course, some of the changes are necessary owing to the many changes in recent years in methods of manufacturing materials, changes in fabrication methods, and, in large part also, the increased use of some of the formerly rare metals into common industrial use. An example of this is the high-strength steels and the increased use of magnesium, titanium, and similar metals. They all have made quite an impact upon the industrial equipment field.

Should you have any question concerning any section of the Code, it is only necessary for you to contact the ASME Boiler and Pressure Vessel Committee, whose address is 345 E. 47th St., New York, NY, 10017, and they will consider all inquiries. Or they may have you contact one of the members living nearest you who is on the committee, who will give you his opinion and advice.

No set of rules or codes will be effective unless properly set into operation and accepted by the industry it is designed to protect. The same is true of the Boiler Code. Fortunately, the Code has been very well accepted, and it is the standard followed by the majority of the boiler inspectors and insurance agencies throughout the United States and Canada.

The ASME Boiler Code is administered mostly by means of the numerous boiler inspectors employed by insurance agencies, and federal, state, and municipal authorities. The best of these inspectors belong to a group known as the National Board of Boiler and Pressure Vessel Inspectors, or will carry a commission issued by the Board. This is your assurance that the inspector commissioned by the Board is fully qualified to inspect your boiler and pass upon its merits and conditions. He cannot, of course, be held accountable for improper operation and maintenance of the boiler.

The following Section will describe the National Board of Boiler and Pressure Vessel Inspectors and its functions.

C. National Board of Boiler and Pressure Vessel Inspectors

Rather than attempt to give you our impression of the purpose and activities of the National Board, as it is generally called, we shall reproduce here portions of their "Information Booklet," which is NB-21, rev. 1, dated 1980.

PREAMBLE

The National Board of Boiler and Pressure Vessel Inspectors is an organization comprised of chief inspectors in states and cities of the United States and provinces of Canada, and is organized for the purpose of promoting greater safety to life and property by securing concerted action and maintaining uniformity in the construction, installation, inspection, and repair of boilers and other pressure vessels and their appurtenances, thereby ensuring acceptance and interchangeability among jurisdictional authorities responsible for the administration and enforcement of the various sections of the American Society of Mechanical Engineers (ASME) Boiler and Pressure Vessel Code.

Q What is the National Board of Boiler and Pressure Vessel Inspectors?

A In general, the National Board of Boiler and Pressure Vessel Inspectors can be described as representing the enforcement agencies empowered to ensure adherence to provisions of the American Society of Mechanical Engineers Boiler and Pressure Vessel Code.

Q Who are the members of the National Board?

A Its members are the chief inspectors or other jurisdictional authorities who administer the boiler and pressure vessel safety laws in the various jurisdictions of the United States and provinces of Canada.

Q Does the National Board have a permanent staff?

A Yes. Its central headquarters is located at 1055 Crupper Avenue, Columbus, OH 43229, where an office force is maintained under the executive director to administer the National Board programs.

An Executive Committee of eight members and six Advisory Committee members meet periodically to establish policy for the National Board and conduct such other business as may be necessary.

Q Is there any liaison with boiler or vessel manufacturers and authorized inspection agencies that insure and inspect boilers and pressure vessels in use?

A Absolutely. An advisory committee provides a representative of boiler manufacturers, a representative of the welding industry, a representative of insurers of boilers and pressure vessels who are also authorized inspection agencies, a representative of pressure vessel manufacturers and two representatives of boiler and pressure vessel users. All the above are deeply involved and highly interested in boiler and pressure vessel safety, the first and foremost concern of the National Board. Close liaison is maintained with these groups. The National Board also has liaison with the ASME Boiler and Pressure Vessel Code Committee, and all of the principal ASME subcommittees as well as other organizations interested in safe construction and use of boilers and pressure vessels.

Q What are the objectives of the National Board?

A The objectives are stated in the National Board constitution:

(a) to promote uniform administration and enforcement of boiler and pressure vessel laws, rules, and regulations;

(b) to promote standards for acceptance of boilers, pressure vessels, parts, and appurtenances to ensure safe operation;

(c) to promote one uniform code and one standard stamp to be placed on all registered boilers, pressure vessels, parts, and other objects constructed in accordance with the requirements of that code;

(d) to promote one standard of qualifications and examinations for inspectors who are to enforce the requirements of said code;

(e) to gather and make available information and statistics useful to the members, inspectors, and others interested in boiler and pressure vessel safety.

Q How is this done?

A Uniform enforcement of boiler and pressure vessel safety laws, rules, and regulations, as contained in the laws of the states and municipalities of the United States and provinces of Canada, is required so that boilers and pressure vessels constructed in accordance with the ASME Code, stamped with the applicable ASME symbol stamp, and registered with the National Board can be accepted across jurisdictional boundaries without difficulty. These jurisdictions, which have accepted the ASME Boiler and Pressure Vessel Code as the basis of their safety laws, rules, and regulations, are thus ensured that vessels so stamped and registered have been constructed in accordance with the ASME Code and inspected during construction by a commissioned National Board inspector for compliance with the Code.

Q Who are these commissioned National Board inspectors?

A National Board commissioned inspectors are highly qualified personnel experienced in fabrication, installation, and maintenance of boilers and pressure vessels. They have proven their familiarity with the technical and practical applications of the ASME Code by successfully passing a written examination covering that standard.

Q Does the National Board inspect the manufacturers too?

A ASME will normally refer a manufacturer's application for use of the Code symbol stamp to the jurisdiction in which the plant is located. Where a jurisdiction does not employ sufficient personnel to make the evaluation of the manufacturer's quality control program as related to his ability to control, design, and construct to ASME Code requirements, or where the jurisdiction for other reasons does not choose to make this evaluation, the National Board will perform these duties in its behalf.

Q Authorized inspectors, Code inspectors, commissioned National Board inspectors: who are they and for whom do they work?

A Authorized inspectors must work for an authorized inspection agency that, by definition under the National Board by-laws and ASME Code, is one designated as such by the appropriate legal authority of a state or municipality of the United States or a province of Canada. The authorized inspection agency employs the authorized inspectors, who hold National Board commissions, to perform shop or field inspection required by the Code and laws of the various jurisdictions. The agency may be a state or municipality of the United States or province of Canada or an insurance company authorized to write boiler and pressure vessel insurance in the various states.

Q What's this about registration of vessels with the National Board?

A The National Board has duplicate files, maintained in two widely separated locations for security reasons, for over 12,300,000 vessels. Copies are available by supplying the manufacturer's name and the National Board registration number. That information is easily obtained because it is permanently stamped on the vessel or nameplate.

Q But who needs these copies of data reports for boiler or vessels possibly ten or more years old?

A A manufacturer's records may be destroyed by fire or flood and it can obtain virtually instant replacement. An owner may require extensive repairs to a boiler or vessel and need to know the composition of materials used in original construction, size and type of tubes, wall thicknesses, or other information required to make effective repairs or alterations. If a used boiler or vessel is to be sold or reinstalled at another location, the original manufacturer's data report or a true copy is a mandatory requirement of the jurisdiction involved.

Q If a manufacturer violates Code requirements, does the National Board get involved?

A Violations of ASME Code requirements are usually reported by jurisdictional authorities when they inspect a boiler or pressure vessel received in their jurisdiction for installation. However, violations may be reported by owners, users, inspectors, or others. Such reported violations are all thoroughly investigated.

Q Any other areas of National Board we haven't mentioned?

A Yes. A highly important one is the re-

sponsibility of administering the capacity certification for safety valves and safety relief valves constructed in accordance with ASME Boiler and Pressure Vessel Code requirements. The National Board testing laboratory located in Columbus, Ohio is one of the most extensive facilities in existence for pressure and capacity testing of steam, water, and air type safety valves and safety relief valves of various sizes.

The National Board also makes available several publications, principal of which is the National Board *BULLETIN*, dedicated to the promotion of the organization's objectives previously discussed. It is a source of information to members and other interested parties. *The National Board Inspection Code* is an indispensable reference book for the field and shop inspector. Various other publications such as *Safety Valve and Safety Relief Valve Relieving Capacities* are available from the National board of boiler and Pressure Vessel Inspectors, 1055 Crupper Avenue, Columbus, Ohio 43229.

Q Are there any uniform rules or regulations available for repair or alterations of boilers and pressure vessels?

A Yes. The National Board Inspection Code contains rules for repair and alteration of boilers and pressure vessels.

Q Does the organization making such repairs or alterations require authorization from ASME to make such repairs or alterations to ASME boilers and pressure vessels?

A No. The ASME rules are for new construction. However, such organizations may obtain a certificate of authorization and Repair symbol stamp from the National Board.

Q What is the purpose of the National Board Repair Stamp ("R"), and how is it obtained?

A The "R" symbol stamp is a certification that repairs performed on an ASME Code boiler or pressure vessel have met requirements of the National Board Inspection Code and of the state, municipal, or provincial jurisdiction concerned.

Boilers and/or pressure vessel repair organizations not holding valid ASME certificate(s) of authorization may obtain the National Board stamp ("R") by an application to National Board. Basic requirements are that all welding procedures and welders be qualified and documented in accordance with Section IX of the ASME Code, a written description of how ASME, National board, and jurisdictional requirements will be fulfilled and how documentation of repairs will be accomplished.

Q Does the National Board have any other symbol stamps?

A Yes. National Board issues two other stamps. The Valve Repair stamp ("VR") is used for the repair of safety valves and safety relief valves and the Nuclear Repair ("NR") is necessary for repair and/ or replacement of nuclear components.

D. Boiler Explosions

Unfortunately, in spite of all the precautions that have been taken by our safety and regulatory agencies, we still have an appalling number of boiler explosions. Why is this so, when so many laws and regulations have been enacted to prevent it?

The answer is that not all inspectors, insurance agents, and other people concerned with boiler operation are able to be present in each and every boiler plant for long enough periods of time to spot unsafe practices. There are so many plants using steam, that it would be impossible to police the industry that thoroughly. Like a great deal of other conditions in our life, someone in authority, or with the background and experience to be able to advise others, can only use his powers of persuasion to induce changes for the better. The threat of lawsuits or the threat of higher insurance rates does go a long ways toward cleaning up bad conditions, however.

There is still a feeling among some plant owners and managers that low-pressure boilers are not as dangerous as high-pressure boilers. That may be true in regards to steam drum failures, but there are other sources of explosions in boilers that do not depend on the steam pressures for their explosive force. Many of the explosions are caused by accumulation of explosive gases in the furnaces or firing chambers of the boiler, and these are not dependent on the steam pressures in the drum. Therefore, you will find quite a lot of attention has been given lately to preventing such dangerous accumulations in the boiler furnaces. The complicated programming panels being installed on many boilers today are a result of just such concern for preventing furnace explosions. The purge, light, monitoring, purge, and light again cycles are there for a very good reason, so we do not advise any boiler plant operator or maintenance man to tinker with them unless he has received proper instruction.

When boilers started coming out of the manufacturer's shops with the automatic controls installed on them, too many of the makers were selling them with the idea that they required very little attention. Fortunately, that condition has been fairly well cleaned up, and now you will very seldom find a boiler salesman attempting to sell his product on the idea that you can install it, push the button, and walk away from it. Like any piece of mechanical equipment, the more automatic controls or gadgets you put on it, the more maintenance the gadgets require, and the more susceptible to trouble the equipment becomes. In adding more controls and automatic devices to a piece of equipment, you can easily reach the point of diminishing returns, where the reliability decreases, instead of increases.

The proper balance of automatic controls of a reliable make along with proper supervision and maintenance by competent operators will produce the safest and most efficient boiler plant, providing the basic equipment is properly designed and installed.

We still find a propensity for some plant owners or managers to buy a boiler overly endowed with automatic controls with the idea that he can save money by hiring cheaper labor to operate it. He can, for a short period, until trouble develops.

E. Prevention of Boiler Explosions

We stated earlier in this book that furnace explosions are as dangerous, or even more so, than rupturing of the steam drum or tubes. In fact, in the case of a water tube boiler, furnace explosions are as much to be feared and prevented as the possibility of bursting the steam drum or splitting of the steam generating tubes.

Furthermore, the damage from furnace explosions can be very devastating regardless of the pressure being generated in the steam boiler, since the amount of energy contained in the fuel being released into the furnace is not a function directly of the actual pressure on the water side of the boiler.

To illustrate the importance of accident prevention in the boiler plant, we reproduce here the partial contents of Bulletin 139 by the National Board of Boiler and Pressure Vessel Inspectors.

PREVENTATIVE MAINTENANCE ON POWER BOILERS

by Richard E. Jagger

In spite of some of the most rigorous, well-conceived safety rules and procedures ever put together, boiler and pressure vessel accidents continue to happen. In 1980, for example, 1,972 boiler and pressure vessel accidents occurred causing 108 serious injuries and 22 deaths as reported to The National Board of Boiler and Pressure Vessel Inspectors.

Relative to the huge numbers of boilers and pressure vessels in service, accidents fortunately are quite rare. But they still do occur. The National Board recently conducted a survey to help find out why.

Essentially, the survey asked the question: "Based on your experience, how likely are you to find each of the following conditions during a typical accident investigation or during an in-service inspection and in what type of vessel?" Frequency was tabulated as either "frequent," "sometimes," or "rare." Location was either "firetube," "watertube," or "pressure vessel."

Poor Maintenance Leads

Of nine general problem areas, "poor maintenance and testing of controls," "scale and sludge," and "operator error" rank well ahead of most other categories. Thus the areas linked to human error outrank problems due to mechnical deficiencies and environmental causes. Only "miscellaneous problem areas" (of which code violations are a part) and "corrosion" approach the frequency of the human-related problem areas.

As for location, the results suggest problems are seen most often in firetube boilers, next most often in watertube boilers, and least often in pressure vessels. But in cases where the potential for the particular problem is equally common to all three types—such as acidic or galvanic corrosion—the three locations rank about the same.

Some Cures for the Causes

Knowing which problem areas are most common is obviously only a first step. The second is knowing what corrective and preventive measures to take for these potential accident causes. The lists at the end of this section present "cures" that the National Board recommends for each of the 55 problem areas in the survey. The cause and cure charts also list where the primary responsibility for each cause and cure lies.

Boiler and pressure vessel safety experience shows that safety and longevitey can be achieved in three major ways:

proper design and construction;

proper maintenance and inspection;

proper operator performance and vessel operation.

A close look at the survey results shows that 69% of the causes and their cures depend on the human element—proper operator performance and vessel operation. The other two categories, "design and construction" and "maintenance and inspection," account about equally for the remaining 31% (16% and 15%, respectively).

Design and construction cures are dependent on the formulation and adoption of good construction and installation codes and standards. Equally important is regular third party inspection by National Board Commissioned Inspectors to ensure that ASME Code standards are met. Such conformance to Code can be ensured, of course, by purchasing ASME Code boilers and pressure vessels that have been registered with the National Board.

Owners and users have only limited control over vessel design and construction other than to order Code vessels and participate on the various ASME Code committees. Owners and operators can have a very direct impact, however, on cures in the maintenance and inspection category. These cures require frequent inspection, both internal and external by plant maintenance personnel as well as regular inspections by a National Board Commissioned Inspector. National Board Commissioned Inspectors are trained to locate potential Code violations and possible accident producing conditions.

Thus they are an important part of any in-service boiler or pressure vessel inspection plan.

The key to a good maintenance program is a good record keeping system, diligently used. Records show the owner and user when the vessel was last inspected or tested, and offer both guidance and reminders as to when inspections and/or tests should be made.

Attitudes and Practices Behind Human Error

Proper operator performance and vessel operation, the last category, differs from the first two in that it rests totally with the owner and operator. As the data presented earlier show, the great majority of unsafe conditions can be traced to human error on the part of the owner and/or operator. Why do such conditions occur again and again?

National Board Commissioned Inspectors have noticed attitudes and practices which, over the years, have consistently been associated with unsafe boiler and pressure vessel conditions and the accidents they cause:

Too much reliance is put on automatic controls, sometimes even to the point of substituting qualified personnel with automatic devices.

Boiler and pressure vessel safety and operating standards sometimes are lowered or completely forgotten.

Owner/operator knowledge and training are insufficient.

Operators are not provided with adequate operating, maintenance, and testing standards, procedures, and equipment.

Owners fail to apply sound employment practices in selecting and hiring operators.

Regular testing programs for controls and safety devices are not followed.

Carelessness or apathy are allowed to exist around boilers and pressure vessels.

Some Guidelines to Combat Human Error

National Board experience points to human error as the primary factor in most unsafe boiler and pressure vessel conditions. Cures for these unsafe conditions and problems have already been presented. But to keep most of these problems from ever occurring in the first place, owner and operator attitudes and practices must first be oriented to safety. Here then, is a list of guidelines to combat the conditions that lead to human error and consequently accidents of boilers and pressure vessels:

Use only skilled, trained personnel to operate or maintain a boiler or pressure vessel.

Never assume someone else has opened or closed a valve or switch. Check personally. Use signs, locks, written performance logs.

Always give operators complete, clear instructions and orders.

Look for potential hazards. A good operator studies his plant every day for possible safety problems.

Provide a contingency plan for your boiler should an emergency occur. Although it can't prevent emergencies, such a plan can help bring an accident under control and mitigate its consequences.

Operate a boiler with a good set of safety rules; make sure everyone is familiar with those rules.

Stick to established rules and procedures.

Ensure that any operator placed in charge of a boiler or pressure vessel has been thoroughly trained and tested in all applicable safety and operational rules.

Provide continuing education and training to help the operator stay abreast of new equipment, procedures, operating conditions, and hazards.

Never take short cuts in maintenance, repairs, or operational or safety guidelines.

One of the best ways to follow these guidelines is to set up an internal training program. Such a program might help prevent a serious boiler or pressure vessel accident saving the tragedy of injury or death to an employee, and possibly a large property damage loss as well.

Poor Maintenance and Testing of Controls

Problem or Cause	Cure
Safety valves	Set up a testing procedure
Low-water cut-offs	Test regularly—dismantle annually
Pressure gauge	Check regularly against a calibrated gauge
Water column or water glass	Keep them clean
Feedwater supply system	Maintain pumps and feedwater regulators
Condensate return system	Save and reuse more condensate
Indicating and recording instruments	Instruments and recorders require regular specialized maintenance
Pressure regulators	Have a testing procedure and follow it to ensure operability

Scale and Sludge

Problem or Cause	Cure
Overheated tubes	Keep internal surfaces clean
Overheated other parts	Pay close attention to feedwater chemistry
Blockage of circulation to controls	Utilize regular test schedules for all controls

Operator Error

Problem or Cause	Cure
Blocked or by-passed controls	
Improper or inadequate blowdowns	
Thermal shocks—fast startups	
Improper feedwater treatment	
Improper operation of soot blowers	Operator training in all cases
Inadequate repair, maintenance, or operational records	
Operating with low water	
Allowing oil or contaminants into waterside	

Miscellaneous Problem Areas

Problem or Cause	Cure
Manhole and handhole plates	Keep gaskets in good condition
Stays and staybolts	Inspect annually for defects
Tube rupture	Prevent circulation blockages and hot spots
Shell rupture	Make certain boiler is never overpressured
Leakage at any pressure part	All repairs should be performed by National Board stamp holders
Rivets or rivetted seams	Check rivet heads and buttstraps regularly for deterioration
Code violations	Use only ASME stamped and National Board registered vessels

Corrosion

Problems or Cause	Cure
Oxygen pitting internally	Proper feedwater treatment
Galvanic corrosion	Avoid dissimilar metals—maintain proper pH
Acidic corrosion	Maintain proper pH in feedwater
Slag corrosion	Keep fireside surfaces free of slag
Soot or external corrosion	Eliminate moisture and sulfur combinations
Intergranular corrosion	Match metals to environment service
Crevice corrosion	Eliminate leakage crevices

Piping Failures

Problems or Cause	Cure
Design, material and installation	Provide strength and stability, drainage to prevent water damage; follow Code Rules
Freeze ups	Avoid careless routing of lines; water still freezes at 32°F
Operation	Proper operator training
Vibration	Use proper supports, scrubbers, and shock absorbers

Combustion Control Equipment

Problem or Cause	Cure
Overfiring or overheating	Don't overload or over fire
Firebox explosions	Make sure the purge cycle is in good order at all times
Flarebacks	Keep combustion control equipment well maintained; always follow firing instructions
Carbon monoxide poisoning	Maintain proper fuel/air ratio
Flame impingment	Never overfire
Fire checks	Avoid extended surfaces into fireside
Insufficient draft	Over-design rather than under-design
Lack of fresh oxygen supply	Complete combustion requires an adequate supply of oxygen

Mechanical Deterioration

Problem or Cause	Cure
Improper design material or installation	Design to ASME, register National Board
Baffles and refactory	Keep free of cracks and holes that could allow short circuiting
Cracks in pressure parts	Reduce thermal cycling
Bulging or blistering of tubes	Don't overfire—guard against thermal shock
Bulging or blistering of other parts	Watch for flame impingment
Supports, setting, casing, foundation, or stack	Heavy constructional loads require constant maintenance
Outside forces or external mechanical shocking	Frequent inspections, with proper maintenance

Erosion

Problem or Cause	Cure
Fireside due to slag	Maintain close watch on coal quality
Tube fireside due to bridging or channelling by fly ash	Follow established soot blowing procedure
Improper soot blower alignment	Install properly

F. Fire Prevention and Protection

This is another subject about which a complete volume could be written. It is such an important item in today's industrial plant that we will not attempt to give complete coverage to it in this book. We will, however, make a few suggestions, which we sincerely trust you will follow.

If you have any questions concerning your fire protection, there are several agencies which may be approached for help. One of the better known ones is the National Board of Fire Underwriters, commonly known as the NBFU. They have offices in most large cities, and will either help you directly, or refer you to one of their member agencies. They publish several helpful pamphlets, which are handy guides to proper fire protection in the industrial plant.

In the case of fire hazards applying to the electrical equipment, the National Electrical Code®, together with the Underwriter's Laboratories, have formulated a classification system for designing the degree of hazard for various areas. It is an excellent guide for the plant manager or owner in helping to choose the proper equipment.

The National Fire Protection Association is another agency that has published many codes and guides for use by fire underwriters and insurance agencies. Contact them, if one is conveniently at hand.

There are also various state, county, and city agencies that may be called in to offer help. The local building and safety department, or its equivalent, will have its own set of codes and standards that will have to be followed. Also, the local fire department probably has someone on its staff who is a specialist in spotting potential danger areas and who will know which agency to refer you to for further help.

There is one thing we will strongly urge right here, and that is that it is a matter of good policy for you to become acquainted with the local fire protection authorities, since it is much better that you cooperate with them than try to avoid them or their rulings. They have seen many more fires than you will ever see, and it is very much to your advantage to make friends with them—or at least don't make enemies of them!

And, above all, do not skip over those fire drills! They are not just "kid stuff."

In most plants, fire protection is under the supervision of either one person or a committee. This does not relieve you, the boiler operator, of responsibility, however. As a practical matter, there are several areas in your boiler plant where you should be constantly alert for the possibility of fire damage, or impediments to fire protection, and we shall list some of them here.

As the fuel supply is probably the main source of danger, the main supply or shut-off valve should be clearly identified and marked, with any tool required to operate it readily available at the valve location; the pathway to the valve should always be free and clear. Ideally, the valve should be far enough from the boiler so as to be in a safe place for access and operation.

The front of the boiler, or the area just in front of the burner, should be free of debris at all times. This includes removing all trash containers from the danger area, wiping up all spilled inflammables, and keeping all rags, loose paper, etc., from the area. The log table, with the records, papers, notes, etc., should be in a relatively danger-free area.

Know where all of the fire extinguishers are located, as well as the fire hoses, hydrants, foam trailers, etc.

Electrical switchgear should be isolated from the rest of the plant, if at all possible, and the correct type of fire extinguisher for fighting electrical fires should be readily at hand. If the switchgear is protected by one of the package, special types of fire fighting systems, such as a CO_2 or Halon system, then this system should be carefully checked at regular intervals, and all debris or plant equipment spares should be kept away from the system. Here, again, there should be a clear pathway to the equipment.

One hint, which may help you in your fire protection scheme of things; steam is an excellent medium for smothering a fire. If you have any highly flammable material in storage tanks, vessels, or vats, give some consideration to piping an emergency steam smothering line from your boiler header over to the storage container, with a lever-operated valve in a handy location on the line, clearly marked, with a lead seal on it that can be easily broken. Be sure, of course, that there is no one in the tank, vessel, or vat before you turn on the steam! And be sure to check local codes.

G. When an Accident Occurs

You have followed all of the prescribed procedures, safety regulations, and maintenance functions required, and still a boiler accident happens. They do happen, even though you feel they shouldn't. So what do you do then? We shall give you a few general steps to be followed to minimize the impact upon the plant, its personnel, and equipment. These are a guide only; you must modify them to suit your particular installation, equipment, and plant policy.

    1. Turn off the igniter, pilot, and fuel supply in that order to the damaged boiler by means of the closest manual valves safely available when you first enter the area after the accident. There should be no question concerning the first steps, as a fire danger is definitely present in any accident involving a boiler being fired.

Please turn to page 358 in the Appendix for a statement that will help to prevent a furnace explosion occurring due to a burner flame-out.

If the accident consists of a furnace explosion, after the above steps have been taken, you should examine the condition and situation of any other boilers on the line to see what needs to be done to save them from damage.

If the accident is a tube failure, it is important to maintain the water level in the boiler to minimize the damage from loss of steam and

water. The steam pressure will drop from loss of steam and the inflow of feedwater, providing there is no steam coming in from other boilers on the line, and the potential damage will diminish with the drop in steam pressure. Do not open the vent valves or the safety valves on the boiler.

    2. In the case of a furnace explosion, turn off all electrical circuits to the boiler, with the exception of the draft fan. If the fan is still in running condition, it is best to leave it on for a few minutes after the fuel has been shut off, to clear excess unburned fuel vapors out of the area. One of the reasons we recommended earlier that the main electric switchgear for the boiler be some distance away is to prevent an explosion of accumulated fuel vapors from the opening of the main switch, which is often accomplished with a spark at electrical contact points.

In the case of a tube failure, it is essential that the feedwater supply be maintained, and the draft shut off, which requires that the electrical control be manipulated accordingly.

    3. In the case of a furnace explosion, turn off the feedwater supply to the damaged boiler, to prevent flooding and possible spreading of fuel oil.

    4. Remove all unauthorized personnel from the area. This includes mostly plant personnel who have gathered to gawk at the carnage, as well as others who may have come in off the street to see what's happening.

    5. Take a quick check for casualities, and call the plant fire marshall, plant fire department or the local fire department, if they are needed. In most cases, it is well to call the local fire department, even if it does not appear to be necessary. They will often have equipment, manpower, and training to lend you professional assistance. Their official report may be needed by your insurance company, also.

6. Do not touch, adjust, or alter any other part of the damaged boiler or its accessories, unless essential to safeguard the area, personnel, or plant.

7. Notify immediately all plant officials necessary, and await their further instructions.

8. Take photographs of all damaged portions or areas, and close-ups of all suspected trouble spots or equipment.

9. Do not attempt to analyze the trouble or find its source at this time. The boiler inspector is the best one to do that, and you may be called upon to assist him, and answer any questions concerning the events leading up to the time of the accident.

10. Make no statements to any outside personnel or unauthorized agencies, firms, etc., until cleared to do so by the plant officials or the insurance underwriter. Damage claims may be involved, and off-the-cuff remarks made to news reporters, gawkers, etc., have a way of getting into the testimony.

Of course, if your plant is properly managed, there will already be a planned procedure in house for guidance should such an emergency occur. You, the boiler plant operator, should be very familiar with it, and, in fact, you should have an active part in formulating it. Following that, you must review it often enough to be sure you know it well, and to be able to update it from time to time, as changes occur in the plant equipment, policies, or personnel.

During times of stable, peaceful, routine operation, the alert operator will occupy the time in preparing for emergencies. By going over in your mind constantly the various troubles and emergencies that could conceivably occur, and having a plan of action worked out in each case, it will be only natural for you to follow the proper actions if that emergency does occur. Your reaction to emergency could very well help to erase any stigma or blame falling on you for the accident.

H. Responding to an Alarm

When an alarm initiated by the boiler monitoring system sounds off, the tendency is for all within hearing of it to stop work, and look for the cause. This can be very disruptive in the plant operation, and could not only cause reduced plant efficiency when not usually warranted, but could also lead to accidents in other portions of the plant.

When an alarm bell rings, it means that something is not right on the boiler, and needs looking into. It may or may not be life-threatening, and if the boiler operator has tended his plant long enough to gain an understanding of its weaknesses and possible trouble sources, he need not spend to much time locating the trouble. If he has followed the advice given in these pages, he will already know where the trouble is most likely to be, and what steps are necessary to correct the problem. However, we shall give some suggestions here that will help the operator maintain control over his plant and his well-being. We shall start by assuming that the operator is in the immediate vicinity of the boiler when the alarm rings.

1. DON'T PANIC! That is always the best and first advice to be followed.

2. Call out to all unauthorized personnel in the boiler plant to leave immediately.

3. Calling upon your mental drills, which you should have consistently practiced for the various emergencies, check all indications from the readings of the various gauges, and make visual checks of the flame to ascertain the source of the trouble.

4. Proceed very carefully, swiftly, but calmly, to correct the problem and restore the boiler to proper condition, or shut it down if the situation requires it.

5. Assess the effect upon the remaining boilers on the line, and take all appropriate action to ensure their safety.

6. If necessary to shut down the faulty boiler,

start up any spare boiler available, and see that it is operating properly to maintain the plant demand.

7. Stay with the boilers until you are satisfied that all boilers are working as required.

8. Fill in the details of the incident in the proper place in the log.

9. Alert the next shift operator as to the difficulties, and ask him to carry on any corrective actions you have started.

Before we proceed any further, some comments on point 2 above are in order.

The boiler plant is a central point of high energy consumption. As such, it is potentially a danger to any person loitering in the vicinity who has no business being there, and who does not have the training or background to appreciate the dangers. For their own safety, therefore, all loitering in the boiler plant should be strongly discouraged, if not absolutely forbidden. It may be necessary to get the cooperation of the Safety Committee to declare the area off limits to unauthorized plant personnel. In some plants, where there is no partition between the boiler plant and the remainder of the premises, and where the boiler area has been used as a gathering place for those wishing to take a few minutes for a smoke, or who like to sit down in a warm place to eat their lunch or have their work break, it may be difficult to make drastic changes, but it must be done.

Most boiler operating codes today permit the boiler plant operator to leave the boiler for periods up to 1 hr, providing he is always within earshot of the boiler control alarms. This is a result of the modern tendency to load the boiler with automatic devices and alarms to meet the insurance and code requirements. Some plants take advantage of this to the maximum extent, by such contrivances as placing alarm bells throughout the plant, wired into the boiler control alarms, thus "authorizing" the operator to stray far from the boiler room on other plant duties. This is stretching the intent of the code, and not to be recommended.

So how do you respond to the alarm from the boiler when you are some distance away in the plant? There are several rules and suggested procedures to follow, and we give some here as an addition to those already given in this section:

1. DON'T PANIC. Keep calm, and walk at a pace a little faster than normal in getting back to the boiler room.

2. Don't do or say anything to cause others in the plant to panic as you make your way back to the boiler. If you do not show undue concern, those in the plant who see you, will feel safe.

3. On your way back, be thinking of what steps are necessary, in the correct sequence to correct the problem, ensure the safety of others in the plant, and protect the remaining boilers on the line.

4. Follow the procedures given previously, steps 2 through 9.

At times like these it pays to know your equipment, know the weak points of the system, and have faith in the safety alarms and shutdowns. This requires a knowledge of the past history of maintenance performed, tests made on the control and alarm systems, and your ability to spot potential trouble spots and anticipate future problems.

As an example, suppose you are quite a distance from the boiler room and the safety valve blows with its high-pitched shriek heard throughout the plant. The first natural reaction is one of extreme danger to the plant and others in the vicinity of the boiler. It may or may not be. A lifting safety valve means that the valve is doing its job, and if it is properly sized, set, maintained, and tested, there should be no immediate danger to the boiler or others.

One final suggestion: Should your plant use a public address system to page personnel, then we suggest they use a code when reporting an emergency, such as, "will the chief engineer please report to room 15." This may help in preventing panic by some timid plant workers.

I. Deliberate Overfiring

The major emphasis in this book has been on getting the maximum efficiency from the industrial boiler, assuming that the boiler plant has sufficient capacity to carry the demand. The time may come, however, when the plant management insists that the boiler operator coax the last pound of steam from the plant that is possible to attain. What are the possible ramifications should such a condition arise?

First, let us explore the reason for this demand on the boiler plant. It obviously comes about because there is a lack of steam available to perform the assigned plant task, either on a temporary or a permanent basis. Before either rejecting or accepting overfiring the present equipment, there are several alternatives to be explored, and we list some of them as follows:

1. Install a rental boiler to handle a temporary increase in load.

2. Install a flash-type boiler either to handle a specific portion of the load or to tie into the supply main.

3. Bring in steam through a temporary line from a neighboring plant.

4. Isolate a portion of the heat load and investigate alternate heat sources, such as direct firing gas or oil burners, circulating hot water, or other heat-transfer medium.

5. Reschedule plant operations to stagger the major loads, and thus reduce the load peaks. See Chapter 20, Section D for a discussion of this method.

6. Install a steam accumulator. See Chapter 20, Section D for a discussion on this item.

Should none of these alternatives be acceptable, then the boiler plant supervisor is faced with a hard decision to make. We urge him to consider very carefully the results to be expected and the dangers to be encountered should he attempt to increase the output of the boiler beyond its rated capacity. Here are some of the results to be expected:

1. Shortening of the boiler's life, due to overheating and possible flame inpingement on the heat transfer surface.

2. Drop in efficiency with increased stack temperature.

3. Possibility of increased atmospheric pollution from soot.

4. Increased maintenance and emergency shutdowns.

5. Increased possibility of furnace and waterside explosions.

6. Increased danger to life and property surrounding the plant.

7. Possibility of any guarantee on the boiler being canceled.

8. Possible cancellation of the insurance on the plant.

9. Possibility of the local inspector or fire marshall condemning the equipment.

10. Expect some changes in operation of the auxiliaries, such as vibrations or pulsations from the deaerator, fan casings and ducts, and boiler breechings.

So now that all the alternatives and warnings have been considered, and the decision is still the same—increase the output, regardless! Where to start?

As the heat-transfer-surface area is fixed, and assuming that the boiler has already been tuned to get the highest efficiency and the most steam possible from its existing design, then the only answer is to increase the firing rate of the burners. This should be combined with an increase in furnace volume, if possible. Years ago, when the horizontal return tubular boiler and similar designs using onsite built-up furnaces were in vogue, this was relatively easy, as the furnace could be enlarged with additional brickwork. Boilers at the time were often conservatively rated at about 10 ft^2/bhp, and some overfiring was possible without serious damage to the boiler. During 1940–1945 it was fairly common to overfire some boilers as much as 200%.

In the case of boilers supplied in more recent times, having furnaces integral with the boiler shell and tube nests, the first approach must be to contact the supplier of the boiler and burner units, and get their suggestion. They should have the first chance at the problem. If the burner was furnished as a separate item, the burner engineer very likely oversized it by at least 5–10%, producing a built-in overcapacity from the boiler. There are often changes that can be made which will increase the burner fuel capacity, such as enlarging burner orifices or increasing the fuel pressure and flow. Additional air supply must be provided, either primary or secondary. See the discussion on seminozzle mix burners elsewhere in this book.

Should the boiler or burner supplier have nothing to offer beyond the preceding changes in their equipment, then we strongly urge the adoption of one of the options given in the second paragraph of this section.

DO NOT ATTEMPT ANY ALTERATIONS IN THE EQUIPMENT NOT SANCTIONED BY THE SUPPLIER, INSURING AGENCY, OR THE INSPECTOR.

REMEMBER: Boiler explosions, whether in the furnace or on the waterside, are often very disastrous affairs.

CHAPTER 20
BOILER TROUBLES: CAUSES AND CURES

A. General

Looking at the list of accident causes in Chapter 19, the cures for some of them are self-evident. However, it will not be amiss if we discuss them here, for purposes of identification and reinforcement.

One of the major causes is unsafe operating practices, and that is one of the chief reasons for this textbook and the course of instruction for which it was produced. We are hopeful that both the book and the course will help to reduce this cause of accidents very substantially in the years ahead. Although we have covered many possible areas of misoperation, there are undoubtedly many that we have missed. The general approach has in most cases been spelled out, permitting you to work out your own safe practices and thus keep you out of difficulties, or at least minimize them to levels that may be handled easily.

In the matter of construction code violations, you no doubt have noted that we continually suggest that you consult local codes for the final word concerning approved methods of installation, maintenance, and operation of the boiler plant. You are responsible for the plant's operation, efficiency, and safety; therefore, it is in your best interest to be sure the local codes and any other codes having authority over your plant are adhered to.

Should you take over an existing plant and find that the plant does not meet existing codes,

then it is still your responsibility to bring that fact to the attention of management. First, however, be sure of your position by checking very thoroughly into all governing codes, then put your findings in a written memo to management, stating the violation, the possible consequences if it is not corrected, and your suggested procedure for correcting the situation. It is very poor management that would ever fault you for taking that step, providing you are correct, sure of your grounds, and use a fair degree of common sense and tact in bringing it to their attention.

The other items on the list of causes of accidents will be covered for the most part in the sections that follow. We shall break the list down into several categories, which should cover them all very well.

B. Overheating

When we speak of overheating a boiler, we mean both general overheating of the entire boiler and localized overheating. Generally, it is localized overheating that is most common and causes the most trouble. When localized overheating occurs, the failure is usually very rapid after the condition is established, but the damage is usually isolated to that particular area. This is mostly true of the watertube boiler overheating situation; but in localized overheating of tubes in a firetube boiler, the result may be disastrous.

In general overheating of the boiler, the result is at least a shortening of the boiler's useful life. However, if there are any weak spots in the boiler that are critical and about to cause trouble, then overheating may cause that weak spot to fail under the stresses generated by the overheating. A boiler may coast along on a moderate load for months, with several weak spots developing, but not severe enough to cause difficulty. But, push the boiler to its maximum load capabilities for any length of time, and those weak spots start to become critical, and trouble comes.

We shall list the most common causes of overheating, and if the cure is not self-evident, we shall cover that subject, also.

1. Scaling of the tubes, which occurs on the water side, usually from improper water treatment, increases the resistance to heat transfer across the tube surface, and thus increases the tube surface temperature on the hot gas side.

   A build-up of scale on the tube surfaces of a watertube boiler will cause the pressure inside the tube to produce a blister on the outer surface, or in the extreme case, a complete rupture,

   Scaling on a tube of a firetube boiler will produce a dent in the wall of the tube, or a complete collapse.

   An accumulation of scale and sludge on the bottom of a firetube boiler may produce a bulge on the outside bottom of the drum.

2. Oversized burner capabilities, not matched to the boiler heat-transfer-surface capabilities, luring the plant personnel into demanding too much from the boiler. This forcing of the boiler to produce more than it was designed for increases the gas temperature throughout the boiler, increases the flue gas discharge temperature, and lowers the efficiency. It also shortens the life of the boiler, and may hasten the failure of those weak spots described previously.

3. Too much draft, pulling the flame beyond the normal designed point in the boiler, may cause direct flame impingement on tubes not meant to handle such temperatures and thus result in overheating. Should there be a layer of accumulated soot (unburned carbon) on the breeching and base of the stack, the flame may ignite this layer of soot, and burn out the breeching, destroying it and spreading sparks over the countryside. The cause of the high draft is usually due to the draft system getting out of adjustment or some form of damage to the equipment—loosened cams, burned out dampers, bent damper rods, etc.

4. Poor water circulation in any of the waterside areas may cause localized overheating. Water on one side of the steaming tubes acts as a cooling medium, carrying away the heat as it is transferred into the water from the tube wall. If that water does not move fast enough, the heat-transfer method changes from nucleate boiling to film boiling, thus increasing the tube wall temperature on the gas side of the tube. (See Figs. 3-4 and 3-5).

5. Internal damage to the baffles or refractory work may cause an upset or change in the gas or water passages through the boiler, which may interfere with the distribution of the heat transfer. Each section of the boiler, each pass, is designed to handle the flows and temperatures to be expected at that location. Any change in conditions may cause local damage or overheating. Any foreign object left in the boiler during shutdown may produce the same effect.

We should expand upon the subject of poor water circulation, since the cause of it is not always under the control of the boiler operator. Generally, in the well-designed boiler, the things that can cause an interruption in the circulation pattern are debris left in the boiler, an abnormal build-up of scale sufficient to choke off a tube, and making a temporary repair to a leaking tube by plugging it at the ends, which can change the water circulation pattern. This last item is done in an emergency only, and should not be done to too many tubes before the boiler is closed down completely for retubing. It is a case of a temporary expedient leading to possibility of more damage and more expense.

That fairly well completes the list of things that are most apt to go wrong with the boiler and cause overheating. As you can see, most of the problems are a result of misoperation, which places the responsibility directly in the hands of the boiler operator.

As you probably noted, the net result of overheating of the boiler can result in a great deal of expensive repairs being required.

C. Corrosion

There are a number of different chemical reactions that come under the general heading of "corrosion," but the ones we are most concerned with here are those involving chemical attack on the metals used in the boiler and its accessories. This subject is being touched upon at other places in this book, and we shall go into it a little more deeply here.

Metals are produced from ores in what is known as a reduction process. Once purified in the metallic state, there is a tendency for most metals to revert to the original state—that of its ore—in what is known as an oxidation process. Conditions for this reversion vary with the different metals, but oxygen must be present, along with other conditions, such as moisture and heat in varying degrees.

Putting it another way, metal is produced from its ore by removing the oxygen from the ore. Putting the oxygen back into the metal causes the metal to revert to an ore. This is greatly simplified, of course, but the principle is all we are attempting to bring out here, for the purpose of this discussion. We are concerned only with the process as it applies to steel, the most common metal used in boiler construction.

The three main ingredients required to cause corrosion of steel are the presence of oxygen, heat, and moisture, all of which are quite prevalent in the boiler, or in the steam system. It is obvious that if we can eliminate one of the three ingredients, we can cut down or eliminate corrosion.

The only one of the three ingredients that we can eliminate is oxygen, since it does not enter into the steam production process in any way on the water side. And the methods used to eliminate, or at least control, the amount of oxygen in the water side are described in Chapters 9 and 10, plus another subject germane to the situation, which will be covered in a later section of this chapter—the presence of air in the steam spaces.

There still remains the problem of corrosion of the steel surfaces on the hot-gas side of the boiler, such as the furnace, firetubes, uptake, breeching, and similar portions in the hot-flue-gas system. The problem may be partially solved by the use of stainless steel, but this is expensive, and in those areas where we desire a high heat transfer, it is not generally economical to use it, since it has a greatly reduced heat-transfer ability compared to carbon steel. Some breechings are being made of stainless steels, where the fuel being burned as waste products produce an abnormal amount of corrosive exhaust gas.

In the chapters on combustion, we pointed out that excess air is needed for complete combustion. Later in this book, we shall go into the methods of controlling the amount of excess air in order to keep it at a minimum. The emphasis there will be on increasing the efficiency of the combustion process, whereas here we are interested in the effect of that excess air on the corrosion of the steel surfaces as it is carried along through the flue gas passages.

The combustion process produces large amounts of water vapor, which is carried along with the hot gas. Thus, with excess air being swept along with hot gas and the water vapor, we have all three elements present to corrode the steel surfaces of the boiler and the flue passages. This explains one of the reasons why it is advantageous to keep the excess air for combustion to an absolute minimum.

The presence of excess air in the flame produces what is known as an oxidizing flame, since it is capable of attacking, or oxidizing, metal surfaces, as previously explained.

The composition of the fuel has a direct bearing on the amount of corrosion to be expected from the flue gas. Most fuel oils have varying amounts of sulfur in them, and this is a source of corrosion, since the sulfur produces oxides of sulfur, which, when absorbed by the water vapor in the flue gas, results in sulfuric or sulfurous acid, both of which are highly corrosive to steel. This is usually only a problem when the flue gases are cooled sufficiently to cause the water vapor to condense on the steel surfaces of the boiler breeching and stack. This subject will be covered in more detail elsewhere in this book.

There is one more element present in the flue gas, and that is nitrogen, most of which comes from the air for combustion. This nitrogen produces various combinations of oxides of nitrogen. They are not normally a problem in the boiler and flue gas passages. The main problem we have with these nitrous oxides is in their effect on the surrounding atmosphere, which can be a considerable nuisance.

Fortunately, there are steels on the market containing alloys that tend to resist the hot corrosive flue gases, and these steels are used quite extensively, at least by the better boiler manufacturers. This does not remove the problem entirely, but only tends to mitigate it. It is still to your best interest to keep the excess air to a minimum, and burn fuel which is low in sulfur.

There remains one other location where corrosion is a constant problem in the boiler plant, and that is in the return lines carrying condensate back to the boiler room. This section of the plant is one of the continual problem areas, since in the average steam plant the condensate lines usually eat through from corrosion much faster than the steam lines.

The major source of this corrosion is carbon dioxide carried over with the steam, which is left in the gas when the steam condenses, then readily dissolves into condensate, forming carbonic acid. The presence of this carbonic acid may be detected by a distinctive reddish brown color of the condensate, and grooving of the bottom of the condensate line. The joints usually eat through first, since they are under higher stress than the remainder of the piping.

Carbon dioxide is a by-product of the boiler water treatment system, when zeolite-softened water is fed into the boiler without being properly deaerated. The result is release of large quantities of carbon dioxide, and we refer you to Chapters 4 and 10 for further discussion on this matter.

Another common source of corrosion in the return lines is oxygen, which has a tendency to form pits in the top portion of the return lines, as well as adding to the corrosion of the iron in the lines. The source of this oxygen, which is often in the free state (not associated with other elements) is the water treatment and the air in the system from the previous shut-down.

Some specific industries and some practices connected with their operations produce other corrosion problems. For instance, steam coils used to heat corrosive products often spring leaks, and the corrosive solution is pulled into the heating coil when the system is shut down, and a vacuum formed inside the coils. The solution is either to make the heating coil of corrosion resisting material or discharge the condensate from those coils to waste, after reclaiming as much heat as possible from it. In this regard, we refer you to Chapter 26.

The presence of corrosion in the steam system is a never-ending problem, and boiler plant operators may as well get used to the idea that they are going to be fighting it all their working lives. We have told of only some of the more common types, but there are others, and the operator had better be prepared to meet the challenge.

D. Common Water-Side Problems

Aside from the problem of corrosion on the water side of the boiler plant system, which was just discussed, there still remains other problems of an operational nature that are encountered quite frequently on the water side. These we shall cover now, as listed below, along with short descriptions of their causes and cures.

1. Low water level is probably the most common source of accidents, and a complete chapter could be devoted to that one problem. However, we shall cover it sufficiently here to permit you to understand the reasons for the low-water conditions to be expected in your boiler from time to time, assuming that your boiler is fed from an automatically controlled system.

The major causes of low water level may be traced to:

Faulty level controllers.

Faulty feedwater regulating valve and its controlling mechanism.

Clogged strainer or closed manual shut-off valve.

Worn pump impeller, or clogged pump.

Drop in feedwater supply pressure entering the pump.

Increase in the feedwater temperature, causing pump cavitation.

Open blowdown valve to the blowdown tank.

Electrical system inoperative to the feedwater system.

2. High-water level is not normally such a problem and is usually not as dangerous as the low-water-level condition. The result is usually a priming and carryover condition, which is annoying and troublesome in the process end of the plant. The sources of the troubles are usually found in:

Leaking or collapsed level controller float.

Corroded or coated electrodes on electrode-operated level controls.

Feedwater regulating valve stuck open.

3. A surging water level, one which never

seems to stabilize in one place on the gauge glass, has been covered previously in this book, and we shall not go into it any further here.

4. Various troubles that may be traced directly to the water treatment methods used, such as scaling of the heat-transfer surfaces, foaming of the boiler water, priming, and carryover of the boiler water into the steam spaces. These have all been covered in the chapters on water treatment, and will be repeated here.

The troubles due to failure of the system in maintaining a proper water level in the boiler are important enough to require the operator to be ever on the alert for their occurrence. To emphasize their importance, we shall summarize them here, as a reminder and a check list for your use:

Causes of Low Water Level:

Faulty level control—dirt, scale, grease, etc.

Boiler feed pump inadequate or faulty.

Scaling or obstruction in feedwater line.

Blowdown system leaking or open slightly.

Faulty feedwater valve.

Clogged strainer in boiler feed line.

Causes of High Water Level:

Faulty level control system.

Faulty feedwater valve.

Leaking float on level control.

Causes of Surging Water Level:

Fast load changes.

Level controller improper for load pattern.

Poor drum internal design.

Causes of Scaling

Improper water treatment.

Improper blowdown procedures.

Poor water circulation.

Causes of Foaming:

Improper water treatment.

Oil in boiler water.

Rapid load changes.

High boiler water pH.

Causes of Priming and Carryover:

Rapid load changes.

Improper drum internal design.

Pushing boiler output beyond its rating.

Total dissolved solids in boiler water too high.

Boiler water pH too high.

The cures for most of these troubles are self-explanatory.

There is not much you can do to correct an improper drum internal design, short of ordering a new boiler and insisting that the supplier show proof of proper design. Once the boiler is installed, unless you can prove conclusively that the drum internals are at fault, you stand little chance of getting any relief from the faulty design. Fortunately, this does not happen too often—very rarely, in fact.

Regarding the causes listed as "Rapid load changes" or "Pushing boiler output beyond its rating," there is something that the plant operator can do to solve those problems, and we refer you to Fig. 20-1 for two simple methods of limiting the steam output from the boiler. The orifice is the simplest solution; however, sizing it properly takes some fairly complicated calculations, and you may have to ask the advice of the boiler supplier. If the steam line where the orifice is installed is horizontal, there should be a steam trapping station just

ahead of it. An alternative to the steam trap is a small hole drilled through the orifice plate at the lower edge of the plate, in line with the bottom of the steam main.

In the case of the back-pressure regulator, the correct selection is a fairly routine matter, and any good supplier of this type of pressure regulator will be able to make the right selection for you.

The back-pressure regulator should be adjusted to shut off at a boiler pressure just under the low setting of the main steam pressure controller for the boiler. For instance, if the steam pressure controller cycles the boiler between 110 and 125 psig, the back pressure regulator should shut off at about 105 psig and remain open at all pressures above 105 psig. The result should be a stable boiler operating condition with fairly dry steam to the process, but at a low pressure and temperature.

The rapid load changes referred to usually come about as a result of the plant operators out in the production department suddenly opening the steam supply valve to the process too rapidly, or too many steam-using processes are put on the line at one time. This is particularly apt to happen when the plant has been expanded faster than the steam supply system's capability to handle the increased load. By installing a load limiter on the steam supply header off the boiler, the production personnel can open all the steam valves they wish, but the steam they get will be at a very low pressure, forcing them to change their procedures.

The correct solution, of course, is a properly sized boiler plant capable of riding out the sudden demand without a large drop in steam pressure.

There is another solution to this problem. If the sudden surge in plant steam demand is due to several processes being instituted at once, such as all retorts in a battery being started at once, then perhaps the plant production department would agree to stagger their start-up times, thus keeping the instantaneous load on the boiler at a more moderate figure. By staggering their loads, there may be an actual decrease in total overall process time, since the

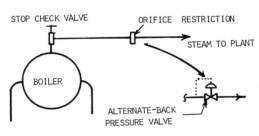

Fig. 20-1.

average steam pressure and temperature being supplied to the process will be higher.

Still another solution to the undersized boiler problem, which has been used in the past, is the steam accumulator vessel. This consists of a pressure vessel, built to Section VIII of the ASME Boiler Code, containing water, into which steam is directed during periods of low loads. The vessel absorbs the excess steam, and builds up a reservoir of water at saturation temperature and pressure, which is floating on the plant load. When the surge demand comes, this high-temperature water flashes off into steam upon drop in pressure in the steam main, thus providing the extra steam for the demand.

The accumulator must be designed by an engineer well versed in this procedure, as the design is critical.

We do not recommend this solution for installation in crowded or built-up locations, as the storage of high-pressure and -temperature water presents a safety hazard. Should it burst, the result would be disastrous.

E. Gas-Side Problems

The problems to be expected from the air and gas side of the boiler system fall into five general areas: dampers, ducts, fans, baffles, and stacks. We shall cover each in turn. Each of them can cause its share of headaches to the boiler plant operator, as often each difficulty will be an occasional or one-time occurrence, which can cause a lot of consternation until the trouble is isolated.

Draft fans, being mechanical and rotating in nature, are subject to the usual problems of this class of equipment. They are often very lightly built, and subject to damage very easily. The things to check are:

Broken and loose fan belts, or belts improperly installed.

Fan casing damaged and rubbing between the fan and casing.

Fan wheel damaged and running out of balance.

Fan wheel balancing clips missing.

Build up of dirt on the fan, causing vibration.

Fan casing pulsating due to improper design or selection, or draft system out of control.

Bent fan wheel shaft.

Normal electrical-side problems.

Dampers are equally lightly built, and easily damaged. Following are a few of the things to watch in obtaining good service from your dampers:

Bent leaves from carelessness and transient conditions.

Presence of rodents, rags, paper, etc., in the dampers.

Bent, damaged, or broken damper rods or arms.

Damper motors inoperative or operating badly.

Manually operated or set damper arms moving out of adjustment.

Damper cams coming loose on the shaft or out of adjustment.

The ductwork is also subject to damage, from collisions, overheating, corrosion, extremes in weather conditions, and vibrations caused when pushing the boiler at maximum loads beyond its rating.

Baffles, when used, must be kept in good condition, since should they become damaged, altered, or dislodged, drastic changes in gas flow may occur, causing overheating or hot spots in the boiler. They may become damaged during maintenance operations and not be detected prior to closing up the boiler. Overheating of the boiler from a high flame may damage them, and in some cases they have been known to disappear entirely, usually from too high a draft situation.

The stack is merely an extension of the ductwork, and is subject to the same sources of damage. There is one additional trouble, which must be guarded against, and that is high winds which could topple the stack and cause property damage. The solution to this is, of course, proper guying with cables or structural steel members.

In a later chapter, we discuss the effects of operating with too low a stack-gas temperature, and we shall not repeat it here. The main problem with small boilers is usually one of too high a stack-gas temperature, which means low boiler efficiency, and seldom will the solution to that situation lower the stack temperature sufficiently to cause trouble from corrosion, providing reasonable care is exercised, as explained in a later chapter.

There is one other possible source of trouble that may be experienced on the gas side of the boiler, and one which may be very elusive in pinpointing.

In Chapter 4 we gave the amounts of combustion air required for the boiler burning either gas or oil. Any boiler operating for long periods of time will require large quantities of air, and as a result, the draft fan handles not only the air for combustion, but also any other stray fumes, vapors, or pollutants that may be floating around in the general area of the fan inlet. All of this goes through the combustion process with the air, and the result may be completely unpredictable, sometimes showing up as an odd deposit or corrosion on the surfaces of the breeching or the gas side of the tubes. The two most common culprits are ammonia and Freon from the plant's refrigeration system.

The location for the boiler plant is generally not based on the availability of clean air for the boiler, but only on the convenience to the use of the steam in the plant.

F. Air and Noncondensables

When the boiler is shut down and cools off, the steam condenses, and if you have performed your work properly, the vents have been opened, allowing air to enter the steam system to prevent a vacuum. Even if you do not open the vents, the condensing steam and the resultant vacuum will draw air into the system through the many leaks around valve stems, traps, connections, etc.

Lighting off the boiler and closing the vents when the pressure has started to rise in the steam spaces will drive out most of that air, but not all of it. Some of it will remain, and will circulate throughout the system, providing oxygen for the corrosive attack upon the steel in the boiler and piping.

In addition, gases from the water treatment, consisting of carbon dioxide, ammonia, and free oxygen, often are released in sufficient quantities to cause their own brand of corrosion problems. The ammonia has an attraction for zinc and its alloys, carbon dioxide forms carbonic acid in solution with water to attack steel, and free oxygen causes localized pitting in the steel portions of the steam system. Thus, you may often discover that your condensate lines are grooved along the bottom from the carbonic acid attack, while the top of the condensate line contains numerous pits or pock marks from the free oxygen in the line. The chapters on water treatment go into this subject also, and we will not pursue it further here. However, there is one other aspect of the air pollution of steam, which we shall go into here. But first, a few more comments on the overall problem, its development, some of the results, and some solutions.

The pressure gauges used in the average steam plant measure the pressure existing inside the bourdon tube, in pounds per square inch above atmospheric pressure surrounding the tube. The gauge is unable to tell you what is causing that pressure.

Dalton's law of partial pressures tells us that in a mixture of gas and/or vapor, the total pressure is the sum of the pressures exerted by the individual components. Thus, if we have a mixture of steam and air, the total pressure of the mixture is the total of the pressures exerted by the air and by the steam individually.

The bourdon tube pressure gauge measures and indicates the total pressure of the mixture inside the tube. But the temperature of the mixture will be that of the saturated steam pressure of the steam in the mixture. For example, let us suppose that the gauge on a boiler registers 100 psig. The steam tables tell us that the saturation temperature at that pressure is 338°F. Now suppose further that we measure the temperature of the steam at the point of the pressure gauge, and discover it is only 320°F. What does this mean? Check the steam tables, and you will find that the temperature of 320°F corresponds to a steam pressure of 75 psig. This means that of the 100 psig on the gauge, only 75 psig of that is steam, and the rest is some other substance. As the only reason to carry a certain pressure is to get sufficient temperature to perform the desired process, it means that 25 psig excess pressure is being carried, and that excess pressure is made up of gases that are probably corroding the system.

Table 20-1 illustrates what will happen should various amounts of air and noncondensables become mixed with the steam. The total steam pressure referred to in the first column is that indicated by the steam pressure gauge. The table covers only low-pressure conditions, but the same principle applies at all pressures.

From the previous discussion on corrosion, it is obvious that the gases making up the excess pressure are probably air, oxygen, ammonia, or carbon dioxide, or a mixture of them. For the sake of simplicity, in discussions of this type, we usually lump them all together under the general term of "air." This is what we shall do in the following paragraphs. They all have one thing in common, in that they are considered to be noncondensable.

It is obviously to your advantage to get rid of this air before it causes trouble in the system. But how to do it, that is the problem.

The best solution is to prevent its formation in the boiler, which means a very carefully worked out program of testing and water

treating. Even then, the results may be disappointing.

Another solution is to get rid of the air and noncondensables in the system where they are released from the steam, and before they can dissolve into the hot condensate.

When the mixture of steam and air comes in contact with the heat transfer surfaces in the process, the steam condenses, and the air and noncondensables are left in pockets, ready to do harm to your system.

Perhaps you have heard the debate concerning what happens to the air in a mixture of steam and air—does it settle to the bottom, or does it rise to the top of the mixture? In reality, it is purely irrelevant, since the mixture is usually in so highly an agitated state that the mixture does not get a chance to stratify. The only thing that can be said with any degree of certainty is that the steam upon entering the cold spaces of the heat exchanger will condense and leave the air above its surface. The condensate will flow along the path of least resistance to the trap opening, and this condensate will tend to produce isolated pockets where the air may accumulate, unless the design of the heat exchanger is such as to prevent this. The problem then, is one of examining each case, and determining where the isolated pockets of air might be located.

Once you have determined where those pockets are likely to be, it then is a matter of tapping a hole into those pockets, and installing an automatic thermostatic vent to bleed off the air. These vents are similar to thermostatic traps, and in some cases, thermostatic traps may be used for that purpose. One warning, here, however. On high-pressure service, when the traps discharge, be careful where the discharge is piped, since the blast can be dangerous.

Table 20-1. Effect of Air on Temperature of a Steam–Air Mixture

Total Pressure (psig)	Pure Steam (°F)	5% Air (°F)	10% Air (°F)	15% Air (°F)
2	219	216	213	210
5	227	225	222	219
10	239	237	233	230
20	259	256	252	249

G. Unburned Oil in the Furnace

Should you be so unlucky as to ever find yourself with spilled or unburned oil discharged into the furnace, prepare yourself for a dirty and sometimes difficult task of cleaning it up. If the furnace is entirely of steel, with no refractory present to become saturated with the fuel oil, then it is simply a matter of shutting down the boiler, waiting for it to cool, and isolating it properly from any other boilers on the common steam header or stack. Then it is a matter of cleaning up the oil with rags, and lighting off the boiler again. You will find that the draft must be increased for some time, since the remaining spilled oil in burning off requires more air than that being supplied with the burner by the combustion control system.

If the furnace has refractory material in the vicinity of the burner, and should it become saturated with the fuel oil, then it is a much more costly and time-consuming project to get the boiler back on the line. The refractory, being saturated with the oil, will be completely ruined if you attempt to burn the oil out of it, so it is recommended that you not even consider it. The only solution is to replace all oil-contaminated refractory, before lighting off again.

To those of you attending an oil-burning boiler, remember to keep the possibility in mind that you may be faced with the messy job of performing one of the above cleanups, and you should be prepared for the worst.

CHAPTER 21
SHUT-DOWN FOR INSPECTION

A. Preparation for Shut Down

In Chapter 17 we described the procedure for shutting down the boiler for maintenance purposes. To a certain extent, the procedure is the same regardless of the purpose of shutting down. However, there are some steps to be taken prior to shutting down, if it is intended to open the boiler for inspection or repair.

One of the first things you should do is to coordinate that shut-down period with the remainder of the plant. To do this it is necessary that you have some idea of the amount of work which must be done on the boiler and the rest of the plant equipment. This may be approximated by studying the boiler manual, if you have not had enough experience with the boiler to know what to expect when you open it up. Even then, you may be surprised when the boiler is opened, and for this reason you should be prepared for a shut-down period longer than the minimum required to perform the operations you have definitely scheduled.

In scheduling your work for the shut-down period, it is a good idea to list the items to be done by priority, with the items having the highest priority at the top of the list, and adding the other items in descending order of priority. You really do not know for sure what will have to be done, or how much time it will take, until the boiler has been opened, and inspected and the repair work assessed.

If it is time for the annual inspection by a qualified Board inspector, then you must arrange for his appearance, and schedule your work around his schedule. The inspection agency usually allows some flexibility in their schedules to accommodate the plant schedule, within certain limits.

Next, it is a good idea to prepare a plan of action to be used as a guide when the boiler is cool and clean and ready for internal work. To do this it will probably be necessary to obtain a proposal from the contract firm who will be doing the major work, if any is planned. Keep in mind that the contracting firm engaged to perform your maintenance work is often too optimistic as to the amount of time it will take to do the work. Allow yourself some extra time in the schedule for this.

If any replacement parts are required, be sure they are on hand and are the right ones. Do not rely on the word of the supplier that he can get parts you need overnight. It is recommended that you have the parts you know will be needed right in your shop, where you can check them to be sure they are the correct ones, and know just where they can be found when they are needed. It is very discouraging and embarrassing to have the boiler open, waiting days for the parts to arrive. It is also very expensive, since the down time for the plant may be building up to a lot of lost revenue for all concerned.

B. The Annual Inspection

The previous section, along with Chapter 17, has prepared you for this step, but there are still a few things that have to be done in preparation for arrival of the boiler inspector.

The boiler must be cleaned out by removing all loose scale and sludge that may have settled in the bottom of the water spaces. This should be done first with a broom, and then followed up with a thorough washdown with a water hose. Do not remove any of the scale clinging to the tubes or heat-transfer surfaces, since the inspector will want to see the condition of the inside under approximate steaming condition.

In washing out the boiler, it is a good idea to save any loose scale or dirt that you have removed, for his examination. He will probably want to get an idea of your operating methods by looking at the scale and sludge that have accumulated.

In addition to this cleaning procedure, follow any previous instructions that the inspector may have given you in preparing the boiler for his brief stay in your plant.

The combustion gas side of the boiler passages must be washed out, also, and cleaned and dried for the inspector. The external surfaces of the boiler must be cleaned, and all excess equipment, tools, hoses, etc., removed from the vicinity to permit the inspector to gain access to all portions of the boiler. One of the things the inspector will note is how well the premises are being kept in order. Each inspector has his own ideas on how a plant should be maintained and operated, and it is well for you to pay very close attention to his comments. He has seen all types of boiler conditions, and yours is probably typical of many of them. Do not attempt to hoodwink him or influence his decisions. Remember, he is a professional at his trade, and expects to be treated as such. Treat him well and respect-

fully, and he can do a lot toward keeping you out of trouble in your operations, and perhaps save your life and that of others.

C. Water-Side Inspection

Quite frequently the boiler will be shut down for an in-house inspection, especially if there has been some indication of trouble, and you wish to find the cause of it, before it develops into a major problem. In this case, the procedure for shutting down, cleaning, and opening the boiler is much the same as given previously in preparation for the annual inspection.

Once the boiler has been cooled down, and opened up, the first thing to look for is any accumulation of sludge and loose scale on the bottom of the drum or water passages. Examine as much of the tube surface as you can to see how much scale has been deposited on the heat-transfer surface, and the scale's location, thickness, and appearance. These considerations will often tell you how the boiler is steaming, how the circulation of the water is being produced, and may even tell you where there is a possibility of future problems. For instance, a uniform build up of scale in one spot on a bank of tubes, followed by a relatively clean area, may tell you of hot spots or circulation and turbulence problems. This will depend on the design of your particular boiler, and is one of the things your inspector may be able to help you with.

Check all water-side surfaces for evidence of bulging, blistering, or overheating. Such abnormalities increase the stresses in the steel, which could lead to further corrosion problems, because corrosion attacks stressed portions of steel more rapidly than other areas.

Inspect all internal baffling for damage and distortion. It is not uncommon to find that they have disappeared, or are broken off and are lying in the bottom of the drum or nearby. Moisture scrubbers installed in the top portion of some drums on watertube boilers may come loose, may get damaged from tools used inside the drum, or may get clogged with scale or sludge. Sudden changes in steam demand may even rip them loose, if they are corroded at the fastening.

Look at all openings into or out of the drum to be sure they are free and clear. Scale or sludge may lodge in the openings and cause trouble. Blow-down connections into the drums must be kept clear and fully open. Also, the drum surface around these various openings must be examined for signs of erosion or stress corrosion.

All gasketed surfaces, such as around manholes or handholes, must be free of dents, gouges, corrosion, or erosion. Remove all old gaskets to clean metal for inspection.

Look for evidence of erosion and corrosion on all water surfaces. Do this by sighting along the surfaces with a low-voltage "safe" drop light or flashlight at varying angles to obtain a relief view across the surface. Then look at every square inch of the surfaces from up close directly normal, or at right angles, to the surface. Inspect every pit or dent or irregularity that appears in the surface.

Remember, if the wetted metal surfaces inside the boiler are bright, clean, and shiny, this is not good, since it indicates that there is a strong possibility that the metal surface is being eaten away from chemical action. Your water surfaces may actually be too clean. Usually, the inspector is looking for a slight coating of rust or scale on the wetted surfaces. If he finds evidence of chemical attack, such as the shiny surface mentioned above, he may demand a hydrotest of the boiler to check for integrity at full designed pressures. In this case, after performing the hydrotest you had better contact your water-treatment service and discuss the proper solution to prevent further chemical action on the metal.

D. Gas-Side Inspection

Inspection of all hot-gas surfaces and seams is similar to that described for the water side. On the hot-gas side, those seams and surfaces that are subject to atmospheric pressure out-

side may be more susceptible to distortion from the temperature differences across the metal casings or tubes.

Probably the most important place to check is the area directly in the path of the flame, as we have already mentioned in previous chapters what dangers are associated with direct flame impingement upon any of the metal surfaces inside the furnace of the hot-gas passages.

The furnace area in the direct path of the flame should also be given a close inspection for signs of burning away or of corroding away of steel surfaces. Such indications are a sign of an oxidizing flame or of too much excess air being supplied to the burner.

If the fire side of the boiler contains an excessive deposit of soot on the surfaces, this is a sign of insufficient air for combustion, or it could mean poor mixing of the air and fuel mixture. This soot must be cleaned off to permit an inspection of the surface underneath. It should be soft enough to be washed off with a water hose and nozzle. If not, then you have trouble. It is wise at this stage to call in a service representative from your fuel supplier. This condition usually is found only when burning heavy fuel oils with additives. In fact, if the soot deposit appears at all suspicious, it is well to call him anyway, just to be sure of the right solution.

The internal hot gas-side baffles and refractories are subject to high temperatures, and are often burned out completely, or distorted to the point where they no longer perform their function properly. They should all be inspected and repaired where necessary.

Internal brickwork should be inspected for signs of loose bricks, mortar, or anchors. All repairs should be made in the manner prescribed by the brick and mortar supplier, and allowed to cure for the required time before putting the boiler back on the line. Usually, the firebox or furnace has to be heated slowly at low firing rate for an extended time to first dry the repaired mortar and then to cure it. Failure to do this properly can result in a ruined repair job.

Right angle joints in the metal surfaces, where

welding has been used to make the pressure-tight joint, should be given very careful attention in both the water and the hot-gas sides. These joints are usually under great stress, and thus are subject to more corrosion damage than flat surface joints, such as along the flat seams. One of the typical locations for this type of seam is where staybolts are used to shore up flat surfaces, and the ends of the staybolts project through the steel plate to the hot-gas side, then are sealed and strength welded on both sides. Everyone of these joints should be given very careful inspection.

All gasketed joints should be scraped clean and inspected for signs of corrosion. It may be necessary to dress off the sealing surface to obtain a smooth surface suitable for gasketing. Leakage of hot flue gas from the furnace to the boiler room can be very annoying, and at times even dangerous. On those furnaces operating under induced draft, any leaks in the hot-gas or furnace joints from the outside will result in excess uncontrolled air being drawn into the gas passages. This will upset the boiler efficiency as well as your readings when you check for combustion and excess air, as we shall explain later in the book.

E. Boiler Exterior Inspection

The exterior surfaces and appurtenances of the boiler require periodic maintenance, also. Nothing looks more disgusting than to see a boiler with sagging insulation, ripped or torn jacketing, bent small-bore piping to the accessories, such as on the water column, or piping that has obviously been assembled from odds and ends of nipples and fittings. Any inspector who notes such conditions as he enters the plant is apt to make a mental note of it, and no doubt it will influence his findings during the inspection of the boiler's condition.

As a guide, and a guide only, we shall give you a few ideas of what to look for when you inspect the exterior surfaces and piping on the boiler. You should put much careful thought into this portion of the maintenance procedure, and in time you will know your equipment well enough to know where to look for sore spots.

1. Always be on the watch for obvious signs of flue gas or water leaks, such as rust spots around joints, dirt accumulations, etc. Fuel oil leaks will be very obvious, more so probably than water leaks.

2. Keep the insulation material on the boiler casing in place, complete, and well jacketed. It sometimes has a tendency to bunch up under the casing, leaving voids, which lead to hot spots on the jacketing. These may be dangerous, as well as expensive in energy loss. The average boiler insulation is designed to obtain a surface temperature of the jacketing of about 140–150°F, which is not sufficient to cause a bad burn on the skin unless the skin is deliberately held against the surface. The average reaction time for a person touching a hot surface is such that noticeable burn will be produced if a surface at 160°F or higher is touched. The normal reaction is to remove the arm, leg, etc., at once.

3. Watch for any bulging or sagging surfaces, which would indicate trouble internally. This is in addition to the sagging mentioned in the preceding item on the insulation and jacketing.

4. Repair all damaged hinges, latches, or yokes on handholes, manholes, inspection doors, covers, etc. Damaged hinges and latches can usually lead to leaking joints, since it is very diffitult for the door, cover, etc., to close tightly if it is not hung properly and the latching device is not in good order.

5. Valve stems have a habit of becoming coated with scale from leaks, which shorten the life of the stem and damages the packing around it. Therefore, as a minimum, when the boiler is down for repairs, the valves should be backed off their

seats to expose the maximum amount of stem, then the stem should be cleaned with steelwool or emery cloth. It takes but a few minutes to clean each valve in this manner, and it is much easier and quicker than repairing the entire valve or repacking it. Often, cleaning the valve stem and tightening up slightly on the packing is all that is required to stop a leak and prepare the valve for another steaming period.

6. Instrument piping, tubing, and wiring should be checked very carefully and repaired or replaced if necessary. Loose connections in the wiring, leaks in tubing joints, can all cause headaches to an operator attempting to keep his boiler on the line without annoying trip-outs.

7. Check all other electrical connections and controls, attempting to anticipate difficulties that may arise due to loose connections, arced contacts, burned spots on wiring insulation, etc. Around the hot areas of the boiler watch for electrical wiring that may be in contact with those hot spots, as the insulation on the wiring may be damaged or burned.

When the above items have been checked off the list, then there still remains the accessories in the boiler plant, such as feed pumps, hot-well tanks, receivers, water treatment systems, etc., which must be given their fair share of the budgeted maintenance hours. It is at this time that the instruments should be removed, cleaned, checked, and calibrated.

This is all we shall cover on the subject of scheduled shut-downs for maintenance, in this chapter.

C H A P T E R 2 2
MAINTENANCE POLICIES

A. Types of Maintenance Policies

By maintenance policies, we mean the general approach to plant maintenance as set down by management. Of course, we naturally assume that all boiler plant operators will ascertain the management's wishes, and follow them diligently.

But what types of policies are there?

The maintenance methods being followed by industrial plants and other organizations with large amounts of mechanical equipment, which must be kept operating, generally follow one of two systems, preventive maintenance or emergency maintenance.

Preventive maintenance merely means that a system of equipment checking and replacement of parts is followed on a periodic basis. The theory behind this method is that it is best to inspect the equipment and repair or replace parts before they fail in service, which is usually at a very inconvenient time, and thus very expensive in down time for the plant's production schedule. The regular vacation time for the production personnel is often devoted to maintenance of the plant equipment.

In practice, the boiler plant is usually based on the preventive maintenance system, with the goal being to reduce to as near zero as possible the number of emergency shutdowns for repair. The number of emergency shutdowns is usually taken as a measure of the effectiveness of the maintenance program.

Emergency maintenance programs are generally based on the theory that if a piece of equipment is operating satisfactorily, leave it alone until it breaks down, then repair it. The implementers of this policy do not usually call it "emergency maintenance," but will attach some other more acceptable term to it.

Emergency maintenance methods require more spare parts to be kept on hand, more expertise in their maintenance personnel, and the willingness on the part of the maintenance personnel to pitch in when the emergency comes and stay with the job until the equipment is back on the line. This policy is usually found in high-speed, high-volume expensive production lines of a highly complex nature. It is not easy to predict where this class of equipment is going to break down, unless it has been in service for years and an accurate history of its breakdowns has been accumulated.

We are not attempting to recommend one policy over the other, but the mode of operation of most plants seems to fit the preventive maintenance pattern very well, and it is usually the one recommended by the boiler manufacturer. Of course, in reality, even the preventive maintenance program is interrupted at times by serious breakdowns.

The preventive maintenance policy lends itself very well also to the use of contract maintenance firms, as covered in the next section.

See Section D for a suggested procedure for setting up a preventive maintenance program, if your firm does not have one.

B. Contract Maintenance

So far we have discussed only those normal maintenance problems that crop up frequently during the operating life of an ordinary boiler. These are items that the boiler plant operators are expected to be able to handle by themselves. They probably amount to about 75% of the total maintenance cost of the boiler plant over a period of years.

There are a number of major repair jobs that must be done on boilers from time to time, the frequency depending on how well the operators have done their job, coupled with a fair amount of luck. Some boilers go through their entire life without having to have major repairs done on them, while others may have to be torn into and major repairs made every few years.

The type of repairs we are speaking of are major retubing, replacing damaged tubes, replacing portions of tube sheets, crown sheets, staybolts, and sometimes entire furnaces or boiler breechings. These are best left to one of the many contracting firms that make a specialty of this type of repair work. They do an excellent job, usually, since they have over the years of business made repairs to many boilers and of the same order of magnitude as yours. To expect the average boiler plant staff to be able to effect a quality repair on the major pressure parts of a boiler, and do it cheaply and fast, is not practical. The average plant staff does not have the experience to do it, unless there is ample time for them to study the procedure and take their time learning the methods. In the long run it is more economical to call in an outside contractor to perform the work, and get a guarantee of the quality in the bargain. It all reduces to the fact that the contracting firm are probably specialists, whereas the plant staff are amateurs at the more complex repair jobs.

Another occasional maintenance job that should not be tackled by the small boiler operator is acid cleaning of the waterside surfaces. This requires special equipment and knowledge of water chemistry and metallurgy, which the contracting firms have mastered over the years. It is true that you can probably rent the equipment in most areas, and possibly the water treatment service firm you are using may be able to assist you in the cleaning process. However, there is a fair amount of danger involved in the proper handling and disposal of the acids and neutralizers used, and there is always the possibility of accidents. This involves the matter of liability for personnel burns and disabilities. It is far better to engage a responsible firm who is equipped and insured to do the job safely and fast and with a guarantee of final results.

There are several other maintenance areas that could be assigned to contract firms. The degree to which this is done depends on the size and complexity of your equipment, and the composition of the staff in the plant.

We will mention only a few of the most obvious and common ones:

Refractory repair and replacement.

Electrical repairs and rewiring.

Control system repairs, replacements, calibration, and adjustment.

As always, it is a good idea to use competitive bidding to obtain the best job for the money paid out. Some firms, after accepting bids, will select the one most qualified, then continue to rehire them for subsequent jobs, without further bidding. This has the advantage of saving money only if the plant equipment they are working on is of a specialized nature unique to the plant or the process. Engaging the same firm time after time thus saves the expense of the learning curve being required each time a new firm works on the job. However, if the job and the equipment are purely routine, run-of-the-mill projects, there is no sense in not taking bids for each job. Taking competitive bids helps to ensure that the firm is getting the best price, and tends to keep all bidders on an equal level.

It also reduces the possibility of under the table payments and gouging on the part of the contractor.

C. Cleaver-Brooks Maintenance Manual Approach

The following brief reprint is from a manual used by the Cleaver-Brooks firm in their training seminars given throughout the country. It will give you a general idea of what is expected of the boiler operator and what responsibilities he is expected to assume. It is an extension of material presented in Chapter 16, Section B, and covers firetube package boilers as supplied by that firm. Other firetube boilers of the Scotch type are similar in operating characteristics, and the same general instructions apply.

ANNUAL MAINTENANCE

This should be coordinated with the annual pressure vessel inspection performed by insurance or government groups. Establish a firm procedure with all outside inspection groups so that your equipment will be in a proper state of readiness. If your equipment is open, clean, and cool, your inspector will be a happier man. After all, he is a professional looking after your best interests to secure maximum equipment life and safety. As a matter of routine, whenever a boiler is taken off the line, disconnect the main power supply and lock switch in "off" position. Whenever there is more than one boiler connected to a common header, establish routine procedure of locking the header valve on any unit that is down for cleaning or inspection.

1. Clean fireside surfaces by brush or water washing. Use a powerful vacuum cleaner to remove soot. After the cleaning process and if boiler is to be left open, it is advisable to spray all fireside surfaces with waste oil or some type of corrosion preventive.

2. Remove all handhole and manway plates, inspection plugs from water column tees and crosses, float assemblies from water columns and thoroughly wash all waterside surfaces.

3. It is recommended that all pipe plugs be eliminated from inspection openings and be replaced with either a solid stock plug or a nipple and cap assembly that can be easily removed.

4. Waterside cleaning is best accomplished as soon as the boiler is cooled enough to work on comfortably. Use a high-pressure hose to flush out sludge, scale, etc., washing from top to bottom. If the boiler is not going to be used for a period of time, it is advisable to dry out the inside with heaters or fans. Humidity conditions will dictate what method of control must be used to prevent internal deterioration if the boiler is left open. If boiler is laid up wet, then it should be fired to drive off oxygen, treatment added to prevent corrosion, and left filled to top of shell. This procedure is considered so important that certain state boiler codes spell out this recommendation very carefully.

5. Immediately, upon opening fireside areas, give the refractories an inspection and start repair as soon as possible. Area of repair should be carefully prepared and built up according to instructions of refractory supplier. The majority of failures involving refractory repairs can usually be traced to the failure to read and follow instructions for that particular material. After repairs are completed, give entire refractory area several water-thin wash coatings of high-temperature cement material. This seals refractory surfaces and lessens deterioration.

6. Oil storage tanks should be inspected or checked annually for sludge and water

accumulation. Keep tank filled with oil to prevent condensation during summertime.

7. After a cleaning, the entire combustion process should be carefully checked, CO_2 readings taken, and necessary burner adjustments made. Make certain that readings are recorded and used as a basis of comparison for future tests.

8. Check electronic controls. It is recommended that vacuum tubes be replaced annually.

9. Mercury switches on all types of controls should be inspected and replaced at first sign of deterioration.

10. Check fluid levels on all hydraulic valves—if any leakage is apparent, take positive corrective action immediately.

11. Check oil preheaters by removing the heating element and inspect for sludge or scale. It is most important that heat-transfer surfaces be kept absolutely clean.

12. Check all filter elements, clean or replace as needed. On all so-called self-cleaning filters, make certain that impurities are flushed or discharged from the filter body.

13. Check gauge glass for possible replacement. If internal erosion at water level is noted, replace with new glass and rubber gaskets. On all unattended boilers, the gauge glass mounting should be of the safety style with stop-checks in case of glass breakage.

14. Remove safety valves and have them reconditioned by an authorized safety valve facility. This one most important device possibly receives less attention than any other item on a boiler. Why not have a spare set of safety valves so that they may be changed annually? How can anyone estimate loss or justify savings by ignoring safety valve maintenance!

15. If oil fuels are used, check on the condition of the fuel pump. These items wear

out and the annual inspection is the opportune time to rebuild or replace this item.

16. Boiler feed pumps and strainers should be reconditioned. Feed pump elements wear and must be replaced. Sometimes a review of the condensate return system and chemical feed arrangement will reveal causes of short pump life.

17. Condensate receivers should be emptied and washed out internally. Make an internal inspection if possible, since there are usually several openings available to accomplish this. If the receiver has a make-up valve mounted, this item should be overhauled and checked for proper operation.

18. Chemical feed systems for water treatment should be completely emptied, flushed, and reconditioned. Metering valves or pumps should be reconditioned at this time.

19. All electrical terminals should be checked for tightness particularly on starters and movable relays.

D. Establishing a Preventive Maintenance Program

The steam plant equipment is very well suited to conventional preventive maintenance procedures. With few exceptions, it is usually well established in a particular location, is confined to specific functions of a repetitive nature, and is usually crucial to the plant's processes and purpose.

Before launching into the procedure for organizing a plant preventive maintenance (PM) program, some remarks of a general nature are in order.

First, assuming that there has been no such program in the past, the extent to which one is to be implemented depends on several very important features. There is a lot at stake, since the set-up costs may be higher than management anticipates, which is usually the case when something new is proposed by those who are not directly involved in the implementation.

Probably the first decision to be settled on, after the decision to start a PM program is made, is the determination of maintenance functions for which the plant staff is to be responsible, and which ones are to be handled by outside contractors, as mentioned in Section B. This decision may have to wait until well into the organization of the PM program, since some very startling shortages in maintenance items and functions will no doubt be discovered as the organizing proceeds. This could lead to a reevaluation of the entire program, as could other unforeseen requirements.

Another budget-buster could be the need for special instruments and tools, since the plant may have been limping along for months without adequate provisions for proper maintenance.

And, finally, the manpower requirements may change, or the skills required may not match the available personnel. This could require retraining, at the least. The end result is that the plant's staff, from Management down, should be prepared for some changes. Obviously, Management must be sold completely on the need for the PM program, as must also those who are to implement it.

What are the goals, and how are they to be achieved? The purpose is, in brief:

   a. Maintain the boiler plant equipment in such condition as to ensure uninterrupted operations for as long as possible, without disastrous plant shutdowns for repairs.

   b. Maintain the equipment in such condition that it will always operate at the highest possible efficiency.

Basically, the PM program accomplishes this by the following general procedures, performed on an established, recommended frequency:

   a. Equipment inspection.

   b. Cleaning the equipment.

   c. Tightening of all loose joints and connections.

   d. Adjusting and lubricating all moving parts.

   e. Testing the critical portions of the systems involved.

   f. Correcting any defiencies discovered.

   g. Recording and signing the PM Record sheets in the proper manner, adding all pertinent comments.

   h. Updating the plant equipment and spare parts records as required.

We shall now present one method of organizing a PM program for the average industrial plant. We do not intend that it be considered the only method, but it should at least give the enterprising plant maintenance worker the basic procedures, which may be altered to suit the individual plant size, complexity, and manpower availability.

To start, assemble all available literature covering the manufacturer's recommended maintenance procedures for the equipment. Add to this a supply of standard size ruled, three-hole punched notebook paper, with a suitable binder.

Assigning one complete sheet to each item of equipment to be maintained, place the name of the equipment at the top, along with the following data:

   a. Plant equipment number. If numbers have not already been assigned this is the time to start this very important function.

   b. Location of the equipment, by floor, column number, etc.

   c. From the maintenance manuals, list all items requiring service, with the frequency (weekly, monthly, etc.) listed on the right-hand side of the sheet.

   d. After each maintenance function listed in (c), list the tools required, including instruments, the manpower, and the estimated time required at the site. The tools and instruments may be separated by those normally carried on the belt, in hand-carried tool boxes, or on service carts.

The next step is to obtain a supply of standard 3 × 5 index cards, and make out one card for each maintenance item and frequency combination from the previously described data sheets. The card should be headed by the equipment name and the frequency of service required. It should then be followed by those items (a) through (d) above, and with any spare parts data from other plant files, as described in Chapter 18. This card file may be incorporated in the plant equipment files, depending on location and availability. See Fig. 22-1 for an example.

Once these index cards have been completely filled out as described, they should be placed in an ordinary file box, with tabbed index dividers, separating the cards by frequency of service, such as weekly, monthly, etc. If the size and number of maintenance functions under each frequency heading permits it, they may be further subdivided by manpower requirements or tool kits required for the function.

If the total plant equipment maintenance requirements are minor, then the card file step may be omitted, and the PM Record sheets, described next can be produced directly from the basic work sheets.

Finally, the last paper form to be developed

is the "Preventive Maintenance Record" sheet. This should be on standard notebook paper, three-hole punched, and similar in form to Fig. 22-2.

In performing their regular PM functions, the crew should have a three-ring binder with them, containing the following material:

a. A schedule or routing sheet, giving the locations and sequence of the maintenance functions to be performed under the existing frequency schedule being followed.

b. The PM Record sheets previously described, in the same order as shown on the routing sheet.

c. Photocopies of the manufacturer's maintenance instructions for the function to be performed.

As the PM program is placed into operation, it is to be expected that some alterations to the first approach may have to be made. As the crews involved progress along the learning curve, the times and schedules may have to be adjusted, and the paper work, records, and plant files altered to bring the entire program up to date.

In Fig. 22-2, note the space for Remarks at the bottom of the page. This is for insertion of comments, keyed by number, (1), (2), etc., to the entries in the columns above or to items in the heading.

```
Preventive Maintenance
    Data                          Weekly
(Equipment Name)

(Equipment Number)

(Location)

(Maintenance Function Required)

(Tool Kit and Instruments Required)

(Manpower Required)

(Time at Jobsite)

(Spare Parts Required)
```

Fig. 22-1. Data file card, preventive maintenance program.

E. Other Sources of Help

The boiler plant operator is not alone in his problems, for there are many outside sources to which he may turn for advice and help. Some of this help may even be free advice. In this field, that of steam boiler plant operation, there are few, if any, trade secrets. Owing to some plants having patents on their processes or products, the plant personnel may be under strict secrecy rules regarding those patents and secrets. This does not usually prohibit the boiler plant personnel from sharing their problems

PREVENTIVE MAINTENANCE RECORD

(Equipment Description)   (Location)

Equipment Number)   (Frequency)

(Maintenance to be Performed)   (Manpower)

(Tools, Instruments Required)   (Spare Parts)

Date	By	Date	By	Date	By	Date	By	Date	By	Date	By

Remarks

Fig. 22-2. Service record sheet, preventive maintenance program.

with others, as long as no plant secrets are disclosed. In this case, absolute discretion is called for.

We shall list here some of those outside sources of help we mentioned, although some of them may have already been referred to in this book.

The boiler inspector and his employer will often give you good advice and literature to back it up. It is to their advantage as well as yours for them to do this, so do not be afraid to request their help.

The manufacturer, or his representative of your equipment, including the boiler, its accessories, and the other plant equipment all want to see their product working properly and for that reason will usually be very cooperative.

Boiler plant service firms, if you are a good customer, will often give free advice at times.

Your water treatment service specialist may be in a position to offer good advice. If he does not know it personally, he may be able to call upon a trained specialist in his organization. Do not hesitate to make full use of that service.

Burner and control representatives may be in a position to help you. This is especially true if the burner and controls were furnished separately through the local source. If not, then often the local representatives have an arrangement with the factory to be reimbursed for any service rendered on their product shipped from another territory into their sales area.

Other trades, such as the electricians, either in-house or from the outside, who are reg-

ularly doing work on your equipment, may be willing to assist you in a problem solving capacity.

The national code organizations, such as the ASME, UL, NFPA, and similar groups dedicated to reducing losses from boiler accidents.

Other members of the various trade unions and professional associations, such as the National Association of Power Engineers, and the Engineers and Operators Union.

So remember, don't be bashful. Ask for help when you need it.

C H A P T E R 2 3

BACK IN SERVICE

A. Before Closing the Boiler

Now that the maintenance work has been completed, it is time to close the boiler and get it ready for service again. This will be performed by the contractor's force, if they were hired to work on the boiler. If not, then it is the responsibility of the plant's maintenance crew to get it ready for another stretch of duty. For that reason we will give you a few hints here of what has to be done, mostly as a guide. The details will have to be supplied by those doing the organizing for the plant's boiler crew.

The first thing that should be done before closing up the boiler is to thoroughly inspect all available parts of the boiler's interior, both the water side and the fire side. Make sure there are no tools, rags, or other debris lying around inside that could cause trouble. This happens too frequently, and the result can be damaging to the boiler, and to someone's reputation.

If there are tubes or areas that are not easily visible, then it is necessary to resort to other methods of testing for the presence of unwanted items. Tubes can be checked by running a plumber's snake through them. Nooks and crannies are a more difficult matter, and will tax the entire ingenuity of the staff in improvising approaches to check them. One method is to keep close account of all tools used on the work, then make an accounting of them before closing up the boiler. This is very difficult to do, but it may catch some of the larger

and heavier tools. Small tools, such as hand chisels, screwdrivers, and scrapers, are more difficult to account for, and they may come up missing without anyone being aware of it. Of course, not all of these small items will cause trouble if left inside the boiler, but if one should find its way into a tube, and reduce the circulation through that tube, then the boiler's efficiency will suffer.

All loose dirt and debris should be swept out, being careful not to sweep it into the tubes, drains, or blow-down connections.

Check all internal piping, hangers, brackets, scrubbers, etc., to be sure they have not been broken loose from the maintenance work.

Flush out all bent tubes with a water hose, to be sure they are clear and free. Open the drains to be sure they are free, and will drain out all water used for flushing. Close all drains after checking to be sure all water has been drained out.

All sealing surfaces should be given a final check for dents, scratches, and gouges. Water joints under pressure are very critical, and may have to be scraped, then dressed off flat with a carborundum block, or similar abrasive surface. Replace all gaskets on the water side, since the cost of doing so is minor compared to the nuisance and inconvenience of having to shut down, drain the boiler, and replace the gaskets after starting up. Do it now, and save your headaches for elsewhere.

Gasketed joints on the gas side are not as

critical, but still require close inspection and cleaning. They are not under as much pressure as the water side, but must still be fairly gas tight. Replacing the gasket will depend on your inspection and decision, since some of them are quite large and expensive. Also, it is not as difficult to replace them during normal steaming periods; often shutting down at night will give sufficient time to do it.

B. Testing the Casing for Tightness

Fortunately, it is much simpler to check the boiler casing surrounding the hot flue gas passages for tightness than it is for testing the water side. This is necessary, since all boilers have pressure differentials in some degree between the fire side and the atmosphere.

The method we describe is only one way to test the casing. You may have others, but the principle will be the same, and the tools and the equipment may differ.

Assuming the boiler is of the forced draft style, the draft fan may be used to place the casing under a pressure high enough for a fairly good test. To do this, it is well to have a manometer installed for the very first time this method is used. The manometer should be installed to give the pressure inside the furnace or firebox, relative to the atmosphere. By closing off the outlet damper from the firebox breeching, and running the draft fan at its normal speed, you can determine the amount of internal pressure the fan is capable of producing. Once this is done, and the conditions and results are recorded for future use, the manometer may be returned to the firm from which you borrowed or rented it.

Now that the draft fan is running, and the casing or firebox is under a slight operating pressure, leaks in the joints between the fireside and the atmosphere may be detected by passing a lighted candle along every inch of the joints being tested. Any leaks of air through the joint will readily be shown by the effects upon the flame.

Another method is to use a soap suds mixed in a can or pan, and applied with a paint brush. Use only suds, not soapy water. The joint will have to be wiped dry after the test is completed, mostly for preserving the appearance of the boiler surface and protecting the insulation.

For induced draft boilers, the same procedure may be followed, but this case calls for the flame behaving in a slightly different manner. The inward suction of air through a leaking joint under vacuum will pull the flame into the leak, whereas in the case of forced draft boilers, the flame is blown away from the leak.

Boilers under natural draft are to be tested in the same manner as induced draft boilers.

There is another, more sophisticated method of testing the casing, which involves the use of Freon gas, as used in refrigeration and air conditioning systems. By feeding a slight amount of this gas into the suction of the draft fan, for forced draft boilers, the presence of this Freon gas may be detected leaking through the joints by running a halide torch along every inch of the joints.

This method does not work, of course, for induced draft boilers. Nor does it work too well to pressurize the boiler furnace and gas passages on an induced draft boiler, then test for leaks as in the method described for forced draft boilers. This is often difficult to accomplish, and also, a joint may leak in one direction but not in the reverse direction.

C. Placing Back in Service

Now that all preservice maintenance and inspections have been carried out and the boiler is buttoned up and ready for service, it is recommended that the boiler be given a short test run under full steaming conditions before placing full reliance in it on the line. Your particular situation will of course determine to what extent this is feasible. Keep in mind that after extensive overhaul and repair on the boiler and its accessories or control circuits, it is unwise to immediately put it on the line without this test run. For example, would you take your family on an extended long vacation in the family car, the day after it came out of the shop, where it had a major overhaul?

Chapter 15 gives in detail the steps for lighting off under several different equipment situations and conditions. We shall not repeat those steps here. However, we add the following suggestions and admonitions:

1. Observe all safety rules very carefully while going through the initial start-up and into the full test run. Almost anything can—and often does—happen at this stage.

2. Keep all unnecessary personnel out of the area.

3. Be prepared to observe and record all steps taken and the results obtained for future reference in case of callback to the service contractor, if any were involved. This is also an excellent idea for your own use.

4. If outside contractors were involved in the servicing, insist that they have a serviceman on the job during the test run. It would be best that this be included in the service contract.

5. The boiler should be put through several cycles of full to reduced firing rate after each adjustment is made in the controls.

6. Before signing off any completion notice from the contractor, be sure you are satisfied with the performance and the work. If there are any reservations in the work, be sure they are recorded on the sign-off sheet, with a stipulation that the items will be corrected within a definite time. Do not leave anything to personal goodwill.

7. When the boiler is finally back on the line under full service, be sure and keep an eye on its performance for several days, until it has proved its worth.

C H A P T E R 2 4
ENERGY CONSERVATION BASICS

A. The Goal

In several places in this book we have referred to the efficiencies to be expected from the small to average industrial boilers, and they were discouragingly low. This is due to the usual practice in the past, of not putting as much design effort and expense into increasing the efficiency of these units. This picture is now changing, owing to the rapid increase in fuel costs in recent years. However, there is a definite limit as to what can be achieved from them.

Table 24-1 illustrates this point well, and in a general manner. Note the increase in efficiencies to be obtained as the boilers increase in size. Part of this increase is due to the inclusion of heat reclaiming equipment, which are often standard items with the larger boilers.

Note also the difference in efficiencies among the three classes of fuels, which may cause you to question why gas is so popular today, with fuel costs so high. The answer is in the ease and cleanliness with which gas is burned and handled. Part of the loss in efficiency is offset by the decreased maintenance costs obtained with fuel gas. Another possible reason is the relative cost of the fuel on a British Thermal unit basis in some areas of the country. We shall cover this subject in Chapter 25.

The boiler's efficiency is a direct function of heat losses from the boiler and the combustion processes. This is shown in Fig. 24-1, in which the various heat losses are plotted against the air/fuel ratio for the burner. As you can see, the total heat losses are the sum of the radiation losses from the boiler surface, the loss due to incomplete combustion in the furnace, and the losses going up the stack in the flue gas. The lowest point on the resultant "Total Losses" curve is the desired operating point, as shown by the vertical dotted line.

Note that the radiation losses are a constant value throughout almost the entire boiler output, so there is very little that can be done by the boiler plant personnel to reduce this loss, other than to maintain the insulation and covering on the boiler in good condition. Fortunately, this loss constitutes only about 1% of the total losses at rated output, and about $2\frac{1}{2}$% at 50% of rated output.

The losses from incomplete combustion drop rapidly as the air for combustion is increased. However, this is offset by the rapid increase in the flue gas losses as the air supplied increases.

We ask you to note, also, the difference in total losses caused by the result of poor air/fuel mixing, as shown by the gap between the solid "Total Losses" curve and the dotted curve at the vertical dotted optimum point line. This brings out the advantage of obtaining as near perfect mixing as is possible, within economic reason.

How much can you, the boiler plant operator expect to increase the efficiency of your

Table 24-1. Maximum Economically Achievable Efficiency Levels[a]

Fuel	Rated Capacity Range (million BTU/hr)		
	10–16	16–100	100–250
Gas	80.1	81.7	84.0
Oil	84.1	86.7	88.3
Coal			
Stoker	81.00	83.9	85.5
Pulverized	83.3	86.8	88.8

[a]33,475 BTU/hr = 1 bhp.

boiler? Table 24-2 will give you a good idea of what is obtainable in the best situations. We do not believe that all boilers will be able to produce the gains shown in this table, and you most certainly will not be able to make the best use of all of them. But, the goal is there for you to attempt to reach, and by so doing, the net result can only be a decided improvement in your boiler plant efficiency, with a dramatic drop in fuel costs.

Several items in Table 24-2 involve fuel and steam saving measures that do not affect the boiler efficiency directly, but should still result in fuel savings, if carried out as far as possible. Chapter 26 contains steam saving suggestions, as does also pages 382 through 385 in the Appendix.

The principles involved in the energy-conservation processes that we shall describe in

Table 24-2

Corrective Action	Possible Increase in Boiler Efficiency
Change burner operation from on–off to modulating modes	10% (or more)
Raising feedwater temperature	8%
Reduce excess air to a minimum	5%
Reclaim heat from flue gas	3%
Reduce scale and deposits on heat-transfer surfaces	2%
Select fuel on cost/therm and basic efficiency basis	2%
Operate each boiler in a battery at peak efficiency point	2%
Reduce quantity of blow-down	1%
Recover heat in blow-down	1%
Reduce boiler operating pressure	1%
Preheat combustion air	1%
On oil burners, use air in place of steam for atomization	1%
Reduce trap and steam line leaks	1% (or more)

the following pages are generally applicable to the average industrial boiler installation, if there is such a thing as an average installation. There are so many different boiler–burner–fuel combinations in service, that any approach can only be considered general in nature. Consequently, the plant operator who conscientiously follows the energy-conservation procedures given here should not be surprised if the results obtained do not match his calculations exactly. If the operator is consistent and careful in his procedures, satisfactory results should be obtained.

The purpose of the remainder of the chapter is to point you in the right direction toward achieving the goals given in the table. We shall go into sufficient detail, including the simple mathematics involved, to permit you to correct some of the inefficient designs and operating procedures, and to calculate the cost savings for management. The equipment, instruments, and methods will be relatively cheap and simple, so you need not flinch at the

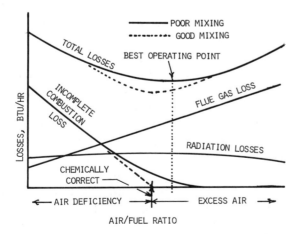

Fig. 24-1. Boiler heat losses. (Courtesy of University of Kentucky, Engineering Library.)

work. It will be interesting, educational, and highly rewarding.

B. Analyzing Boiler Flue Gases

In maintaining the boiler operation at the optimum efficiency point, it is essential that the operator become proficient in analyzing the flue gases, and that is the purpose of this section. Not only is it necessary to analyze the flue gas for setting the burner fuel and air ratio, but it is also necessary that the flue gas be periodically and regularly checked to ensure continuing high efficiency. Therefore, it is very much to your advantage to become proficient in the methods that you choose to use for your particular installation.

The oldest and best known method is by means of the Orsat Flue Gas Analyzer, shown in Fig. 24-2. It is well known in the steam boiler field, having been in use for years by the industrial, utility, and marine steam plant installations. Consequently, it is the standard by which all others are compared and calibrated.

To operate the Orsat Analyzer, and others that

Fig. 24-2. Orsat Flue Gas Analyzer.

we shall cover, it is necessary that a great deal of patience and attention to detail be mastered by the operator. Even then, the results will seldom be consistent, since the flue gas composition changes frequently during the normal firing sequence in the boiler.

One of the first things that must be provided is an opening in the stack by which a flue gas sample may be drawn with consistent accuracy and convenience. A $\frac{1}{2}$ in. pipe nipple welded into the stack, with a pipe plug or cap to close it off when not in use, will be satisfactory. In inserting the sample probe into the pipe nipple, care should be taken to have the tip of the probe approximately in the center of the stack or breeching, so that consistent, well-mixed samples will result. You can experiment with the instrument in taking samples and running tests with the probe at different locations in the flue gas stream, and satisfy yourself that you are getting the best possible sample. Then mark the probe so that your subsequent tests will all be taken at the same location inside the stack or breeching.

The sampling point should be downstream of the economizer or air preheater, since the flue gas temperature will be lower.

Should you find that the stack or breeching where you have made the sampling connection is under a negative pressure, then a slight alteration may have to be made in the probe. This would be indicated by an inrush of air when the plug or cap is removed from the sampling connection. If this should happen, then it is only necessary to make a sealing disk of thick neoprene, leather, or some other soft plastic, which fits tight around the probe; this will serve as a stop and a seal when the probe is pushed into the opening right up to the plastic disk. The negative pressure inside will pull the disk up tight around the sampling connection and effectively seal it off.

The reason this procedure is necessary is to prevent an inrush of air from affecting the reading you are attempting to take. The air will give inaccurate results when mixed with the flue gas.

The operation of the Orsat Analyzer is simply a matter of following the detailed instructions very carefully. We shall not go into them here, except to state that the general method is as follows:

1. A sample of flue gas is pumped into the instrument, and is cleaned, cooled, and measured.

2. The sample is passed through a chemical bath several times, which absorbs the carbon dioxide in the flue gas.

3. The amount of carbon dioxide in the gas is read directly off a scale on a glass tube or reaction bottle.

4. The remainder of the flue gas is passed through bottles of chemicals that absorb, in turn, the oxygen and the carbon monoxide, in the same manner as for the carbon dioxide.

5. The readings taken from the scales on the bottles or tubes are used to determine the percentages of the three constituents, by taking the differences in the scale reading after each test.

A much simpler arrangement for accomplishing the same result is shown in Fig. 24-3. It consists of separate reagent bottles, holding

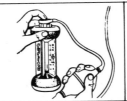
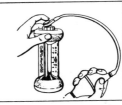
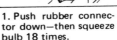

1. Push rubber connector down—then squeeze bulb 18 times.

2. Lift finger from rubber connector—this seals Fyrite.

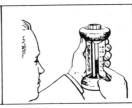
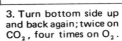

3. Turn bottom side up and back again; twice on CO_2, four times on O_2.

4. Read fluid level on scale for percentage of gas.

Fig. 24-4. Operating the Fyrite Analyzer. (Courtesy of Bacharach Istrument Co.)

the chemical absorbents required to separate out the carbon dioxide or oxygen. Figure 24-4 illustrates the simple steps needed to perform the tests and take the readings. It is much easier, quicker, and just as accurate a method as the Orsat apparatus. The results are comparable, depending on the care and expertise of the person making the tests.

Should you be fortunate enough to have a management that is willing to go to the expense of more sophisticated equipment, there are now several types of instruments available that will give direct readings within a few seconds after inserting the probe into the gas stream. These range from small, portable hand instruments, to large cabinet units, with probes installed permanently in the breeching or stack, with continuous readouts.

All these methods are to be used with charts or tables based on each of the known fuels—natural gas, fuel oils of all grades, or other fuels. By the use of these guides, and by determining the amount of carbon dioxide in the flue gas, it is possible to determine the combustion efficiency in the boiler.

Fig. 24-3. Basic Fyrite equipment. (Courtesy of Bacharach Instrument Co.)

C. Combustion Efficiency Charts and Graphs

The graphs referred to in the previous section will now be explained, and shown.

Figures 24-5 through 24-12 are made for the common grades of natural gas; No. 2, No. 4, and No. 6 fuel oils; coal; wood; and baggasse fuels. The graphs are based on taking the carbon dioxide content of the flue gas, and the flue gas temperature along with the temperature of the combustion air entering the burner, which is often the boiler room air temperature. If an air preheater is used, then the air temperature should be that ahead of the air preheater. By projecting a vertical line for the reading of the stack gas temperature minus the combustion air temperature along the bottom of the graph, up to the % carbon dioxide reading, then horizontally across to the left scale, the combustion efficiency will be determined for that fuel. Our first example in this exercise is already drawn in on Fig. 24-5.

A quick example will illustrate their use.

We shall assume that you have very carefully taken the carbon dioxide content of the flue gas three times, and that you have averaged your readings, and get an average content of 9.0% carbon dioxide, and that the gas used is natural gas. At the same time, the reading from the stack thermometer is 450°F. For this example, and for all succeeding cases, we shall assume that the combustion air temperature is 70°F.

Laying these coordinates out on Fig. 24-5, we find that the combustion efficiency is 79.9%. This means that 20.1% of the heat content available in the fuel gas is going up the stack and is wasted. This is not a good situation, and we shall soon help you correct, or at least mitigate, the high loss of heat, with a resultant saving in money spent in fuel gas.

Figures 24-13 through 24-18 show percentage of excess air, oxygen, and carbon dioxide equivalents to be expected for the more common fuels. In addition to being required to adjust the fuel/air ratio, they are handy for your convenience as a check of the accuracy of your readings and for converting one reading to another. Should you only have facilities for taking the oxygen content of the flue gas, these graphs will permit you to convert that reading to the equivalent carbon dioxide reading for use in determining the combustion efficiencies in Fig. 24-5 through 24-12.

Figures 24-13 through 24-18 cover excess air percentages between 50% and 70%, which is obviously much higher than good practices dictate. Should the initial readings indicate excess air percentages higher than these amounts, then it is the first and primary goal to adjust the fuel/air ratio to bring the readings down onto the graph as a starting point.

Remember: the goal is to reduce the excess air as much as reasonably and safely possible to produce the best combustion efficiency for the boiler.

The discussions contained in this book pertain basically only to the industrial boiler field. There are numerous designs of burners in use in industry in process ovens and furnaces, in which the flue gas composition is a part of the chemical process being carried out within the oven or furnace. In these cases, efficiency and the fuel/air ratios are adjusted to different goals than is done for boiler service.

Little has been mentioned so far about taking the stack temperature. This is simply a matter of installing a dial thermometer in the stack or breeching, near the point where the gas samples are taken. Be sure the thermometer has a sturdy, stainless-steel sensing bulb, and is projected far enough into the gas stream to get a true reading, at least one-quarter of the way across the breeching or stack diameter. The dial should have large enough numbers to give a very clear reading.

So far we have not given any indication as to what to expect when taking the stack gas temperature. This will vary with the particular design and firing rate of the boiler at the time the temperature is observed. Most of the average industrial boilers in use today have been designed to produce an exit temperature (leaving the boiler and entering the stack) of about 150°F above the design steam temperature at

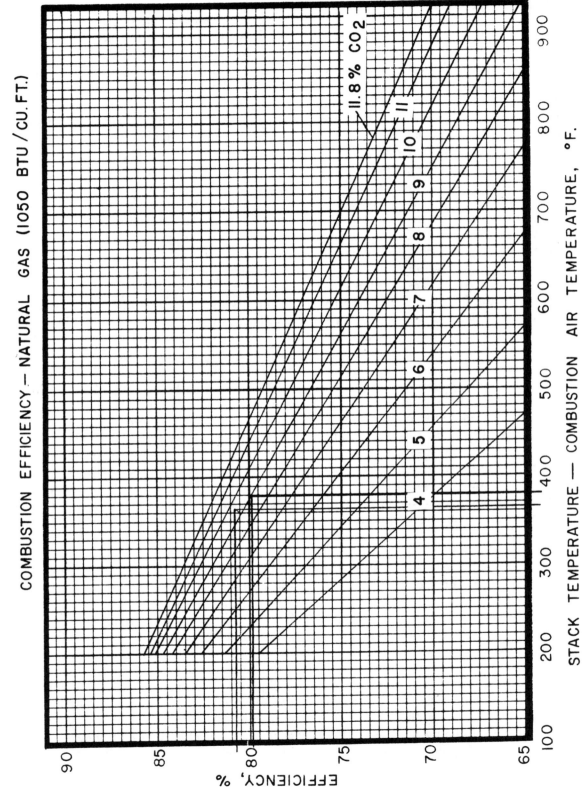

Fig. 24-5. (Courtesy of the Boiler Efficiency Institute.)

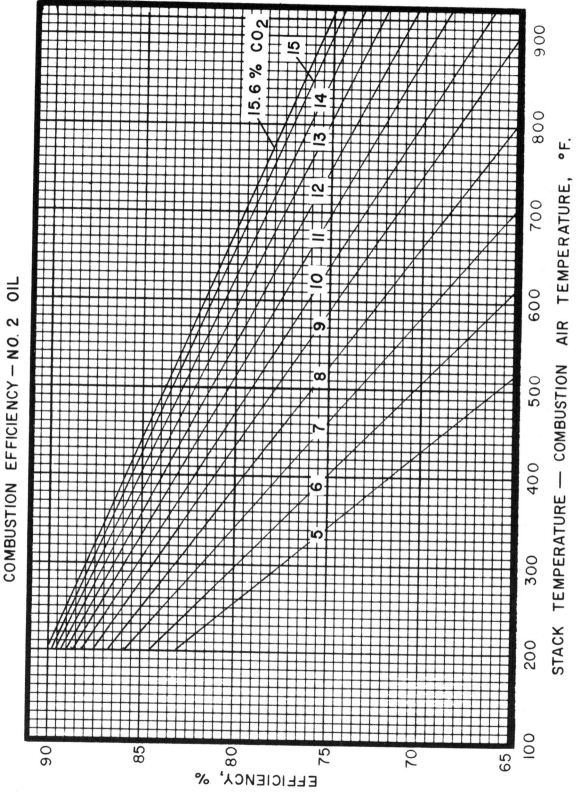

COMBUSTION EFFICIENCY — NO. 2 OIL

STACK TEMPERATURE — COMBUSTION AIR TEMPERATURE, °F.

EFFICIENCY, %

Fig. 24-6. (Courtesy of the Boiler Efficiency Institute.)

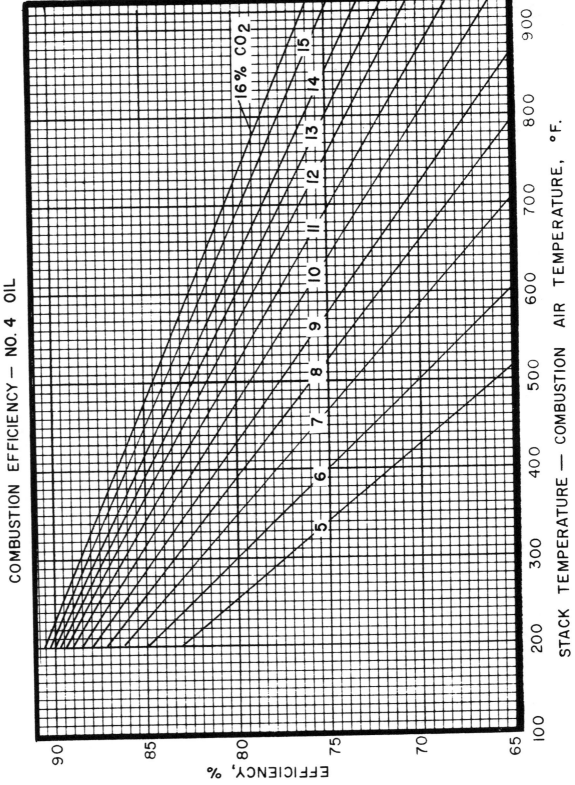

COMBUSTION EFFICIENCY — NO. 4 OIL

STACK TEMPERATURE — COMBUSTION AIR TEMPERATURE, °F.

Fig. 24-7. (Courtesy of the Boiler Efficiency Institute.)

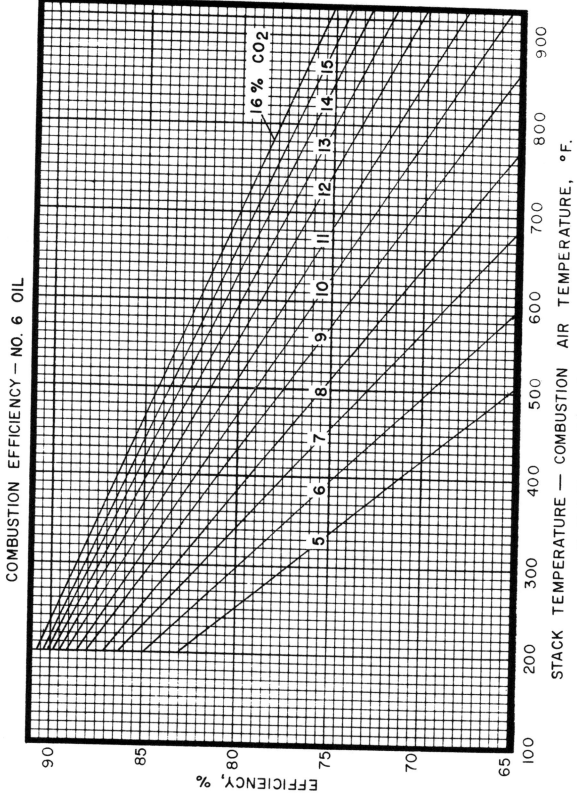

Fig. 24-8. (Courtesy of the Boiler Efficiency Institute.)

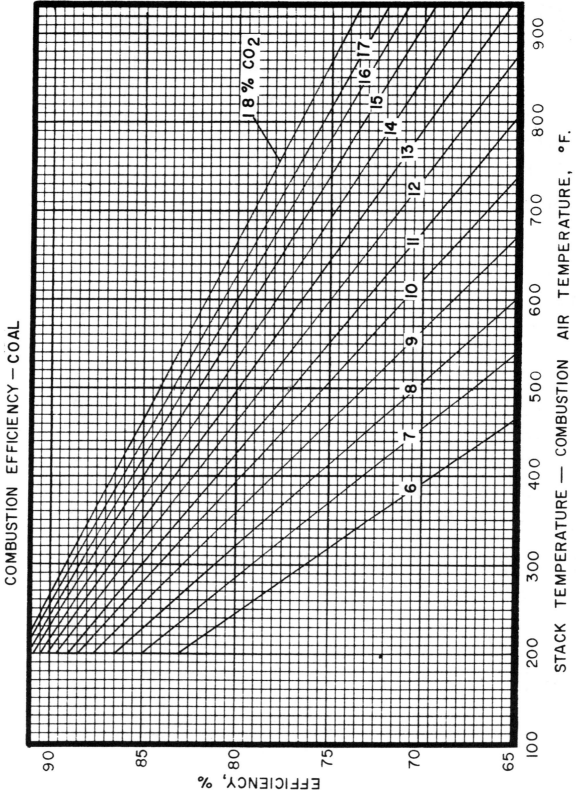

Fig. 24-9. (Courtesy of the Boiler Efficiency Institute.)

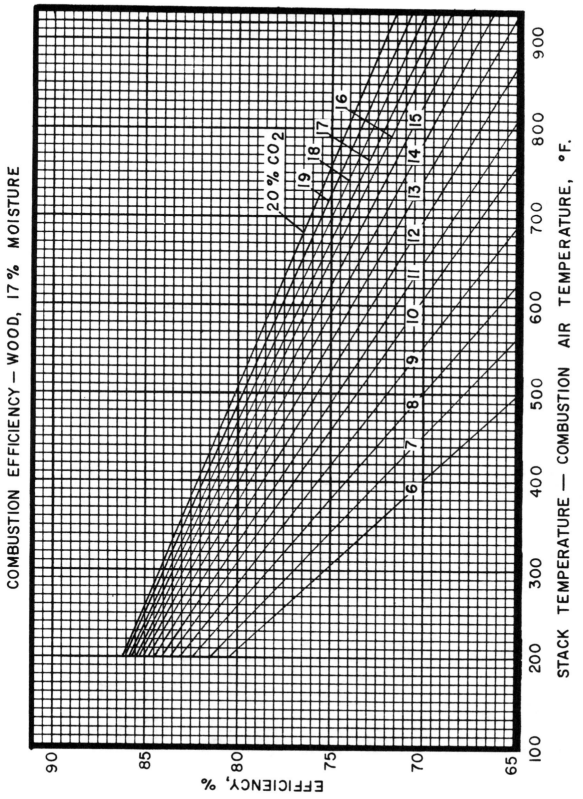

COMBUSTION EFFICIENCY — WOOD, 17% MOISTURE

STACK TEMPERATURE — COMBUSTION AIR TEMPERATURE, °F.

EFFICIENCY, %

Fig. 24-10. (Courtesy of the Boiler Efficiency Institute.)

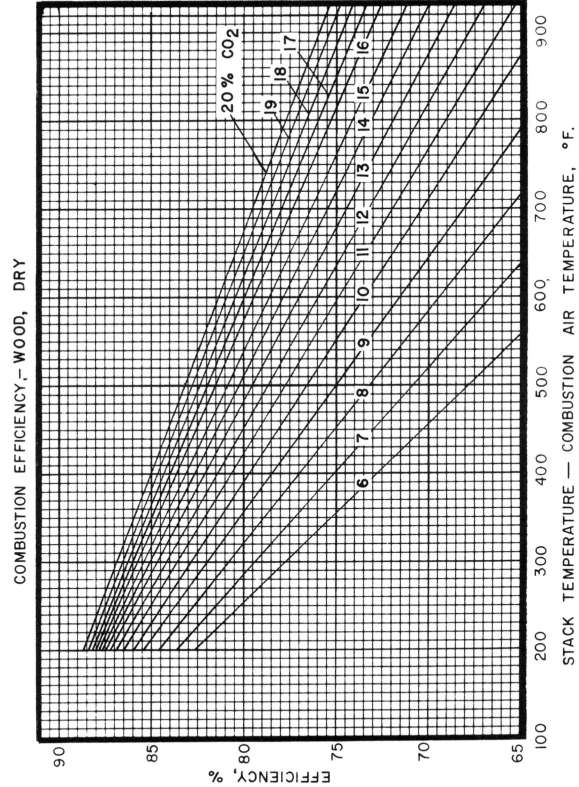

COMBUSTION EFFICIENCY—WOOD, DRY

Fig. 24-11. (Courtesy of the Boiler Efficiency Institute.)

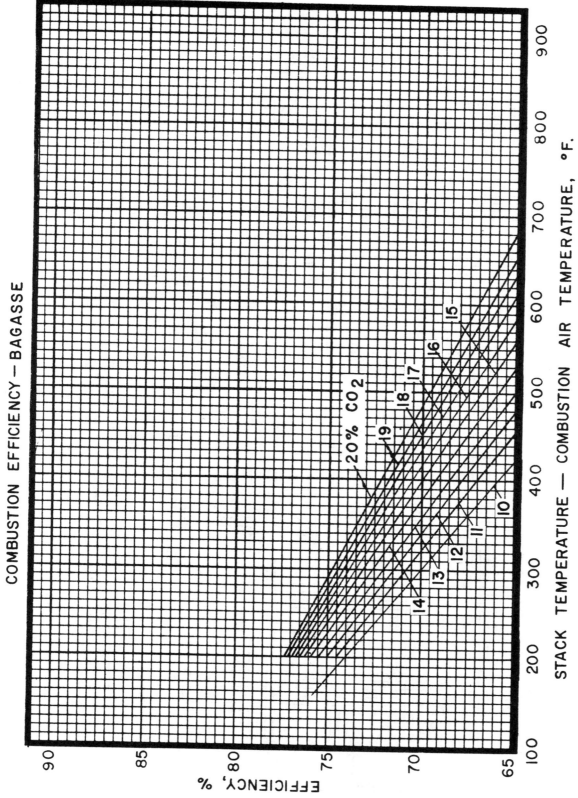

COMBUSTION EFFICIENCY — BAGASSE

Fig. 24-12. (Courtesy of the Boiler Efficiency Institute.)

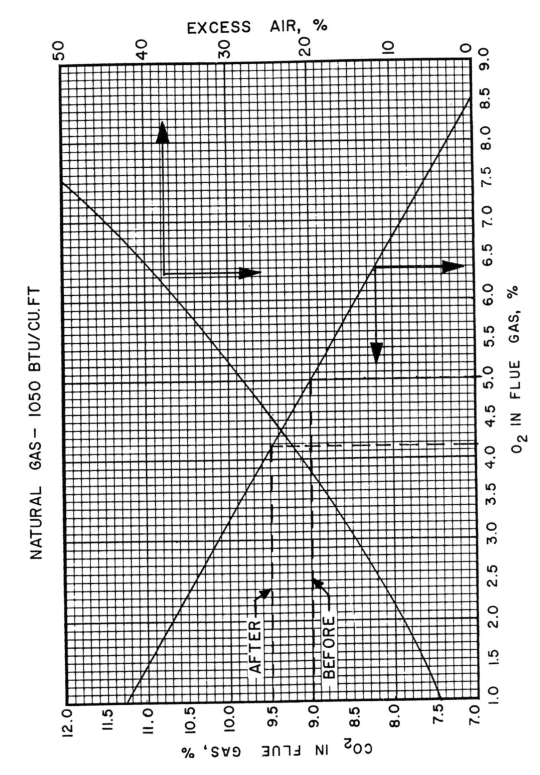

Fig. 24-13. Flue gas equivalents, natural gas firing. (Courtesy of the Boiler Efficiency Institute.)

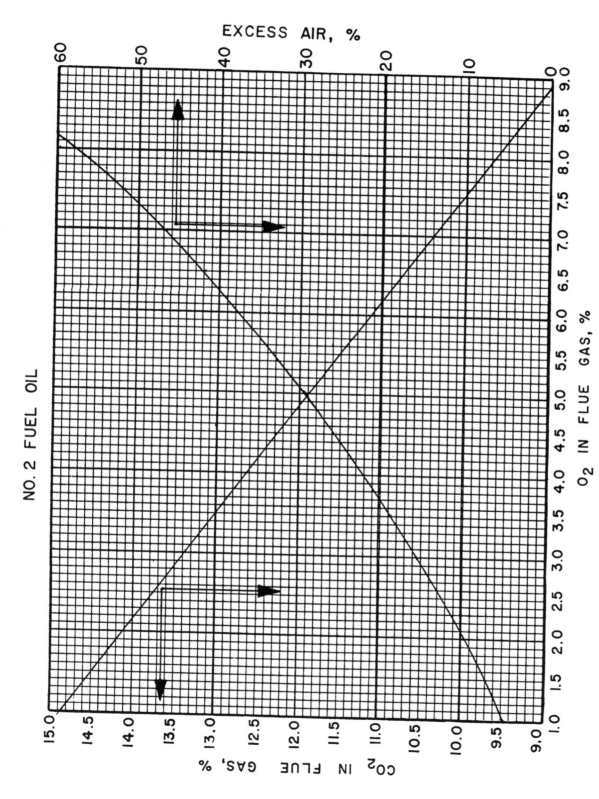

Fig. 24-14. Flue gas equivalents, No. 2 fuel oil firing. (Courtesy of the Boiler Efficiency Institute.)

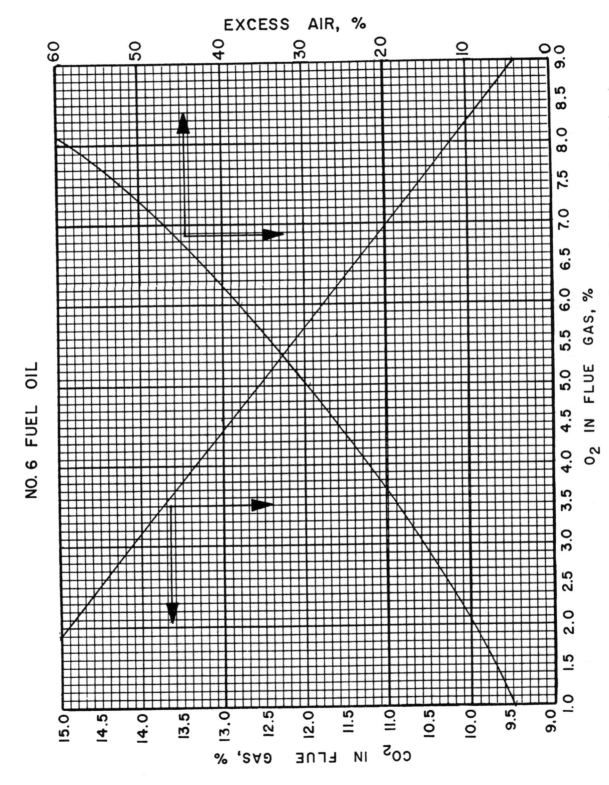

Fig. 24-15. Flue gas equivalents, No. 4 fuel oil firing. (Courtesy of the Boiler Efficiency Institute.)

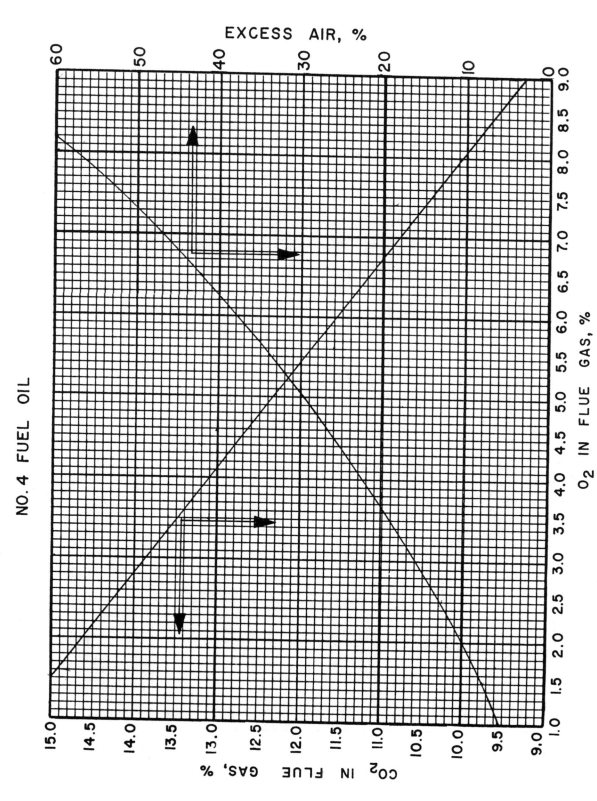

Fig. 24-16. Flue gas equivalents, No. 6 fuel oil firing. (Courtesy of the Boiler Efficiency Institute.)

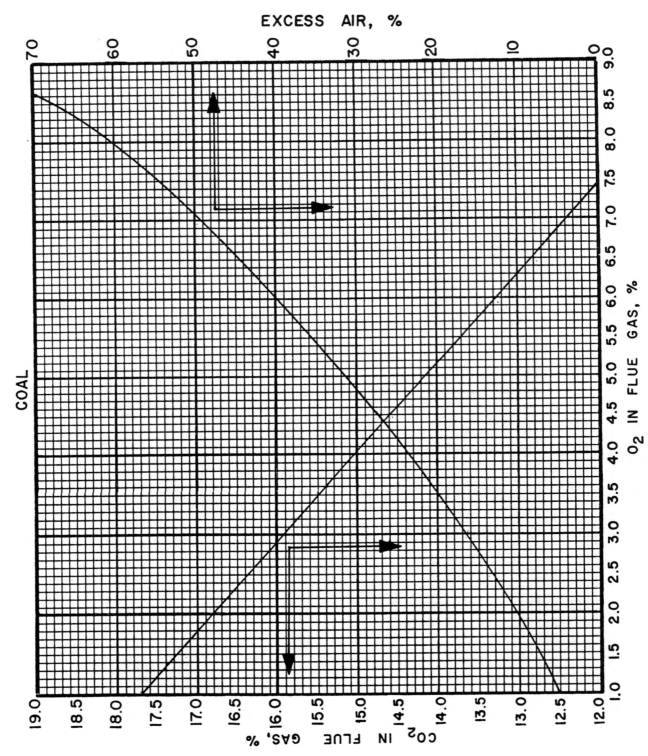

Fig. 24-17. Flue gas equivalents, coal firing. (Courtesy of the Boiler Efficiency Institute.)

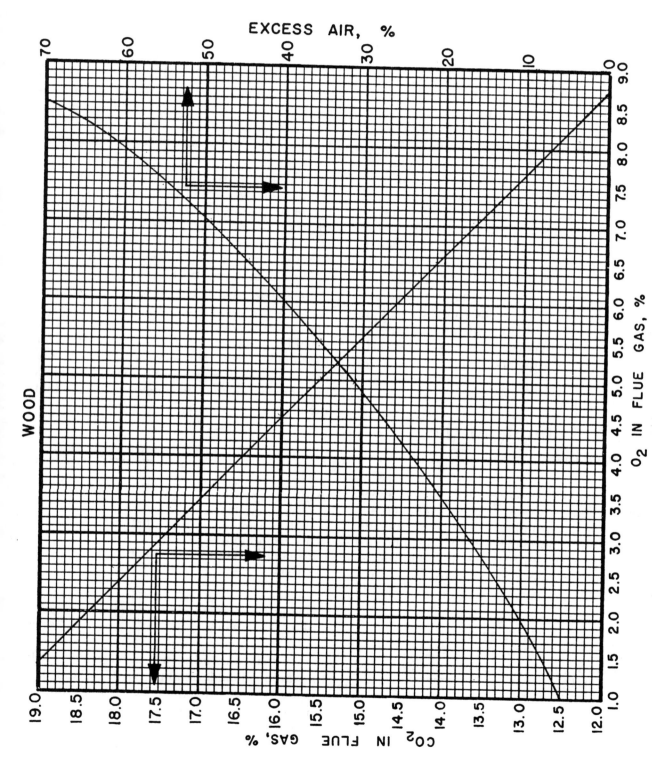

Fig. 24-18. Flue gas equivalents, wood firing. (Courtesy of the Boiler Efficiency Institute.)

boiler rating. In the plant, this may be altered to meet your particular fuel conditions, but there should be little change from the temperature difference as stated by the boiler manufacturer.

The reason 150°F is chosen as a differential is that it is not economical to build the extra heat-exchange surface required to reduce that outlet temperature differential. Therefore, the boiler designers have left it up to the individual plant owner/manager to decide whether it is worth the money expenditure to add the extra surface as an adjunct to the basic boiler. That is the subject of Section C in Chapter 25.

For your general guidance, we give here in Table 24-3, the expected stack temperatures above the design operating steam temperatures for the various output ratings of a typical industrial boiler. As an example, if your boiler is rated at 125 psig saturated steam temperature, and is operating at 75% of rating, you should expect the approximate exit gas temperature from the boiler, entering the stack, to be = 353 + 115 = 468°F. This applies only to the conventional style boilers, and some variations should be expected.

D. An Example of the Method

This section will be devoted to a detailed walk-through of the testing and calculations involved in determining the effects of the adjustments in improving the combustion efficiency. From section A, we learned that this is one of the first approaches toward improving the efficiency and cutting down on the fuel bill.

Table 24-3

Boiler Output Rating (%)	Exit Gas Temperature above Design Steam Temperature (°F)
100	150
75	115
50	80
25	50

Furthermore, it is relatively simple, as long as the boiler operator understands the combustion control mechanism on his burner, and he should by now.

For the purposes of this example, we shall assume the following basic equipment and conditions: The plant's steam is being supplied by a 100 hp boiler, operating around the clock, all year. It is gas fired, with natural gas, costing $6.00/1000 ft³, and the boiler output averages 60% of its maximum rating over the year. This is known as the Use Factor.

Before you do anything to change the air/fuel ratio, it is necessary for you to take at least three samples of the flue gas and analyze them for the carbon dioxide and oxygen contents, using any of the methods you have available. Following that, the ratioing device on your combustion control system should be adjusted, very slightly, reducing the combustion air intake, while watching the effect on the flame through the peephole. Then it is time to take another set of flue gas samples and test them to determine what changes, if any, have resulted. This should be done after waiting a few minutes for the flame to stabilize under the new set of conditions. We shall now assume, for the purposes of this example, that the following are the total tabulations of the samples analyzed:

	Before Adjustment	After Adjustment
Stack gas temperature	450°F	435°F
Boiler room temperature into the draft system	70°F	70°F
Stack temperature— room temperature	380°F	365°F
% carbon dioxide, average of three best readings	9.0%	9.5%
% oxygen, average of three best readings	5.0%	4.2%

Note that your adjustments have increased the carbon dioxide reading by 0.5%, and decreased the oxygen content by 0.8%. This may appear to be discouraging, and hardly worth the effort. But, before you get disgusted, let us carry this through to the ultimate conclusion.

Throughout the remainder of this book, on the subject of energy conservation, we shall use this same basic boiler in any examples given, to see what is possible to attain in the way of fuel savings.

We now refer you to Figs. 24-13 through 24-18 for the next step. These graphs contain the equivalent percentages of excess air, carbon dioxide, and oxygen to be expected in the flue gas when firing the more common fuels.

Thus, if you take a reading for the amount of carbon dioxide in the flue gas, the oxygen content and the amount of the excess air present can readily be obtained. Furthermore, if you take a sample of the oxygen and the oxygen content is not within ±0.1% of the corresponding figure in the graph, then the sampling was not correct, the test was inaccurate, or the fuel is different than that for which the table was designed.

Another comment on the use of the graphs applicable to the accuracy in taking the oxygen and carbon dioxide readings is appropriate here. The composition of all fuels varies considerably throughout the country, so some variation between your readings and those in Figs. 24-5 through 24-18 should be expected. Do not let this deter you, however, since all of the fuel saving calculations are based on the *difference between readings* on the same graph used as a base. Thus, Fig. 24-5 or 24-13 may be used for natural gas throughout the country, with sufficient accuracy to assist you in your goal of reducing fuel consumption.

The same comment is true for the other fuels in this discussion, with the possible exception of wood. Most wood being burned in boilers contains some moisture, and is seldom really dry. Therefore, we suggest you use the graph for 17% moisture in your wood fuel, unless the wood is actually dried before burning.

We have drawn in with dotted lines the readings taken for our sample case, on Fig. 24-13. The graph indicates an "after" reading for the oxygen as being about 4.16%, but we have rounded it off to 4.2% for the purposes of this example. This is legitimate, as it is not practical to expect an accuracy better than ±0.1 with the normal equipment in use today.

You are now in a position to be able to commence tabulating the results as follows:

Reading	Oxygen	Carbon Dioxide	Excess Air	Stack Loss (%)
"Before"	5.0%	9.0%	28.1%	20.1
"After"	4.2%	9.5%	22.2%	19.3
		Difference in stack losses =		0.80

In the preceding tabulation of reading for this example, to obtain the "before" stack loss in percent, it is first necessary to obtain the combustion efficiency. This is done by projecting a line vertically up to 380°F temperature difference line on Fig. 24-5 as shown, to the diagonal 9.0% carbon dioxide line, then horizontally to the combustion efficiency scale on the left. The "after" reading for the 9.5% value is drawn in the same manner, to obtain a combustion efficiency of 80.7%, using 365°F temperature difference. The stack losses in both cases are then

$$\text{Stack loss} = 100 - \text{Combustion efficiency}$$

$$\text{Stack loss "Before"} = 100 - 79.9 = 20.1\%$$

$$\text{Stack loss "After"} = 100 - 80.7 = 19.3\%$$

You have just been shown the basic method in determining the condition of your combustion equipment, with the exception of one important point; the output of the boiler at the time of the test sampling. This may be rather difficult on modulating-style boilers, and may tax the operator's ingenuity, but it is important that some means be found to ascertain the approximate firing rate at the time the test samples are taken. Timing the fuel meter, if one is installed, noting the position of the jackshaft positioning lever or the feedwater regulating valve stem are examples. Another one is to record the inlet and outlet pressures on the feedwater pump, and plot the pressure rise across the pump on the pump head-capacity curve. Some help from your boiler or feedwater equipment supplier may be necessary at this point.

This reinforces the need for a flow meter on the boiler feedwater line, and we strongly urge you to consider adding one if there is none installed on your system. The simple multiple dial positive displacement meter is sufficient, requiring the use of a watch with a second hand to time the rate at which the water is entering the boiler. Better yet is one that indicates actual instantaneous rate of flow.

The ideal situation is to be able to set the boiler firing rate at any desired point, then take your set of flue gas readings for at least the 25%, 50%, 75%, and 100% boiler output points. Then you may plot the boiler efficiencies, draw a curve through them, and compare your present operation with the efficiency curve supplied by the boiler supplier. This will only be meaningful if all conditions are as stipulated by the boiler supplier when his curve was produced. Such a curve will enable you to compare your own results from time to time, which will be of some help in your quest for the ultimate operating efficiency.

With the existing stack temperature given in the example, you will note that in order to obtain higher efficiencies, it is necessary to reduce the excess air considerably to lower the stack loss. In fact, depending on the size of the boiler, the sophistication of the combustion control system, and the operator's patience, it may not be worthwhile attempting to increase the efficiency any further by reducing the excess air supply. There are other ways to increase the efficiency, which we shall bring into the picture in the next chapter.

Most boilers are probably at any one moment operating with more excess air than is really necessary. Most boilers in the size range we are discussing here are probably set up in the beginning to operate with about 20% excess air. On natural gas there should be no difficulty in reducing that amount to about 15%, or perhaps slightly lower. Larger boilers are being operated below 10% excess air on natural gas, but they have continuous monitoring of the stack conditions, and are being watched much more closely than the smaller plant can afford to do. The larger industrial and utility boilers obviously have more at stake in saving fuel than the smaller plants, and in those cases, a fraction of a percentage saving in fuel consumption results in large savings in dollars.

We should warn you at this point, that there is a limit in the amount to which you can safely reduce the excess air supplied to the burner. Starving the combustion process of air will result in smoking, soot accumulation on the gas side of the tubes, a reduction in efficiency, and an increased possibility of a furnace explosion.

As a guide to the minimum limit of excess air that can be expected for industrial boilers without sophisticated continual monitoring instrumentation, the following is offered;

Fuel	Excess Air
Natural gas	10%
No. 2 oil	12%
No. 6 oil	15%

We suggest that whenever the carbon dioxide and oxygen readings are taken from the boiler stack, that the readings be checked against those on Figs. 24-13 through 24-18, whichever are applicable. If the two readings consistently fall off the diagonal line, this indicates that the fuel is probably slightly different than that for which the graph was made. In this case, we refer you to the discussion on accuracy of flue gas composition readings in Section D.

It would be well, also, to check your readings against Fig. 24-5 through 24-12, as a constant check on the boiler efficiency. Logging the results will be a help in monitoring the condition of the boiler heat-exchange surfaces. Keep in mind, however, that the efficiency changes with the load, so that will have to be considered in evaluating the results.

E. The Calculations

Now that the groundwork has been laid, it is time to go into the detailed calculations to determine what savings can be expected in the annual fuel bill. The "bottom line" is the important thing.

It is necessary to determine the approximate amount of fuel that our 100 hp boiler requires, and to do this we refer you to page 307 in the Appendix. We have marked in the appropriate value on the table for natural gas consumption, and find that the 100 hp boiler requires about 4200 ft³/hr, at 79.7% efficiency, which figure is close enough for our use.

As a result of the foregoing, we can now set up the procedure for calculating the annual savings to be expected from your seemingly insignificant adjustments:

Line No.	Item	Value
1.	% of fuel wasted up the stack, before	20.1%
2.	% of fuel wasted up the stack, after	19.3%
3.	ft³/hr of fuel burned at rating from Appendix, page 307, approximate.	4200*
4.	Use factor, average for the year (See page 237)	0.60
5.	Average hourly fuel consumption, ft³, line 3 × line 4	2520
6.	Hours in operation per day	24
7.	Average fuel consumption per day, line 5 × line 6	60,480
8.	Days operation per month	30
9.	Monthly fuel consumption, ft³, line 7 × line 8	1,814,400
10.	Months per year in operation	12
11.	Yearly fuel consumption, ft³, line 9 × line 10	21,772,800
12.	Improvement in efficiency, line 1 − line 2	0.80%
13.	Line 12, expressed as a decimal, 0.80/100	0.008
14.	Annual reduction in fuel consumption, line 11 × line 13, ft³	174,182
15.	Annual savings in fuel cost, line 14 × $6.00/1000 ft³	$1,045.00

$$*\text{ft}^3 \text{ hr} = \frac{(\text{therm/hr}) \times 100,000}{\text{BTU/ft}^3}.$$

Thus, we see that by simply reducing the amount of combustion air fed to the burner a small amount, you have saved your employer about $1,045.00 a year in the fuel bill. This is assuming, of course, that the burner was using too much air to begin with.

CHAPTER 25
FINE-TUNING THE BOILER

A. General Comments

Now that we have covered the basic principles of testing, adjusting, retesting, and proving the economic worth of the changes, we shall go into the many areas in the operation of the boiler where improvements in efficiency may be possible. Then, after taking advantage of as many of these possibilities as you can, you will be able to say that the boiler is producing at as high an efficiency as it is possible for one that size and design to produce.

First, we shall define what is meant by "efficiency."

So far we have been speaking mostly of combustion efficiency, which simply refers to the degree with which the internal heat energy in the fuel is released in the burner and combustion chamber, and is approximately equal to

Combustion efficiency

$$= \frac{\text{Heating value of fuel input} - \text{Stack losses}}{\text{Heating value of fuel input}}$$

$$\times 100$$

It does not matter what time factor is used in calculating the efficiencies. It can be calculated in heat input per minute, hour, or any other time you wish to use. Normally, the heat input per hour is used, since it is a common unit when speaking of boiler rating.

You can see from the preceding that the condition of the boiler passages and heat-transfer surfaces have a very large effect on the combustion efficiency. The burner only, with its air intake and mixing apparatus, and the combustion pattern within the boiler are not the only things affecting the combustion efficiency.

There is another common efficiency term used, and that is the fuel-to-steam efficiency, sometimes simply called "boiler efficiency." This takes into account many more elements in the operation of the boiler. In large boilers, it becomes a rather complicated problem to determine the true fuel-to-steam efficiency. Fortunately, for the boilers we are dealing with here, the method of determining the fuel-to-steam efficiency (which we shall call by its better-known term "boiler efficiency") has been reduced to relatively simple calculations, as long as the fuels involved are the better-known types of plain fuels, such as natural gas; Nos. 2, 4, or 6 fuel oils; coal; wood; or bagasse. In general, this simplified method reduces to the following procedure:

1. Record the stack temperature of the flue gases.

2. Test the same flue gases for carbon dioxide and oxygen content.

3. By use of the graphs, such as given in Figs. 24-5 through 24-12, determine the combustion efficiency.

4. From tables supplied by the boiler manufacturer obtain the percentage radiation

losses for the boiler. Subtract that figure from the combustion efficiency, and the result is the fuel-to-steam, or boiler, efficiency.

5. The radiation losses may be estimated at 2%.

We hasten to remind you, however, that the preceding method is not applicable to large boilers, since there are many minor items that have been ignored owing to the small amount of heat involved. The result by the methods we have described is close enough for our purposes. In the calculations for determining the payout economics, we shall be dealing only with the differences in efficiencies between two operating conditions, and the minor items referred to will not appreciably change from the one operating condition to the other in making the adjustments recommended. Therefore, our method is valid, and you may use it with confidence.

Now concerning the calculations for the time that the boiler is on the line during the course of a year, known as the "use factor," this is an area where some ingenuity on the part of the boiler operator may be required. Often, only an educated guess is all that can be made. The result will directly affect the payout calculations, of course, so we suggest that some careful thought be given to arriving at the total annual "on" time and the average percentage of maximum rating for the boiler output for the year. If possible, some arrangement of a timer installed across the starting circuit of the burner should be attempted, in order to arrive at a total firing time for the boiler, over a definite time span, such as one week. Failing that, some method of making and recording observations during the same time span at frequent intervals should be done.

We wish to emphasize that anything done to the burner, the controls, or other parts of the boiler should have the approval of the boiler supplier. In fact, it is a good idea to call the supplier in before you make any adjustments, just in case he has some input that may help you achieve your goal. At times like these, with energy becoming more expensive, our boiler and

equipment suppliers are becoming adept at helping their customers squeeze the most efficiency from their equipment. If for some reason you do not want to ask his help, or have asked and are not satisfied with the response, then try one of the commercial boiler plant servicing contractors. They will charge you for their services, of course, but if there is any appreciable savings to be realized, then it could well be worth the cost.

In calculating the expected payout for the contemplated changes, be sure and include any costs incurred for items other than the cost of the fuel burned for the year. This would be limited to costs that would not otherwise be incurred normally. For instance, the wages of the plant personnel doing the adjusting or alterations would not be included in the costing calculations, unless overtime is involved.

Some of the heat and fuel conservation methods being proposed here require alterations in the draft system. We have assumed mechanical draft systems are involved in all of these examples. In those cases where natural draft systems are involved, appropriate adjustments in procedure may be necessary.

It will help to remember that the induced draft stack, however short or long, is merely replacing the fan in an induced-draft system. The stack does impose a definite limit on the alterations possible, unless it is economically feasible to increase the stack height in arriving at the desired boiler efficiency improvement.

B. On–Off vs Modulating Burners

Many older boilers, and some of the newer, but smaller, boilers, have burners operating in the on-and-off modes only. The burners are designed to produce a heat output slightly more than the boiler rating requires, in order to assist the boiler to pick up the load and restore the heat lost in the boiler furnace and casing and the piping, once the burner control has ignited the burner after an "off" cycle. This aggravates the situation, since the burner then cycles much more often than if the burner was able to produce just the boiler rating.

The whole system is designed to attempt to follow a changing load demand curve by a series of on-and-off burner cycles, so as to produce the required total steam over a period of time. Each time the burner control shuts it down, the boiler casing and furnace cool down, and the furnace is purged of hot gas. When the burner starts up again, the furnace is purged again, sweeping out warm air, and the burner then has to replenish all of the wasted heat, and bring the whole system up to operating temperatures again.

Figure 25-1 shows approximately what the result is under this system of burner control. We wish to point out, however, that the average industrial boiler does not produce the 82% efficiency shown, unless auxiliary heat-reclaiming equipment has been installed.

To alleviate the high fuel consumption for this arrangement, we suggest that the boiler manufacturer be called in and the problem discussed with him. If the boiler is an older one, it is possible that the firm who furnished it has since designed a better, more efficient burner and control arrangement. The energy shortage has caused many boiler and burner makers to redesign their products, and modulating burners are now being used where the simpler on-and-off systems were furnished previously.

The problem first is to determine what percentage of the time your on-and-off system is operating to actually produce steam, and what percentage of the time it is off and cooling. This was explained in the previous section, and will

result in the starting point of the program to increase the efficiency of the unit.

There are several styles of burners and controls that may be installed or at least considered to achieve better fuel economy. The ideal arrangement is to convert to complete modulating equipment, in which the burner is capable of putting out maximum rating for the boiler, and will modulate its output down to a well-defined minimum output. If the boiler output requirement drops below this minimum capability, then the burner shuts down. When starting up again under demand for more steam pressure, the burner starts under a low flame condition. As the main flame is proven, the controls then increase the firing rate by a programmed schedule until it is producing its maximum, or until the burner is floating on the load demand. The entire process is much more efficient than the on-and-off cycling arrangement, as Fig. 25-1 indicates.

For example, let us assume that, using the 100 hp boiler in the example in Chapter 24, the boiler has an on-and-off burner control, and that you have determined by test that it operates only about 50% of the time the boiler is on the line. From Fig. 25-1, it is obvious that the boiler can be operating at only about 72% efficiency overall.

The boiler supplier is called in, and he makes a few tests to determine your plant load pattern, and decides that the best he can do for your particular unit and with the load demand, is to convert the burner and control system to a two-step arrangement, one step being 67% of total boiler rating, and the other step 33% of the boiler rating. By making this alteration, the supplier estimates that the burner will be on approximately 75% of the time. From the graph, Fig. 25-1, we see that the expected boiler efficiency is then about 76%, an improvement of 4%.

At this point, it is well to give a little thought to that 50% figure previously arrived at. It is obvious that if the boiler is only operating 50% of the time, then the boiler is actually twice the capacity that the present load requires, since it is able to supply the total demand by being on the line only 50% of the time. However, the

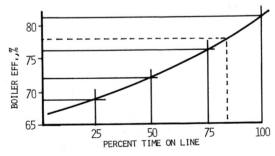

Fig. 25-1. Effect of burner cycling. (Courtesy of Boiler Optimazation Course, BTU Consultants.)

50% figure is not out of line for some installations, so we shall use it for purposes of illustrating our example.

We now have sufficient data to make the following calculations:

1. Present average hourly fuel rate, 4200 × 0.6   2,520 ft³

2. Present annual fuel rate, 2520 × 24 × 365 (actually required to handle the load)   22,075,200 ft³

3. Expected efficiency by converting burner, from Fig. 25-1   76%

4. Present efficiency, from Fig. 25-1   72%

5. Expected improvement from conversion, line 3 − line 4   4%

6. Decimal equivalent, line 5/100 = 4/100 =   0.04

7. Expected annual savings in fuel consumption, line 2 × line 6   883,000 ft³

8. Annual savings in fuel cost, line 7 × $6.00/1000ft³ $5,298

However, the cost of making the conversion must be deducted from the projected savings. The difference then becomes the amount of actual savings to be realized by converting the burner and controls.

Usually, the cost of making the conversion will exceed the amount to be saved in one year in fuel costs. This brings into play the "payout" factor, which we have mentioned but have not defined. The payout factor simply means the number of months or number of years that it takes for any projected change, improvement, or new installation to pay for itself in either savings or increased revenue.

The best method for illustrating this is to carry the present example another step, and arrive at the number of years it will take to write off the total cost of making the conversion. Thus,

$$\text{Payout time} = \frac{\text{Total cost of conversion}}{\text{Annual savings}}$$

The total cost of making the conversion must include the contractor's cost, the plant downtime, and other costs, such as employee overtime, associated with the project. For this example we shall assume that the total costs add up to $11,000. The calculation then becomes

$$\text{Payout time} = \frac{\$11,000}{\$5,298}$$

$$= 2.0 \text{ years, or 24 months}$$

This payout time is feasible for most industrial plants, and we can then assume that the conversion would be well worth the expense.

Now let us carry this example one step further, and assume that the boiler may also be fitted with a modulating style of burner and controls. We have established that the amount of fuel burned annually to handle the total demand is, from line 2, 18,396,000 ft³, more or less, depending on the final combustion efficiency. Starting from there, we shall assume that the contractor submitting his proposal has estimated that the modulating-style conversion will result in the boiler and burner producing steam with an "on" time of about 85% of the total time. The calculations then become

1. Present annual fuel consumption, line 2, previous calculation   22,075,200 ft³

2. Estimated time boiler will be on line, after conversion   85%

3. Expected revised efficiency, Fig. 25-1   78%

4. Present efficiency   72%

5. Expected improvement by converting, line 3 − line 4 = 78 − 72 =   6%

6. Decimal equivalent, line 5/100 = 6/100   0.06

7. Projected fuel savings, annually, line 1 × line 6   1,324,500 ft³

8. Annual fuel cost savings, line 7 × $6.00/1000 ft³ $7,847

You now should be in a position to make the necessary basic calculations in determining the payout rate or the feasibility of any projected changes in the equipment.

C. Reducing Flue Gas Temperatures

We have mentioned several times that a large amount of waste heat usually disappears up the stack in the form of hot flue gases. In fact, this is usually the single largest source of loss, and is, therefore, the first place one should look for effecting a saving in fuel cost.

Figure 25-2 illustrates the amount of increase in boiler efficiency that can be expected by recovering some of the waste heat in the flue gas and using it to raise the feedwater temperature. It is applicable for any method available that will increase the feedwater temperature. In large boiler plants, this may be done by feedwater heaters using waste steam or by passing the feedwater through heat exchangers installed in the boiler breeching and recovering some of the waste heat in the flue gas. In boilers of the size we are discussing, the usual and most feasible method is with the heat exchangers in the boiler breeching. These are called "economizers," since they save on fuel costs.

The savings gained by raising the feedwater temperature or by increasing the temperature of the combustion air as described in Section H can only be credited to the boiler if the heat to accomplish this increase in feedwater temperature is taken from heat that would ordinarily be wasted, such as that in the flue gas. Thus, the savings come from removing some of the waste heat from the flue gas and putting it back into the boiler, as we are doing here. If waste heat from the flue gas is used to heat some process stream, such as water or air being used elsewhere in the plant, there would be an overall savings in fuel cost for the entire plant, but no increase in boiler efficiency.

If the feedwater or combustion air temperatures are raised by using waste heat from other plant processes, the direct result will be an increase in the boiler efficiency, as shown by Fig. 25-2 and 25-11, with a decrease in flue gas temperature brought about with the reduction in fuel consumption. However, in this example we are concerned only with recovering some of the waste heat in the flue gas to raise the feedwater temperature. After going through the procedure given here, the boiler plant operator may wish to examine other possibilities of recovering waste heat from the plant as a whole.

This gain in efficiency is not without some cost and limitation. The heat-transfer equipment and the piping and controls are not cheap to install. Also, the maintenance may be rather high, especially if there is a tendency for soot to form on the heat-exchange surfaces.

We mentioned earlier that there is a considerable amount of water vapor being produced and carried along with the flue gas. This moisture combines with any sulfur compounds in the flue gas when the temperatures drop to certain limits and forms highly corrosive conditions. The temperatures below which trouble may be expected depend on the percentage of sulfur in the fuel. Natural gas usually has, at most, only a trace of sulfur, and, therefore, the flue gas for natural gas may be reduced in temperature to the lowest point for any of the fuels, except for LPG fuels. Fuel oils contain sulfur in varying amounts, and one of the main reasons why fuel oil has risen in price recently is due to the requirement for low-sulfur oil. This is an environmental problem in many areas, and, as a result, many fuel suppliers have had to install expensive sulfur-removal plants to prepare the fuel oil for sale to the boiler plant operators. This was brought out earlier in our discussion on atmospheric contamination.

Table 25-1 will give the approximate lowest

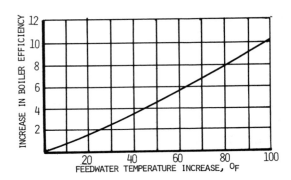

Fig. 25-2. Raising feedwater temperature to increase efficiency. (Courtesy of the Boiler Efficiency Institute.)

Table 25-1. Minimum Stack Gas Temperatures

Fuel	Minimum Temperature (°F)
Natural gas, sulfur-free	220
Fuel oil, No. 2, less than 1.0% sulfur	330
Fuel oil, all grades, more than 1.0% sulfur	390
Subbituminous coal, 0.8% sulfur	230
Bituminous coal, 2.8% sulfur	290

temperatures to which the flue gases may be discharged out the stack without producing corrosive conditions in the breeching and stack. We warn you that it is not a good idea to attempt to install heat-reclaiming apparatus that will reduce the flue gas temperature to those shown, since the result may be a bad case of "acid rain." For this reason, we recommend that you call on your fuel supplier for his recommendation on the lowest safe flue gas temperature.

When considering the addition of heat-reclaiming equipment in the stack uptake or boiler breeching, keep in mind that if the boiler is to be operated for extended periods of time at reduced rating, the flue gas temperature leaving the boiler furnace will be reduced, as shown in Table 24-3. Consequently, whoever designs the heat-reclaiming equipment should be made aware of this limitation.

Figure 25-3 shows one of the commercially available economizers on the market. The feedwater, which is under pressure, is inside the tubes, and the hot flue gas flows around the outside of the tubes. If the economizer is to be used with No. 6 fuel oil with high sulfur content, then the unit must be equipped with a soot blower attachment.

Figures 25-4 and 25-5 illustrate the two general methods of installing the economizer. Figure 25-4 is the one used mostly with low- to medium-pressure boilers, since the tubes in the economizer must withstand the boiler pressure.

Figure 25-5 is the method used mostly for higher-pressure boilers, since the economizer tubing is not subject to the full boiler pressure.

The flow of water through the economizer

Fig. 25-3. Typical economizer. (Courtesy of Energy Conversion Systems.)

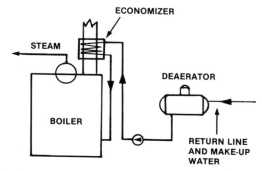

Fig. 25-4. Economizer installation, #1. (Courtesy of Energy Conversion Systems.)

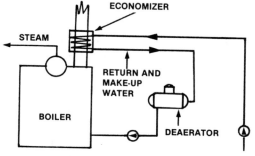

Fig. 25-5. Economizer installation, #2. (Courtesy of Energy Conversion Systems.)

must be controlled in some manner so as to limit the exit gas temperature to the temperature previously determined. Figure 25-6 shows one of the accepted methods for doing this with low-pressure-boiler economizer installations. For high-pressure-boiler installation, the dis-

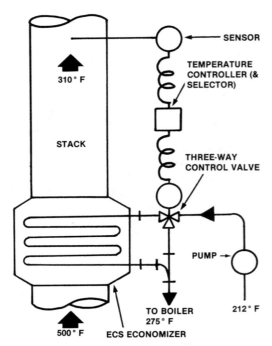

Fig. 25-6. Recommended economizer piping. (Courtesy of Energy Conversion Systems.)

charge would probably be into the deaerator or hotwell instead of into the boiler.

We shall now continue our example of the 100 hp boiler started in Chapter 24, and take it further by utilizing the principles we have explained in this section. The boiler has been adjusted to give the best efficiency possible under the existing conditions, as covered in Section E, Chapter 24. We are going to investigate the installation of an economizer to increase the feedwater temperature from 200 to 240°F, and to reduce the stack gas temperature to 350°F. The burner adjustments have already been made and the expected annual fuel consumption decreased slightly, for an annual savings of $1,045.00.

Of course, this is not the end of the calculation, but it is far enough to show you the method. The cost of installing the economizer, the equipment, piping, and control costs, and all other associated costs must be taken into account to find the payout time, as we explained in the last section.

The calculations are given here for installing an economizer:

1. Present flue gas temperature   450°F

2. Present feedwater temperature   200°F

3. Boiler room temperature, fan inlet 70°F

4. Exit gas temperature from economizer   350°F

5. Feedwater temperature from economizer   240°F

6. Increase in feedwater temperature, line 5 − line 2   40°F

7. Increase in boiler efficiency, Fig. 25-2   3.5%

8. Decimal equivalent, line 7/100   0.035

9. Annual fuel consumption, Chapter 24, Section E, line 11 − line 14   21,598,600 ft³

10. Annual savings in fuel, line 9 × line 8   755,950 ft³

11. Annual savings in fuel costs, line 9 × $6.00/1,000 ft³ $4536

The "breakeven point" for this project, assuming the payout time of three years, would be:

Breakeven point = Total costs × Payout period

= $4,536 × 3 = $13,608

This means that the firm could afford to invest up to $13,662 in making the alteration to the boiler plant by adding the economizer and the associated piping and controls, if their payout policy is to recover the costs in three years or less.

We mentioned earlier that there are limitations to the use of economizers, and one of them is the temperature of the feedwater leaving the economizer. We use 240°F for our example, because that is a reasonable temperature for industrial boilers. At 240°F, sending the feedwater into the deaerator would require the deaerator to be at 10 psig, which is within the pressure rating of the average deaerator. At a temperature of 260°F the deaerator pressure would have to be 20 psig, which is the highest practical deaerating temperature. Figure 9-4 indicates that at temperatures higher than 260°F there is very little increase in carbon dioxide removal and none for oxygen.

It is also important to be aware of possible trouble from local corrosion where the feedwater line enters the economizer. If the entering feedwater is below about 200°F, there may be condensation on the flue gas side of the economizer piping. Also, this condition may be reached if the boiler is fired for any length of time at greatly reduced output.

Any increase in feedwater temperature calls for a check of the adequacy of the feedwater pump to handle the increase. A new pump may be required.

There is a plus feature to be gained from this exercise of increasing the feedwater temperature, in the case of Fig. 25-5 installation. If you will refer to Fig. 9-4, you will see that there is a definite drop in solubility of carbon dioxide when the temperature of the feedwater is raised above 200°F. This will result in less trouble from corrosion in the system, and better heat transfer in any heat exchangers using the steam. These items are difficult to evaluate in dollars and cents, but they are there, nevertheless.

Figure 25-6 shows a typical control system for a well-designed flue gas heat reclaimer, such as an economizer. The temperature controller (and selector) is set to maintain the desired minimum stack temperature, which is decided by the fuel characteristics. In the illustration, the minimum stack temperature is set at 310°F. At temperatures above that, the controller positions the three-way valve to pass all of the boiler feedwater through the economizer. As the stack temperature approaches 310°F, the valve diverts a portion of the feedwater around the economizer, direct to the boiler, as required to maintain the set stack temperature.

We have shown only two of the typical arrangements available when the decision is made to install an economizer. There are other systems that may be applied, depending on the existing equipment, the result desired, and the type and availability of the equipment to be installed. The application is best left up to the chosen supplier, because some detailed engineering is required.

D. Losses due to Scale and Soot

So far in the examples we have assumed that the heat-transfer efficiency was perfect on both the water side and the hot-gas side of the boiler tubes. Experience tells us that this is very seldom possible, and we can expect that there will always be some scale deposits on the water side of the tubes, and some soot on the hot-gas side. It is the goal of every boiler operator to keep these deposits to a minimum. If you have any doubt about that, we refer you to Figs. 25-7 and 25-8. Note the high percentage of heat lost due to a very thin layer of either scale or soot on the tubes.

Looking first at Fig. 25-7, we see that there are three types of scale included. The one marked "Average Scale" is the one usually encountered in the lower-pressure boilers, and the one we shall assume for our example. This scale composition is mostly calcium and magnesium compounds, as found from the ordinary boiler water conditioning process. The other two scales are extreme cases and are often products of higher-pressure service conditions. Obviously, if you do encounter them on your tubes, the problem requires immediate help from the water-treatment-service experts, because the fuel loss can be very high as we shall soon see.

Figure 25-8 covers the same situation for deposits of soot on the tubes. This comes about from improper combustion, usually from

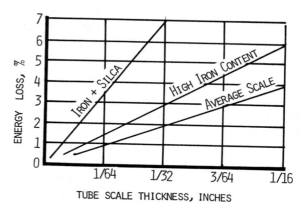

Fig. 25-7. Energy loss from tube scaling. (Courtesy of NIST, Handbook No. 115, Suppl. 1.)

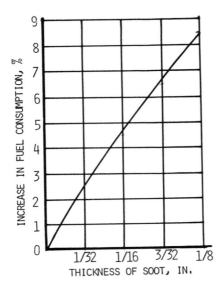

Fig. 25-8. Effect of soot on fuel consumption. (Courtesy of Boiler Optimization Course, BTU Consultants.)

starving the burner for air, or incomplete mixing of the air with the fuel. This situation is often a result of improper air louver adjustment, too low a firing rate for the burner design, or the air/fuel ratioing device getting out of adjustment.

The graphs do not tell the entire story, since they are simplified to a great extent. They are correct for the conditions shown, but in actual practice, the scale and soot deposits do not appear uniformly throughout the tube surfaces, and they do not just build up to a certain amount and stop at that point. Both deposits tend to concentrate upon certain trouble spots, and leave other areas completely free of deposits. Also, once the condition has been reached where they are forming, the deposit continues building up until something is done to stop it.

As an example, we shall illustrate what can happen to the annual fuel costs when deposits are allowed to form. Starting with the 100 hp boiler already being used, we shall assume that the boiler is being operated under indifferent monitoring of the water and combustion conditions. As a result, the boiler is cleaned thoroughly each shutdown period, the deposits form gradually over the succeeding months, and are deposited in the usual selective areas for this

boiler and class of service. The tubes start out clean, but, over the operating period, the deposits build up on both the water and hot-gas side, so that the average thickness for calculation purposes is $1/32$ in. for both of them.

This figure will be used as an average for the entire surface of the tubes, although we know that some areas will be clean, and some areas will have more than $1/32$ in. deposit just prior to shutting down for cleaning. We shall also assume that the scale is of the ordinary calcium and magnesium composition. We shall start our example from the end result of the calculation in the previous section. Thus,

1. Annual fuel consumption, from Section C
   21,598,600 − 755,950   20,842,650 ft³

2. Increase in fuel consumption, from Fig. 25-7, $1/32$ in. thick, average scale   2%

3. Increase in fuel consumption, from Fig. 25-8, $1/32$ in. thick soot deposit   2.5%

4. Total increase in consumption   4.5%

5. Decimal equivalent, 4.5/100   0.045

6. Annual increase in fuel consumption, line 1 × line 5   937,900 ft³

7. Annual cost of scale and soot deposits line 6 × $6.00/1,000 ft³ $5,628

You can now readily see how much a relatively small deposit of scale or soot on your boiler tubes can increase the annual fuel bill. This could wipe out any or all of the savings arrived at by the procedures we are covering here for improving the efficiency of your boiler. Keep in mind that the above amounts of deposits are not uncommon in small boiler operation. Make no mistake about it, scale and soot represent a major loss.

We hasten to repeat here, however, a warning made earlier in this book, that it is possible to open the boiler and find that the tubes are sparkling, bright clean. This also is not good, because it may mean that the water conditioning is etching away the tube metal. Most of the time, the boiler operator's problem will be keeping the scale and soot off the tubes, and it will often prove to be a constant battle.

Now that we know what trouble these deposits can cause, how do we check for their formation while the boiler is in operation, and how do we remove them, once formed?

As the deposits are formed, the boiler efficiency drops off, causing the combustion control to attempt to compensate for the increased fuel requirement. This causes the flue gas temperature to increase, and the burner to remain on high fire more often. The increase in flue gas temperature is very easy to detect, providing the boiler breeching or stack has been equipped with a thermometer permanently installed, as we stated in the discussion in Chapter 24 on setting the fuel/air ratio for best efficiency. By keeping a regular log, as the boiler operator should be doing, and recording the stack temperature along with all of the other readings, it will be very easy to spot any increase in stack temperature over the course of time. This will be a direct indication of the formation of scale and soot, or of the burner controls becoming out of calibration. Either condition calls for immediate action to correct.

To remove these deposits, there are several common methods in use and we list them.

Soot deposits may usually be removed simply by washing with a high-velocity-water nozzle. This must be done with care, since the sludge that results can be very irritating, and even harmful, if the eyes are not protected while performing the operation.

Another method for removing build up of soot is to increase the amount of combustion air. This is best done before the layer of soot has built up too high, since it is possible that burning soot may escape up the stack, and, conceivably, start a fire in the neighborhood. At the least, it could cause a nuisance, and some concern among the neighbors.

Scale deposits may be removed from the inside of tubes in a watertube boiler by mechanical scrapers, driven by water, air, or motor, as the scraping tool is pushed through the tubes.

In firetube boilers, the usual method of removing the scale is by chemical cleaning, best performed by contracting firms who are well qualified and are insured.

There are other methods being used to perform the cleaning jobs, with more or less success. In most cases, special training and expertise is required, and we suggest that the boiler operator would be best advised to engage an outside contracting firm to do it.

Should you decide to attempt the cleaning process yourself, then be sure you are adequately protected with goggles, rubber boots, and clothing, or other such gear required for personal protection.

Remember, if the boiler is not going to be placed into service right after being washed out, it must either be given the wet layup procedure or dried thoroughly and sealed up tight with bags or with trays of moisture absorbants inside.

E. Selecting the Cheapest Fuel

Many boiler plants are lucky enough to be in an area where more than one type of fuel is readily available. This gives the industrial plant more flexibility in operating the boiler and in purchasing the fuel for the plant. It generally leads to lower fuel costs, also, owing to the competition.

In selecting the best and the cheapest fuel to use, there are several factors to be taken into consideration, besides the difference in cost from the supplier. We shall take these items one at a time and give enough pointers on the use of the various fuels to permit a logical and economical decision. Our choices will be limited to the fuel oils and natural gas. The principles are the same, whatever fuel is being evaluated.

Fuel should be evaluated by the lowest common denominator, which is generally taken to be the cost for 100,000 BTU, which is known as a "therm." The fuels are purchased by cubic feet in the case of gas, and by gallons, in the case of liquid fuels.

The formula for reducing all fuels to the cost per therm is as follows:

Cost per therm

$$= \frac{\text{Cost per unit of measurement}}{\frac{\text{BTUs per unit of measurement}}{100,000}}$$

The preceding costs are as delivered to the plant, without any other factors being taken into account. Other factors to be considered are

Combustion efficiency

Cost of handling and storage

Cleanliness from combustion and handling

Effect on the boiler and other equipment

Cost of safety features involved, insurance, etc.

Environmental effects

Reliability

Availability

The first two listed items are usually the easiest to identify and evaluate, but the remaining factors are often only matters of judgment and opinion, and we shall omit them in our example. If you, as the boiler operator called upon to make the decision, have any concrete information to help in regard to the other factors, then we urge you to take them into full account. You and your management are the ones who must be satisfied, and are the ones who will have to justify the decision.

The data for our example will be shown in Table 25–2. The Cost Per Therm values are calculated by the formula given above, and will not be illustrated here.

To convert the Cost Per Therm into the practical value for purposes of selecting the cheapest fuel, we must calculate the practical use value of it in the boiler, by factoring in the Aproximate Combustion Efficiency in the last column, thus:

Cost per therm as utilized

$$= \frac{\text{Delivered cost}}{\text{Combustion efficiency}}$$

This enables us to produce Table 25–3, which tells us the actual cost per therm as burned in the boiler, considering only the delivered cost and the combustion efficiency.

Of course, you must alter the data to fit your particular situation. The example is purely illustrative, and it will take some effort on your part to establish some of the costs involved. But if there is a lot of fuel being burned in your plant over the course of a year, it may well be worth the effort.

What do the results in Table 25–3 tell us?

1. There appears to be a stand-off between natural gas and No. 6 fuel oil as burned.

2. No. 4 fuel oil is an excellent back-up to No. 6, if No. 6 should prove difficult to obtain.

3. No. 2 fuel oil is obviously out of the running for a steady fuel in most boilers, if any of the other fuels are available, de-

Table 25-3

Fuel	Actual Cost Per Therm
Natural gas	$0.76
No. 2 fuel oil	$0.90
No. 4 fuel oil	$0.79
No. 6 fuel oil	$0.76

Table 25-2

Fuel	BTU Content	Unit Cost	Cost Per Therm, Delivered	Approximate Combustion Efficiency[a]
Natural gas	1050/ft^3	$6/1000 ft^3	$0.57	75%
No. 2 fuel oil	141,000 BTU/gal	$0.97/gal	$0.69	77%
No. 4 fuel oil	146,000 BTU/gal	$0.90/gal	$0.62	78%
No. 6 fuel oil	150,000 BTU/gal	$0.88/gal	$0.59	78%

[a]The combustion efficiencies are approximate only, for this example.

pending on the current prices. It may be considered as a back-up for highly intermittent use.

We shall give you a few items covering each fuel in turn, to permit you to make a proper evaluation.

Natural gas—Cleanest fuel to burn, easiest to control, requires the simplest equipment. Not always available, cannot be stored on the plant premises, often may be turned off on short notice.

No. 2 fuel oil (diesel fuel)—Usually easy to obtain, burns readily, easy to ignite, fairly clean burning, easy to store and pump, requires no preheating, fairly safe to handle and store.

No. 4 fuel oil—Easier to handle, store, and burn than No. 6. May need some heating and atomizing by artificial means, but only minor. Often not available, some suppliers obtain it by blending Nos. 2 and 6.

No. 6 fuel oil—Cheapest of the oils, has the highest heating value, usually easy to obtain, safe to store and handle. Most difficult of the fuel oils to store, pump, heat, and burn. Often requires treatment in storage to prevent sludging, usually contains some sulfur, burns the dirtiest of the oils, the most difficult to handle and prepare for burning.

So you have the facts. The rest is up to you.

As an exercise, we shall calculate the annual fuel bill for our 100 hp boiler we have been using in our examples, using the latest fuel consumption from Section D, line 1 from this chapter, thus:

1. Annual gas consumption,
   line 1   20,842,650 ft^3

2. Therms required annually,
   $\dfrac{\text{line 1} \times 1050}{100,000}$ 218,850

3. Annual fuel cost, natural gas,
   line 2 × $0.76 $166,324

4. Annual fuel cost, No. 2 fuel oil,
   218,850 × $0.90 $196,965

5. Annual fuel cost, No. 4 fuel oil,
   218,850 × $0.79 $172,890

6. Annual fuel cost, No. 6 fuel oil,
   same as line 3 $166,324

As you can see, there is a substantial difference, based on a round-the-clock operation of a 100 hp boiler.

F. Reducing Blow-Down Heat Losses

In Chapter 8, Table 8–2 we give the maximum recommended total dissolved solids (TDS) to carry in the boiler water. This is the maximum, and is a general guide. The boiler manufacturer will give his recommendations in the instructions that came with your boiler, and the water-treatment expert may have his own ideas. The end result will depend on your actual experience, using the two listed sources, and the result in your boiler. The water-treatment expert servicing your plant will be the one best guide to follow. Usually you will find that the faster the steaming rate for the size of the boiler, the lower the TDS concentration must be kept. This is because the faster-steaming-rate boilers, which means that the tubes are being forced to produce a higher boiler hp output per square foot of heating surface, produce more agitation at the surface of the water. This agitation is what causes foaming and carryover of water into the steam spaces. The slower the steaming rate, the less agitation on the water surface, and the less foam and moisture will be carried into the steam. This is another good reason for sizing the boiler for operation at less than full rating.

The blow-down piping and procedure described earlier does not include any method of reclaiming the heat in intermittent blow-down methods. The continuous blow-down systems described do contain several heat-reclaiming systems, and we show one package system in Fig. 25-9. The ideal condition is one in which all of the heat being carried out of the boiler by the blow-down water will be picked up by

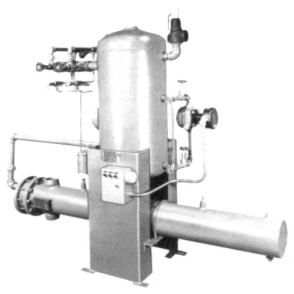

Fig. 25-9. Continuous blow-down heat reclaimer. (Courtesy of Cleaver-Brooks Co.)

the boiler feedwater entering the boiler. This is being achieved in many cases.

The intermittent blow-down system is a different situation. Here the problem is one of removing large quantities of hot flashing water as fast as possible, without damaging the equipment. The speed of removal is important to the process of removing the accumulated sludge in the bottom of the boiler. Normally, it is not recommended that any heat-reclaiming coils be installed in the blow-down tank. The impact upon the coils would cause vibration and erosion and the maintenance problem would probably be prohibitive, for one thing. Therefore, any heat-reclaiming apparatus must be downstream of the first blow-down, or flash, tank. This may be done by installing a tank downstream of the blow-down vessel, through which the blow-down waste must pass, and pause, on its way to the sewer. Inside this second tank, you may place a nest of heat-exchange coils, through which the incoming boiler makeup water is passed before it enters the hotwell, or any other convenient or suitable entrance point. The best point for introducing this hot raw feedwater is usually into the hotwell ahead of the deaerator, or directly into the deaerator.

In Fig. 10-6, we show the blow-down tank as

being large enough to retain one blow down from the boiler between blow-down periods. This is to permit the water to cool to the required temperature for discharging into the sewer. If this blow-down is to be cooled in the manner we have just described, with a nest of coils containing the incoming raw makeup water, then this extra storage capacity need not be in the blow-down tank, but may be in the vessel or tank containing the makeup water coils. The cooling water inlet line and valve shown on Fig. 10-6 will still be required into the cooling tank, just in case the blow-down water needs additional cooling at times. You may be very generous in sizing the nest of coils to be immersed in the blow-down cooling tank, because you want to pick up as much heat as possible. The final decision as to where it should enter the system after picking up the surplus heat from the blow down will depend on the advice of your boiler supplier or water-treatment expert.

Figure 25-10 will give you an idea of how much heat we are working with in this section. From this graph, it is easy to see that the blow-down losses may be very high. The tendency is to work with higher blow-down rates with the smaller boilers, owing to the lower grade of water-treatment methods often used on the smaller boilers. The amount of heat to be saved

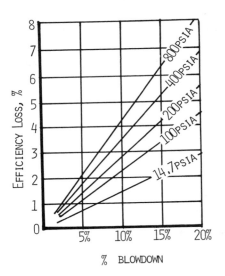

Fig. 25-10. Loss from blow-down. (Courtesy of Boiler Efficiency Institute.)

and the increase in boiler efficiency desired depend on the amount of the annual fuel bill and the desires of the plant management.

We will give you the basic rules for lowering your heat losses from the blow-down, as follows:

1. Carry the TDS in the boiler water as high as practical without causing foaming and carryover.

2. Treat the boiler feedwater to as high a degree of purity as practical for the size of your installation, since the amount of blow-down is a direct function of the purity of the water; the higher the purity of your feedwater, the lower the blow-down rate, and the lower the blow-down losses will be.

3. The goal should be: the discharged blow-down water should be as close as possible to the temperature of the raw water entering the treatment system.

There is one more reason why the blow-down should be reduced to an absolute minimum, regardless of the heat reclaimed. That is the cost of the treatment chemicals dissolved in the blow-down water. This cost is not included in the graph in Fig. 25-10, and it represents a total loss going to the drain in every gallon of water leaving the blow-down system.

We shall not go into an example of the savings to be expected by the use of heat-reclaiming methods in the blow-down apparatus. The method of computing it is similar to those already covered, and you should be well versed in the method at this point in our discussion. You simply start by recording the percentage of blow-down required in the "before" situation, and when the improvements have been accomplished, you calculate the percentage of blowdown "after" the results have been determined. From the two conditions, the graph in Fig. 25-10 will tell you the amount of increase in boiler efficiency to be expected. The calculations are the same as have been shown for the previous examples, based on the annual fuel consumption.

G. Heating Combustion Air

In Section C, we showed what could be done toward reducing the fuel bill by lowering the stack gas temperature by heat reclamation with economizers. In Fig. 24-5 through 24-12, we showed you how the efficiency increases as the difference between the exit flue gas temperature and the boiler room temperature decreases. This was assuming that the boiler room air was the temperature of the air actually entering the combustion system. This may be the case in many boiler rooms, but it is seldom the most efficient method of supplying the air for combustion.

It is commonly understood in the boiler industry that increasing the combustion air temperature will increase the boiler efficiency 1% for each 40°F rise in air temperature. Put into equation form for ease of calculations:

Percentage increase in efficiency

$$= \frac{\text{Increase in combustion air temperature}}{40}$$

There are three general methods for accomplishing the increase in air temperature entering the draft fan.

The hottest point in the boiler room is usually directly over the boiler, or very close to the boiler near the ceiling. Figure 25-11 is a variation of an illustration presented in the first portion of the book, Fig. 4-3. It is shown here to illustrate the point we are attempting to cover; that the best place to take the air for combustion is near the ceiling of the boiler

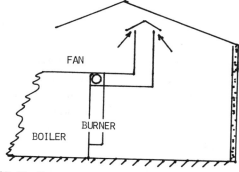

Fig. 25-11. Increasing combustion air temperature.

room. It is not uncommon for the air in that location to be at least 40°F above the temperature of the air at floor level, which means you can count on at least 1% fuel savings by the method shown in Fig. 25-11. The draft fan will have to be checked for sufficient capacity and power if you convert your ducting to this system. Also, the dampers shown in the base of the vertical intake stack in Fig. 4-3 should remain closed, or be omitted.

Another method of increasing the temperature of the combustion air is found only on the larger installations. You may never be in a position to utilize it, but we shall cover it briefly in case you may in the future expand your operations.

In Section C, we show economizers as being one of the preferred methods of reclaiming heat from the flue gas. This method may in some cases be extended to the addition of another heat exchanger between the economizer and the point of exit for the flue gas. This is possible only if the flue gas leaving the economizer is still hot enough to contain sufficient heat for raising the combustion air temperature without being reduced to the dew point of the lower-temperature flue gas. The temperature of the leaving flue gas must be watched very closely to prevent condensation, as we pointed out in Section C.

To use this method for reclaiming additional heat from the flue gas, it is necessary that some engineering be performed, and we recommend that a firm experienced in supply of this type of equipment be called in. The heat exchangers are known as air preheaters in the boiler industry.

As an extension of the exercise in Section C, we refer you to the flue gas outlet temperature from the economizer, which is 350°F. Table 25-1 shows that for gas, the minimum outlet temperature is 220°F. Obviously, we have some extra heat still available in the flue gas, which would be wasted. From the previous discussion we see that it should be possible to pick up at least one or two percentage points in combusion efficiency by installing an air preheater in the flue gas outlet, downstream of the economizer.

The air preheaters in most common use in industry are the fin-tubed and the plate styles. Pressure is very low on both sides of the exchange surfaces, so relatively light construction is possible.

If the fuel burned produces soot, then soot blowers are required, with all of the other attachments that go with them, such as access doors and drains. Again, the draft will be affected, and the fan must be altered to provide a higher pressure to compensate for the loss through the preheater. Also, in all cases, the flue gas temperature drops with decrease in boiler load, so if the boiler operates for extended periods at reduced output, corrosion in the heat-reclaiming apparatus and the stack may result.

The third method of raising the combustion air temperature is to place a steam or hot-water coil ahead of the draft fan intake. This is best done when excess heat in waste water or steam is available, which would normally be wasted. The draft fan will have to be checked for capacity and power, as the increased resistance on the inlet to the fan will affect the fan's performance, as will also the increased air temperature.

The method of calculating the annual fuel bill savings, and the payout time has already been shown, and we shall not go into it.

H. Reducing Steam Pressure

When the boiler plant was originally designed, the generating steam pressure was probably established with an extra safety factor built in to ensure that at all times the process would be served with steam of sufficient pressure to more than satisfy the requirements. In the past, the safety factors used have been rather high and arbitrarily set, by use of typical pressure ranges as used in most steam plants in industry.

A look at the steam tables in the Appendix, page 394 will show, if you do not already know, that as the steam pressure increases, the steam temperature increases. This also means that the heat loss through the covering on the boiler and the steam lines will also increase. Any steam leaks will be aggravated by increased steam pressure, also. In addition, the stack losses will increase with increase in stack temperature.

This whole picture is indicated in Fig. 25-12. As it is rather difficult to predict what the savings will be, we suggest that you make your own tests, and determine by experiment how much the steam-generating pressure can be reduced, without starving the process end of the plant. To determine the savings in fuel consumption, we suggest you turn to Chapter 25, Section C, for a fairly complete procedure.

With the cooperation of the plant production manager, slowly, and in small steps, reduce setting of the pressure controls on the boiler, observing the reaction at the process end to see what effects the reduced pressure has on the plant output. While doing this, keep in mind that as the steam pressure drops, so also does the system capacity, including the lines, control valves, and fittings. This may set the lower limit for the first approach to the pressure-reduction program.

Keep reducing the steam pressure in stages, allowing the plant process to stabilize each time, until the process appears to be suffering or until the plant production manager decides that the safety factor on the process equipment has been lowered far enough, and any further reduction would create problems. When this point has been reached, spend a few hours observing the results on the boiler, and check places in the system where further pressure reduction could be accomplished by making slight changes in the piping or the valving, and store the data

in your mind for future reference and study.

Allow the plant to operate under reduced pressure for at least a month, then check the fuel consumption and calculate the amount of fuel burned during the month. By comparing the fuel consumption with that for the previous month, and making allowances for any changes in the process or in production rates, an estimate of the fuel savings should be possible.

If the test results show that further reduction in pressure may be allowed, then it is time to suggest that some piping and valves may be increased or changed, and further reductions are possible. It may be possible to rearrange the piping and reduce the line pressure losses.

Perhaps, with the cooperation of others in the plant, a rescheduling of some of the plant processes may produce lower steam usage peaks, and thus permit a lower steam main pressure. If two large demands are coming at the same time, requiring high steam pressure to meet those demands, by staggering the two demands it may be possible to lower the steam pressure.

Another advantage to be gained by lowering the steam pressure produced at the boiler is a reduction in scaling and sooting rates on the boiler heat-tranfer surfaces, owing to lower temperatures.

There is still another advantage possible, depending on the rating of the boiler relative to the load demand. If the boiler is greatly oversized originally, then by reducing the steam pressure, the boiler rating will be reduced to give a better match with the demand. This means that the boiler burner will be operating at a more steady condition, with fewer rapid changes in fuel and air flows required. The burner may "float" on the load better than previously.

Of course, as the steam pressure setting is reduced, the combustion efficiency should be checked and readjusted if necessary. Remember, what you are actually doing is to rerate the boiler to a lower output, which will alter the output-efficiency curve, and the new curve should be determined all over again for future operation under the new setting.

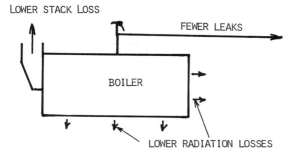

Fig. 25-12. Effect of reducing boiler pressure.

CHAPTER 26
MISCELLANEOUS ENERGY SAVERS

A. General Comments

The previous two chapters on energy conservation and increasing boiler efficiency have primarily been concerned with the boiler only. There are other places in the boiler plant, and other items associated with the boiler operation, which deserve their share of attention, should you embark on a BTU chasing expedition. The boiler does not stand alone, for it needs several very important pieces of equipment to be in operation with it while steam is being produced.

This chapter will focus on several of those areas where some degree of energy conservation may be appropriate. Most of them will be aimed at saving heat energy directly, and some of those will be a direct saving in the use of steam. Either way, there is a return to the firm in cost of fuel saved, although it may in some cases be difficult to predict the projected amount of monetary savings to be expected.

Several of the items covered in this chapter have been taken from a U.S. Government publication, known as the National Bureau of Standards Handbook no. 115, and its Supplement No. 1. The title is, "Energy Conservation Program Guide for Industry And Commerce." This handy booklet contains many more areas of interest beside the steam plant, and it is well worth the cost, which is less than $6.00 at the time of this publication. It is available from the U.S. Government Printing Office, Washington, D.C. 20402. Order by SD catalog No. C13. ii:115 and C13. 11:115/1.

Section I in the Appendix contains several pages of additional heat-conservation hints, including repetition of those considered in these chapters.

B. Heavy-Oil Atomizing Medium

Those of our readers who are burning the heavier oils in the boiler are probably using some of the steam from the boiler for atomizing the oil at the burner gun tip. This has been the usual procedure for years, and it takes about 1% of the boiler's output to do this. The heat in the steam does help to bring the oil up to atomizing temperature, but the water used to produce that steam is a complete loss. In fact, it may increase the amount of trouble due to the condensation of moisture in the flue gas, such as corrosion of the breeching and stack, and any tendency to cause acid rain in the surrounding areas. Obviously, the extra moisture in the flue gas we could very well do without.

Some of the heavy-oil burners using steam for atomization are arranged for starting up on plant air under pressure to provide the atomization until the boiler has produced sufficient steam pressure for the purpose. This is part of the "bootstrap" design built into many boiler burners. Why, then, not continue to use the air for that purpose? There are several advantages in this change:

The air helps in the combustion process, and will reduce the amount of primary air required to be supplied by the draft system.

The cost of the plant air, providing no additional equipment is required, is probably lower than the steam needed for the atomizing.

As previously implied, there is less chance of complaints from the populace in the vicinity of the plant and less corrosion to be expected in the flue gas passages.

If the piping is already installed for the air supply, no additional piping is required. The steam atomizing piping and valves should remain in place, in case the plant air system becomes inoperative.

The air need not be instrument air quality, but reasonable precautions should be taken to be sure the air has all of the water of condensation removed from it before the air reaches the burner.

As usual, we strongly recommend that the burner supplier be contacted and advised of your proposed change. It is possible that he might have an improved design of burner gun for using air as a primary atomizing medium, which may require less air than the present air start-up burner gun uses.

Some further advice concerning the atomization of fuel oil is appropriate at this point.

The final optimum air/fuel ratio depends a great deal on the proper atomization of the oil leaving the tip. Proper atomization in turn depends upon the design of the mixing tip, the oil pressure, and the temperature of the oil. This means that for you to get the most efficiency out of your burner, the temperature and pressure at the tip must be correct. Do not be reluctant to call in your burner representative or the fuel oil supplier and discuss the situation with them. Also, if you change either the supplier or the specification of your fuel oil, you should check the stack gas composition, and the

oil temperature to be sure that the change does not require further changes in your method of handling and burning the new oil.

C. Bare Steam Lines

Regardless of the emphasis in recent years placed on heat conservation, somewhere out there, possibly even in your own plant, there are still many feet of piping, carrying live steam or condensate, without insulation! It may not have been done deliberately, it could very well be that the piping had to be repaired, and there just wasn't time to replace the insulation, so the maintenance crew left it off, planning to come back later and replace it.

Figure 26-1 will show you how much heat is lost for one year for different size lines and at various steam pressures. The loss rapidly builds up to a substantial amount in the course of a year, as you can see. From the cost of producing the heat in the steam from your boiler plant, it is a simple matter of calculating the cost of any bare steam lines in your plant. Perhaps you do not know what the heat in the steam is costing you, but we suggest that it is at least $1.50 per 100,000 BTU. See Chapter 25, Section E for guidance in this respect. Insulation is not 100%

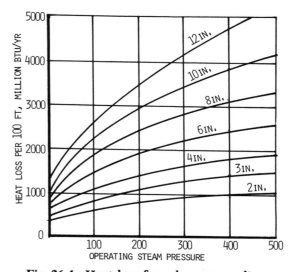

Fig. 26-1. Heat loss from bare steam lines. (Courtesy of NBS Handbook No. 115, Suppl. 1.)

perfect in preventing heat loss, but if properly applied and maintained, it should be at least 95% efficient. Using these basic figures, the formula for calculating the annual cost of heat wasted through bare steam lines is

Annual cost

$$= \frac{\text{Heat lost per year} \times \text{Cost per 100,000 BTU}}{100,000 \times \dfrac{95}{100}}$$

This formula is valid for other evaluation problems, also, providing that the heat lost is given as the total for one year. We shall make use of this formula further in this chapter.

D. Leaking Steam Traps

One of the main sources of steam leaks in the average industrial plant is the steam traps, especially when they are old, improperly selected in the beginning, and subject to indifferent maintenance. One steam trap blowing through continuously can lose large quantities of steam and heat, if not caught soon after the leak starts. Usually the trap hangs open or the seat wire-draws.

Figure 26-2 will show how much heat will be wasted when traps with various size orifices remain open and blow continuously. A simple example will illustrate the point we are attempting to make here.

Annual cost of trap with a 0.125 in. orifice, wide open, under 100 psig steam, discharging

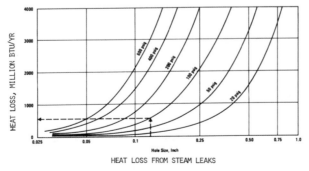

Fig. 26-2. Heat loss from steam leaks.
(Courtesy of NBS Handbook No. 115, Suppl. 1.)

to atmosphere, calculates as follows, using natural gas at a cost of $0.76 per 100,000 BTU, as shown in Table 25-3:

Annual cost

$$= \frac{540 \text{ million BTU} \times (\$0.76 \text{ per } 100,000 \text{ BTU})}{100,000}$$

$$= \$4,104$$

The lesson is obvious; fix that leak! Also, institute a regular program for testing all steam traps and by-passes for leaks and proper operation. In fact, it may be well worth the time and effort to call in the supplier of your steam traps and have him go over your entire system and the method of maintaining the steam traps.

E. Steam and Condensate Leaks

There are many other sources of heat leaks in the steam and condensate system in the industrial steam plant layout. It is a good idea to start a program of "search and correct" for all such leaks, in conjunction with the steam trap patrol. Larger firms have crews assigned to these operations on a regular basis.

We emphasize that the size of leaks, however small, are often deceptive. They can waste enormous amounts of money into the atmosphere if not corrected, as a look at the previous example will prove.

In addition to the heat wastage, there is always the danger of damage to the equipment from the steam and hot condensate, such as corrosion, erosion, and spoilage of insulation. Also, if the steam is highly superheated, there is a very great danger to personnel.

A few of the usual source of leaks to be guarded against and searched out are:

All connections—threaded, welded, and flanged,

All valve stems and packing glands.

All steam trap by-passes.

Reducing-valve diaphragm cages.

Holes in heating coils, both air and other process exchangers.

Condensate and boiler feed pump seals and glands.

Overflows from condensate tanks, heat baths, hotwells, and deaerator storage vessels.

Leaking drain valves to the sewer, and blow-down valves.

These are only a few of the more common ones. Your particular plant system will no doubt have its own unique sources of troublesome leaks.

F. Use of Flash Steam

When condensate is formed from steam at steam temperature, and that condensate is reduced to some pressure lower than the corresponding saturation pressure for that temperature, some of the condensate must flash off into steam again. This is a common occurrence when steam traps discharge condensate at temperatures close to the steam temperature ahead of the trap. Perhaps the best way to explain the process is by an example, and we refer you to Figure 26-3, where the entire process is shown.

The basic heat balance is

Total heat into the orifice

= Total heat out of the orifice

The heat contents given in the figure are taken from the Steam Tables, page 000 in the Appendix, as follows:

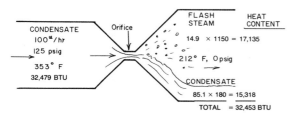

Fig. 26-3. Calculation of flash steam.

At 125 psig and 353°F the condensate contains 324.79 BTU/lb.

At 0 psig, steam contains 1150 BTU/lb.

At 0 psig and 212°F condensate contains 180 BTU/lb.

There is no loss of heat through the orifice, so the formula tells us that the total heat in the condensate ahead of the trap is equal to the total heat in the condensate plus that in the flash steam on the downstream side of the orifice. This is illustrated very well in Fig. 26-3.

Table 26-1 gives the data necessary for determining how much of the hot condensate will flash off into steam when the condensate is passed through an orifice from a higher pressure to a lower pressure. This table is based on the assumption that the hot condensate has just been formed and is at the saturation temperature for its pressure, as determined from the saturated steam tables. For instance, in our example Fig. 26-3, hot condensate is passed through an orifice and reduced in pressure from 125 psig to atmospheric pressure. From the table, under the column headed "Steam Pressure (psig)," find 125 psig, and read across into the first column adjacent to it, and find about 14.9, which is the approximate percentage of flash steam that will be formed when the hot condensate at 125 psig is flashed into a vessel or pipe at atmospheric pressure, as found under the column headed "0," in the "Flash Tank Pressure" section. This 14.9% of flash steam formed tells us that out of the 100 lb of condensate passed through the orifice in the example, about 14.9 lb of flash steam will be formed at 0 psig. The rest of the example is simple mathematics, which we suggest you go through to be sure you understand the principle involved. The values given in Table 26-1 have been rounded off to one decimal place, which is accurate enough to explain the principle involved.

Figure 26-4 shows how this flash steam is put to practical use. In this case, the condensate from the high-pressure-trap returns are all discharge into a flash tank, piped as shown, where the condensate pressure is reduced, and flash

Table 26-1. Reclaiming Heat from Flashing Condensate[a,b]

Steam Pressure (psig)	Percent Flash[a]							
	Flash Tank Pressure (psig)							
	0	2	5	10	15	20	30	40
5	1.7	1.0	0					
10	2.9	2.2	1.4	0				
15	4.0	3.2	2.4	1.1	0			
20	4.9	4.2	3.4	2.1	1.1	0		
30	6.5	5.8	5.0	3.8	2.6	1.7	0	
40	7.8	7.1	6.4	5.1	4.0	3.1	1.3	0
60	10.0	9.3	8.6	7.3	6.3	5.4	3.6	2.2
80	11.7	11.1	10.3	9.0	8.1	7.1	5.5	4.0
100	13.3	12.6	11.8	10.6	9.7	8.8	7.0	5.7
125	14.9	14.2	13.4	12.2	11.3	10.3	8.6	7.4
160	16.8	16.2	15.4	14.1	13.2	12.4	10.6	9.5
200	18.6	18.0	17.3	16.1	15.2	14.3	12.8	11.5
250	20.6	20.0	19.3	18.1	17.2	16.3	14.7	13.6
300	22.7	21.8	21.1	19.9	19.0	18.2	16.7	15.4
350	24.0	23.3	22.6	21.6	20.5	19.8	18.3	17.2
400	25.3	24.7	24.0	22.9	22.0	21.1	19.7	18.5

[a]Percent flash for various initial steam pressures and flash tank pressures.

[b]Courtesy of Spirax/Sarco Co., Inc.

steam is formed. The hot condensate entering the tank will have some flash steam that was formed at the trap discharge, and the reduction in pressure at the flash tank will cause more to be formed, so that in practice, the flash has taken place in two steps. The total result will be the same as if the flashing had all happened in one step, so that the calculations for a one-step process will still be valid.

In Fig. 26-4 we show a pressure-reducing valve feeding steam from a high-pressure source into the flash tank. This arrangement sets the flashing pressure, since the reducing valve is set at the required lower pressure for the end use of the flash steam. Also, when the supply of flash steam drops off for any reason, the re-

ducing valve will open up and feed steam in to supply the demand for the load.

The relief valve is required by code to prevent the flash tank from exploding in case all exits from it should become closed off for any reason. It should be set to open at about 10% below the designed working pressure of the tank.

The size of the flash tank is important, and there is a fair amount of leeway in its size for the amount of condensate passing through it. Most steam trap suppliers will be able to help you size the tank, and they may even have package systems for this service.

There are several advantages to be gained from utilizing flash steam, as follows:

Some of the losses from leaking steam traps will be regained, but this should not be used as an excuse for poor steam trap maintenance.

The insulating of condensate lines is justified, since the heat in the condensate can be retained in the system.

The overall effect will be a saving in the heat energy, fuel consumption, boiler make-up water, treatment chemicals, and reduction in demand on the boiler.

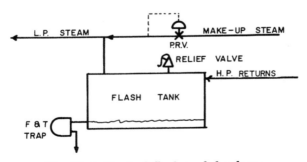

Fig. 26-4. Typical flash tank hook-up.

G. Heating by Direct Injection of Steam

When a fluid, such as water, is heated by a system of tubes in which the steam is condensed and then returned to the boiler, there is always some loss in heat before the condensate gets into the boiler again. In this case, it is well to consider the possibility of heating the water by injecting the steam into the water directly. This is highly advantageous when the distance back to the boiler room is very great.

As an example, we shall take the case of a tank of water at 60°F that must be heated to 160°F in 1 hr, and the tank holds 500 gal. Assume further that the steam pressure available at the tank is 10 psig.

The steam at 10 psig has a total heat content of 1160 BTU/lb. If we inject it into cold water and heat the water to 160°F, the steam will condense and the condensate will then be at 160°F. Therefore, each pound of steam will add its latent heat to the mixture, plus the amount of sensible heat given up by the condensate when cooling from its condensing temperature down to 160°F. This means that each pound of steam will retain 160 − 32 BTU, or 128 BTU, after it is condensed and mixed with the water. If 128 BTU is retained, then each pound of steam has given up 1160 − 128, or 1032 BTU to the mixture.

To find out how much steam is required to heat 500 gal of water from 60 to 160°F by direct injection of 10 psig steam:

Steam required

$$= \frac{500 \text{ gal} \times 8.33 \text{ lb} \times (160 - 60)}{1032 \text{ BTU/lb}} = 404 \text{ lb}$$

If this were a continual process, then it is obvious that some means would have to be used to get rid of the excess hot water, since the continual addition of steam will build up the total mixture until the tank may overflow.

In direct steam injection there is a possibility of producing a final mixture that is contaminated, especially if the steam is far from pure, which is often the case in many industrial steam plants. For instance, some high-pressure plants treat their steam lines with what is called "filming amines." These treatment compounds are poisonous, so should not be used when injecting such treated steam into water that may be used for human consumption.

The steam that is injected into the water is lost and must be made up with additional make-up water. However, for remote intermittent use of hot water, such as for washing-down purposes, the steam injector does an excellent job.

There are several manufactured items on the market designed to mix the steam with water to serve some of the more common applications. Most of them fall into one of the two main classes of applications, either open tank heating or in-line mixing for producing hot water under pressure for hosing-down operations. In either case, one of the main difficulties is obtaining quiet operation, since often the direct mixing causes loud rumbles from the condensing of the steam bubbles hitting the cold water.

The economics of this method is very difficult to perform, and we will not go into it. We suggest that you use it only where there is a definite advantage to be gained, such as in isolated locations, where returning the condensate is difficult.

H. Reclaiming Heat from Waste Streams

Much of what we have covered so far may be lumped under the heading for this section, but we have an idea that will help you get the last bit of heat energy from your waste flows leaving the plant.

Let us assume that you have a considerable amount of hot waste water from the process for your boiler plant that is being discharged to the sewer because it is contaminated. Even though it may be at a relatively low temperature, and at atmospheric pressure, there is still hope of reclaiming some of the heat from it. As long as its temperature is higher than about 20°F above the incoming water to the boiler plant, it is possible to get more heat out of it.

A simple solution is to form a concrete trench or trough in the ground, with the top edge about a foot above the surface, and pipe the waste stream into one end of this trough, and out the other end. The trough forms a holding tank, open to the air. You may cover it if there is a safety problem from its use.

Into this trough you lower a nest of steel coils, carrying the incoming make-up water for the boiler. The flow should be counterflow, with the make-up water coming in the end where the waste stream leaves the trough, to enable you to get the maximum heat from the stream before it goes to the sewer. The larger the trough volume, the longer time is available for transfer of heat into the makeup water coils.

As the process or production department is directly involved in this proposal, we suggest that they be consulted as to the advisability of attempting it. There may be something in the waste stream that could cause problems, such as toxic substances that could contaminate the make-up water if the coils leak. Also, the design of the coils, the holding trough, and the materials required should be decided by the department producing the waste stream.

If in the final analysis, the project is left up to you to design and carry through to completion, then we suggest you call in one of the suppliers of preformed heat exchanger panels. They will no doubt be able to help you design the combination to accomplish the best possible solution.

This section has concentrated on one typical situation, but we do not mean to limit your thinking toward this lone example. It should give you a chance to expand your thinking to other heat-reclaiming horizons throughout your plant.

CHAPTER 27
LOAD SHARING

A. Scope

By load sharing we mean that two or more boilers are used to carry the plant steam load. There are several ways in which this may be done, and we shall cover most of them. The boiler plants we are discussing in this chapter are such as to include the small manually controlled multiunit installation to the use of expensive and sophisticated loading controls on large industrial boilers. Most of the steam boiler plants in industry are handling the loads with one or two boilers only. There are some, it is true, that have more than two boilers on the line, but when that happens, at least one is usually considered a spare. There are exceptions of course.

The problem of load sharing or portioning the demand between the large industrial boilers is usually solved by sophisticated electronic or pneumatic systems, using a central master load distribution panel. This method has built-in control logic requiring inputs from numerous monitoring points in the steam-generating plant, such as steam flow, fuel flow, or combustion air flow, depending on the size of the steam demand. Several boiler control suppliers have developed this field to a high degree recently. There are also a few control specialty firms that have addressed themselves to the medium-sized boiler field, producing compact control systems permitting the operator to distribute the total plant steam load between the available boilers, depending on the efficiency curves of each unit.

Therefore, we shall confine the discussion in the remainder of this section to those smaller units that so far have generally been ignored by the control designers and developers. We feel that the smaller industrial firms with relatively smaller steam demands are also interested in conserving fuel dollars.

We do not mean to imply that the load-sharing methods given in this chapter are the only ones, or the best to be found in industry. From our observations, however, we feel that it represents the most logical first approach for those operators of small to medium-sized boiler plants who have received little or no help from other sources.

B. General Principles

The operator should maintain all boilers in equal condition, and each should receive an equal share of the maintenance time and money. Play no favorites is the rule. Also, the boiler duties should be rotated at regular intervals so that all boilers receive the same amount of use.

The best general rule to follow is as follows: at all times, each boiler in a battery should be operated at as high an efficiency as is possible to attain, regardless of the load being carried, until the load reduces to the point where fewer boilers can handle the load at a higher total efficiency. This means that the operator should avoid falling into the trap of favoring one boiler over any other in the line. When checking and adjusting the boilers for peak efficiency, every boiler should receive equal attention. Only in this way can the highest overall economy in fuel consumption be obtained.

Should boilers have different ratings, then the same general rule still applies; each boiler should be operated as close as possible to its peak efficiency point for the load conditions. The best procedure here is to use the most efficient boiler for the steady, or base, load, and then rely on the less efficient boiler to carry the variable, or swing, load. Unfortunately, this practice too often leads to ignoring the swing load unit, as being not worth the effort to maintain its efficiency, which could cause it to fail just when it is needed the most.

In plants with variable loads, the plant rhythm usually follows much the same pattern from day to day, so that the operator is soon able to predict when the load changes require a change in boiler operation. The load should be watched closely, and when the proper time comes to cut one boiler in or out, the operator should be sure the result will still place the boiler or boilers in the best efficiency point possible for the load. This is going to require some thought and close attention, and it means that the operator must know his equipment very well, and know the efficiency curve for each boiler.

Let us now consider the plant with a steady base load, with one or more short periods during the day when the load runs at a greatly increased rate. This type of load should preferably be handled by more than one boiler. Two boilers at least should be installed, if at all possible. The relative size of the two boilers will depend on the load pattern, and the difference between the various loads during the day. The style of boiler installed to handle the peak load will depend on several factors, such as the speed with which it must be brought on the line, the length of time it will be on, the quality of the steam required, the location in the plant where the increased steam load is needed, the ability

and availability of the operator to bring it on the line and shut it down when it is no longer needed, etc.

What we are attempting to point out in this section is the following:

1. All boilers should, at all times, be operated as close to their peak efficiency point as is possible for the load, regardless of their actual rated capacity.

2. All boilers installed on the line should receive equal attention and care in order to achieve the maximum economy from the plant.

3. The operator should be familiar with the efficiency curve for each boiler under his care.

4. The operator should be in tune to the plant's consumption demand and know its rhythm of load swings.

C. Methodology

The steps required to place the available boilers into a mode of operation which will produce the most efficient and economical fuel consumption will be given here, then discussed in subsequent sections in more detail.

1. The first step is to determine, as accurately as possible, the amount of fuel being consumed by the boiler plant under the existing conditions, for a period of one week. This establishes the basis for calculating results from fuel-saving efforts to follow.

2. The next step is to develop a graph of the load pattern for the plant. This illustrates the steam demand required to satisfy the plant production department and defines the basic problem.

3. The next step is to obtain, if possible, a reasonably accurate efficiency curve for each boiler that will be applied to the system. These curves should then be plotted

on one grid consisting of efficiency versus steam demand similar to Fig. 27-2.

4. The plant load pattern must then be divided into time and load segments that can be handled by one or more of the boilers available. This is done by studying the boiler efficiency curves plotted in Step 3, with the Plant Load Pattern in Step 2, and completing Fig. 27-1.

5. The plant steam-supply pressure control range must then be separated into two or more subranges within that total, one range for each steaming boiler, and the individual boiler steam pressure controls set for the appropriate on and off settings. This is the same philosophy followed for properly setting two steam-pressure-reducing valves in parallel, with the boilers replacing the function of the reducing valves.

6. The plant is then put into operation, and the fuel consumption is monitored for one full week of operation. This total week's consumption is then checked against the original consumption determined in Step 1.

7. Should the cycling of any of the boilers on the line be excessive or the fuel consumption rate not what was expected or the plant demand not be met, then the pressure settings for each boiler involved should be altered, and the results checked for another week.

8. When the results achieved are satisfactory, then the next step is to calculate the savings in dollars, according to the method shown in Chapter 25, Section B. This figure should then be submitted to the management for their edification.

D. Plant Load Pattern

Now that we have set the ground rules and established the methods to be used, we shall illustrate them by means of hypothetical examples.

To determine the results obtained in our efforts at fuel saving, it is essential that for one week before making any changes the fuel meter readings be taken on Monday morning before starting the boiler plant and on Friday evening after shutting down the plant. This is the existing fuel consumption for one typical week and is required for checking the results of the program in fuel reduction.

The next step is to construct a graph, as in Fig. 27-1, showing the daily load pattern for the plant.

This is a simple matter should there be a steam flow meter installed in the supply line to the plant. It then becomes a matter of reading the meter every half hour during the time steam is required, either daytime or 24-hr periods. These half-hour readings are then converted to hourly flow rates, spotted on the graph, and a block load pattern drawn as shown on Fig. 27-1.

In the absence of a steam flow meter, the next possibility for determining the daily load pattern is from fuel meters to the boilers, assuming there are meters for each boiler, or one meter for all boilers on the line. The same procedure is followed as for the steam flow, taking readings every half hour, then calculating the total fuel flow for each half-hour period, converting to hourly flow rates, spotting the readings on the graph as before, and blocking in the load pattern.

The actual steam demand pattern will not be exactly as blocked in on Fig. 27-1, but will be jagged, showing many peaks and valleys. However, the method used here will suffice for the purpose, as long as the readings are taken fairly accurately.

Figure 27-1 has both steam-demand and gas-consumption scales for this example, based on expected efficiencies and gas at a heating value of 1050 BTU/ft^3.

Should the boiler operator be so unfortunate as not to have either a steam flow meter or fuel meters assigned exclusively to the boilers, then it is necessary for the operator to approximate as well as possible a rough load pattern on the

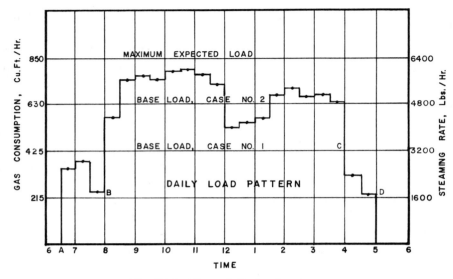

Fig. 27-1. Plant daily load pattern.

graph. Knowing the boiler capacities and the plant operating schedule or rhythm, plus the times in which the boilers cycle on and off, it should be possible to sketch in a rough plant load pattern. From this, it becomes a matter of simply experimenting with various boiler pressure-control settings, as determined from the examples that follow.

Regardless of the accuracy in establishing the plant load pattern, the operator must be willing to experiment and to use a great deal of patience in arriving at the best operating combination from those boilers available.

E. Efficiency Curves

It is highly desirable to have efficiency curves available for the boilers being considered, but it is not entirely necessary. We show them here, Figs. 27-2 and 27-3, for the boilers used in the examples that follow.

If a complete efficiency curve for each boiler in the system has not been established by application of the principles established in the previous chapters, then we suggest that the boiler suppliers be contacted for typical test curves for similar boilers of the type, size, and design being considered in this program.

The examples that follow are all based on two sizes of boilers for Cases 1, 2, and 3. The boilers are assumed to be 100 hp and 150 bhp to han-

dle the load pattern on Fig. 27-1. Their efficiency curves are shown on Figs. 27-2 and 27-3. They are fired with natural gas with a heating value of 1050 BTU/ft^3, and the steam is supplied to the plant between 110 and 125 psig.

Note that where the two efficiency curves cross on Figs. 27-2 and 27-3, a vertical line marked A–A has been drawn. This marks the crossover point for switching boilers as the load changes, in order always to keep the highest efficiency unit on the line. Line A–A represents the ideal situation; with the usual lack of controls on the smaller steam systems, this point is difficult to spot.

Figure 27-2 contains the efficiency curves for the two boilers being considered for Cases 1, 2, and 3. Figure 27-3 contains the curve for a 150-hp boiler, plus the curve for two identical 150-hp boilers operating simultaneously. This is done by first plotting the curve for one boiler, then extending each point, *a, b, c, d,* etc., to the right, an amount equal to the increase in capacity for the additional boiler, as shown by *a', b', c',* and *d'.* Each extended point is at the corresponding efficiency point for the single-boiler curve. This establishes the points on the curve for the combined boilers operating together. The same process may be used for three boilers operating together, if it is desired or necessary.

A boiler will always operate on its curve, regardless of the combination of boilers on the

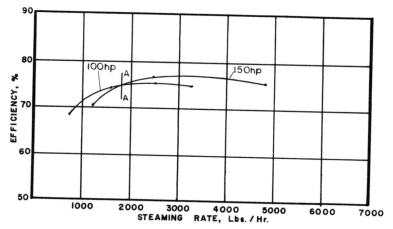

Fig. 27-2. Typical boiler efficiency curves, 100 hp and 150 hp.

line at any one time. Thus, on Fig. 27-2, if the 150-hp boiler is putting out 3,000 lb/hr of steam, it is operating at about 78% efficiency.

The final goal, as stated in Section B, is to have each boiler ride its own efficiency curve at the highest point possible for the time it is operating. This means that the boiler controls should be capable, ideally, of determining when it is time to switch from one boiler to another at higher efficiency. That is the goal toward which we are striving in this chapter, and it is the basis for all control systems designed to achieve the lowest fuel consumption from the boiler plant.

F. Dual-Boiler Operation

Case 1 consists of two 100-hp boilers, identical in design, handling the load pattern in our example, as shown on Fig. 27-1. The base load unit is assigned unit No. 1, and the swing load unit is assigned unit No. 2. From a study of the load pattern, it is easy to assign the units to follow the load pattern and at the same time produce the highest available efficiency. For instance, for the start-up loads from 6:30 AM (point A) to 8:00 AM (point B), and the wash-down loads between 4:00 PM (point C) and 5:00 PM (point D) only one boiler needs to be on the line, as it is capable of 3200 lb/hr of steam. At 8:00 AM (point B) to 4:00 PM (point C) it will require both units to handle the load.

As both boilers are identical, there is no need

to make a hard decision as to which boiler is to be No. 1 and which is to be No. 2. Likewise, it makes no difference which one is used to handle the start-up or the wash-down loads. It is important, however, to alternate services between the two boilers, so that both receive the same amount of use. Of course, if the two boilers are not identical, then we have a Case 2 situation, as described subsequently.

Note, that in this example the load split between the two boilers results in both units operating continuously between 8:00 AM and 4:00 PM, the actual production plant day. This is usually the most desirable situation, since it often results in higher total efficiency over the load split in which one boiler cycles on and off frequently. Experimentation alone will determine if this is the case with your plant.

Assuming that steam must be supplied at pressures between 110 and 125 psig, the first approach would be to set the pressure switch controlling unit No. 1 to operate between 115 and 125 psig. Unit No. 2 would then be set to operate between 110 and 120 psig. Checking the fuel consumption for one week will show what effect the new settings will produce.

After the initial change in pressure settings has been made, and the system placed into operation, observe the boiler operations carefully to ensure that neither unit is cycling on and off excessively, which indicates poor economy, as shown in Chapter 25, Section B. Some experimentation may be necessary to produce the best split in operating pressures.

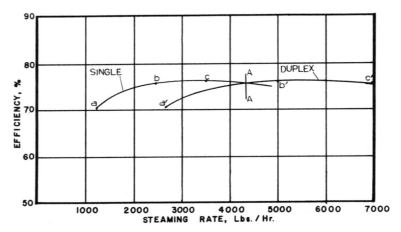

Fig. 27-3. Boiler efficiency curves, dual 150-hp units.

Case 2 is the same as Case 1 except that the base load unit No. 1 is larger than swing load unit No. 2. In this example, unit No. 1 is 150 hp, with a steaming capacity of 4800 lb/hr. The maximum capacity of unit No. 1 automatically becomes the base load, as shown in Fig. 27-1.

The first approach here is the same as for Case 1. As the load is being handled by a different combination of capacities, the split in range between the pressure settings on the boiler control switches may have to be altered, which will be determined by experimentation, as before.

There is a possibility here for another small increase in operating efficiency. Note that the large capacity of unit No. 1 results in the swing unit being forced to cycle frequently, to operate at lower capacity, and thus at lower total efficiency. By consulting with the burner or boiler supplier, it may be possible to alter the burner capacity on unit No. 1 to lock the unit into operation at no higher than about the capacity of the 100-hp unit. The system then becomes Case 1, with a higher total average operating efficiency than the basic Case 2 situation just described.

Case 3 is similar to Case 1, except that both boilers are 150-hp capacity, which means that the first approach to cutting fuel costs is the same as for Case 1. Note, however, that both boilers are too high in capacity for the most efficient combination to be attained. Regardless of which unit is assigned to the base load, the swing load requirements are so small as to

force the swing unit No. 2 into a very poor efficiency range, and to cause it to cycle on and off frequently.

The best solution here is to reduce both boilers to about 3,500 lb/hr of steam, which brings the system into Case 1, and the approach should be altered accordingly.

Another solution, but a poorer one, is to reduce the output of only the swing load unit to about 3500 lb/hr of steam, which produces a system equal to Case 2, first approach.

G. Three or More Boilers

At this point the problems involved are getting more complicated, and will require more ingenuity on the part of the boiler plant operators. This calls for more willingness to experiment, but the payoff in fuel savings will no doubt be worth the effort, since we are dealing here with larger steam-generating plants and larger fuel-consumption rates.

In the cases given in this section, as well as in all of the previous cases, there are two more general concepts to keep in mind, in addition to those already stated in Section B. If all else fails, following these concepts will usually produce favorable results:

1. Once a boiler is brought on the line, it should be operated continuously at outputs as close as possible to its peak efficiency point on the curve.

2. Most industrial boilers should never be operated at below 50% of its rating for any length of time, owing to the resulting poor efficiency.

3. Every time a boiler cuts in and out, it does so with a considerable loss in efficiency. See Chapter 25, Section B for the rationale for this.

Case 4: In this situation, there are three equally sized boilers to handle the plant load, and all three are required during the day's operation. The first approach is to assign one unit half of the total base load, and label it base load unit No. 1. It is expected that this boiler will be on the line at all times during the steaming period.

The second boiler will be assigned as base unit load No. 2, and it will carry the portion of the plant load next in size above unit No. 1. It may cycle occasionally during the day, but not excessively.

The third boiler would then be assigned as swing load unit No. 3, and it is expected to carry all portions of the load that fluctuate above the two base loads.

Obviously, by studying the plant load pattern on Fig. 27-1, the best arrangement would be by the use of three 75-hp boilers, with a capacity of 2400 lb/hr of steam. Also, it is obvious that by the concepts already covered, any boilers larger than 75 hp would result in a system that would be difficult to balance. Boilers of 100-hp capacity could be accommodated, but above that size inefficient operation would result.

The pressure switch settings for each boiler would probably be about as follows, subject to experimentation, of course:

Base Load Unit No. 1	On at 120 psig, off at 125 psig
Base Load Unit No. 2	On at 115 psig, off at 120 psig
Swing Load Unit No. 3	On at 110 psig, off at 115 psig

Applying these three boilers to Fig. 27-1, base load units Nos. 1 and 2 would be on the line from 6:30 AM to 5:00 PM, and swing load unit No. 3 would be on the line from 8:00 AM to 4:00 PM.

An alternative approach would be to apply Case 1 pressure setting by having both base load units Nos. 1 and 2 operate in the 115–125 psig range, and swing load unit No. 3 operate at 110–120 psig. One week's check on fuel consumption should determine which method is best.

Case 5: All boilers are of different sizes, age, condition and efficiencies. The first approach is to determine the best combination of units to handle the plant load, taking into account the principles and cases given so far. If the plant load can be handled by two or three boilers, and this leaves one extra, start with the most efficient, and reserve the least efficient one as a spare. Then attempt to work the chosen boilers into one of the cases covered previously, and use that case as a guide.

The chosen boilers are then assigned their place in the system by setting each pressure control switch in the manner described for Case 4.

Regarding the cold spare unit mentioned, we stress the importance of keeping it in proper condition at all times it is off the line. Also, we recommend that it be operated at least one week out of every four, by working it into the plant boiler battery. This would probably best be done by using it as an alternate to the next larger boiler in the system for purposes of periodic cycling.

During the periods of extended shut-down, the procedure described in Chapter 17, Section D should be followed for the cold spare unit.

If all boilers available are required to handle the peak load, then the problem becomes one of determining the best combination of units to be assigned to the various portions of the plant load pattern. This will require considerable ingenuity and experimentation. It may require additional controls, and, in fact, this situation would be an excellent time and reason to requisition an improvement in the control system to permit one of the more advanced load-sharing control systems to be installed.

CHAPTER 28
COGENERATION

A. Historical Beginnings

In the 1970s there arrived on the industrial scene a concept of producing thermal and electrical energy, known as "cogeneration," for use in the nation's industrial plants. The effects on the boiler plant operator in those plants utilizing cogeneration equipment are profound, and in this chapter we shall explore the ramifications to be expected.

First we shall very briefly explain what cogeneration is, and how it came about. Then we shall briefly describe the basic systems available, the operating and maintenance factors to be encountered, and some of the variations to be found. Our discussion will only be sufficient to give an understanding of each type of installation, and will not be deep enough to make the boiler plant operator an immediate expert.

Finally, we shall express some thoughts on what the present boiler plant operator can expect in his future occupation, as a result of this new concept.

From around 1850 to about 1900, rapid strides were made in the use of power in our factories, most of which were still on a self-contained basis. By this, we mean that fuel and operating supplies were fed into the power plant, from which power was generated or produced only as needed for the immediate purpose or demand, with no excess available for outside distribution. Without realizing it at the time, all of these plant power systems were a form of what we know today as "total energy systems," a term that was not used until the 1960s.

By 1900 the first inkling of what we know today as "cogeneration" existed in some of our large cities and industrial plants, but it was not known by that name. The trend started with the advent of the Edison System of generating two voltages of direct current, and distributing electrical power throughout the building, the firm, and then into close neighbors who needed the electrical power. The prime mover for the generators were reciprocating steam engines, usually exhausting at low pressure into steam mains delivering the low-pressure steam into the heating and process systems. Later, when alternating current equipment was developed, the distribution mains for the electrical power could be extended to great distances, from which evolved the central station power plant and the electrical utility grids that we know today.

Thus was born the first topping cogeneration plants, named from the simultaneous generation of electricity and thermal energy for use in the plant processes. The terminology was not appended to the process until the 1970s, and the term "topping" refers to the fact that electric power is generated by the prime mover as a primary function and the thermal energy discarded by the prime mover is then used for the plant processes. The other arrangement, where the electric power is generated from by-product steam, is known as "bottoming power."

For about 75 years, large plants continued to generate most of their own electrical power, first by reciprocating steam engine, then by steam turbine. The result was excellent overall plant thermal efficiencies, while the central power stations were producing electricity at about 25–30% thermal efficiency delivered to the plant transformer. The energy crisis of 1974 caused a new look to be taken at the total thermal energy situation in the country. This resulted in Congress passing five laws encouraging our industrial plant and public utility managers to cooperate in raising the overall combined efficiency in using our thermal resources. The basic provisions were:

Industries generating surplus electricity have a right to deliver some of it into the local electric utility grid, and receive a fair price for it.

This does not make the industry subject to the laws governing public utilities.

Electric utilities must provide backup power or standby power to the cogenerator at reasonable rates and policies.

Each state was directed to establish the rates and policies under the applicable state Public Utility Commission, or similar body.

B. Justification and Selection

What makes cogeneration so attractive to an industrial plant?

Steam or hot water is produced in the average industrial plant at about 75–85% thermal efficiency. The electricity purchased from the power company is generated and delivered to the plant at about 30% or less, which means the low thermal efficiency is being paid for by the plant and other subscribers. When the plant generates its own electrical power, then uses exhaust or extracted steam to provide the plant thermal requirements, the overall thermal efficiency becomes about 75–85%. Thus, the to-

tal energy cost is much lower for the plant with this arrangement.

It is doubtful that the boiler plant operator will be in a position to help make the decision to convert the plant to cogeneration. The costs involved are such as to require a complete analysis of the plant's energy uses and an evaluation of the costs versus the savings. However, a brief listing of the major factors involved will be listed here:

Favorable rates from the power company for peaking and buy-back electricity.

Reliable, low-cost fuel supply of proper quality.

Proper balance between thermal energy demands from the plant, the electric power being generated, and the control or disposition of the surplus thermal energy and electric power.

Recovery of capital (payout time) in less than five years.

Availability of financing for the project.

The number of shifts the plant is operating; the continuous seven-day, three-shift mode being the most advantageous to the use of cogeneration; the single shift, five-day mode will very seldom prove economical.

As can be seen, selecting the proper arrangement of cogeneration equipment to match the plant load pattern is complicated, and is usually done by a computer. Fortunately, there are several combinations of equipment, options, and basic methods to solve the problem.

Figure 28-1 is a typical daily load pattern for the electrical demands of a prospective plant installation. We shall now examine it as if you were selecting a cogeneration plant. This examination is strictly basic, with no attempt to be definitive or final; it is only an illustration of one of the processes involved in making the decision.

Line A is the base load for the plant. A cogeneration unit sized for this load would run continuously, generating the base load, with all

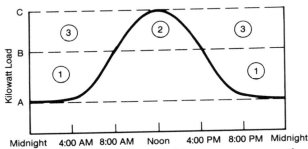

Fig. 28-1. Typical electrical load profile versus cogeneration options.

of the excess load between A and C purchased from the local power company.

Line B is probably the optimum capacity, with the plant selling power to the local power company in areas 1, and buying it back in area 2. An ideal balance would consist of the total of areas 1 being equal to area 2. This would mean that the plant is getting all of its outside power for practically nothing, providing the power company sets favorable rates for selling and buyback power blocks.

A third possibility is to size the cogeneration unit to provide only the peaking demand from B to C, with all power up to point B being purchased from the power company. The justification for this situation would primarily depend on the power company's rate-setting formula for the peaking power between points B and C, should there be no cogeneration installation.

Line C is the maximum the plant requires, in which case a cogeneration unit sized for this amount would provide all of the plant's electrical load, with much excess to sell to the power company, that shown in both areas 1 and 3.

Still a fifth possibility is for the cogeneration unit to be sized to handle the plant load from A to C, with the power company supplying the base load at A.

C. Current Trends

Not all states have seen fit to follow the lead of the Congress in solving our energy crisis. The response from all 50 states has been a complete range from no response to a highly enthusiastic one. The state with the best track record is California, which now has about 45% of the total installations in the country. About a third of the states have done little or nothing to encourage cogeneration.

The response from the electric utilities has been of a similar nature, with some finding it very attractive to cooperate in order to minimize their capital expenditures for new generating capacity. Others are not hurting for capacity, and see little reason to encourage industrial plants to invest in cogeneration installations.

The impetus in those states favorable to the switch-over has produced a wide range of package units with built-in options available to match the individual plant-load demand patterns. There are very few large installations, owing to the large amounts of capital required to make the change. Some package units are being built into new construction, which is the lowest cost method, while other plants have converted existing systems into cogeneration units.

Financing of the new installation is often a problem for the average plant. Large amounts of capital are not always easy to find. This has caused the formation of a number of third party combinations being brought into the picture, with sufficient risk capital available to finance the feasibility study, promotion, construction, operation, and maintenance of the entire cogeneration facility under a contract to supply the electrical and thermal energy to the plant at specified rates. Thus, the savings from using the cogeneration process is split between the plant and the third party combine.

The systems now in general use for cogeneration are:

Steam-turbine prime mover topping units

Gas-turbine topping units

Internal-combustion-engine prime mover topping units

In addition, there is continuing research in solar energy, fuel cells, and geothermal energy; therefore, in the future there may be a larger field to select from in evaluating cogeneration possibilities.

There are variations available in each of the three major systems, which shall be covered briefly under discussion in the following sections.

that is the purpose of the extraction steam. The low-pressure steam can be used for plant heating, water heating, or absorption air-conditioning systems. The low-pressure steam could also be recompressed and fed into the intermediate-pressure line, which will be described later.

The high-pressure steam is generated in a conventional boiler plant, and the condensate system and all auxiliary equipment required for the conventional boiler plant would be essentially the same as for the systems handled in the preceding chapters.

Note the switching station and the control loop governing the output or the infeed of power from the electrical utility grid. This control is described in Section I.

D. Steam-Turbine Topping Cycle

Figure 28-2 shows the basic steam turbine topping cycle for the typical cogeneration installation. The system is one which can very easily be converted from an existing plant where the turbine–generator has been designed to produce only the electric power the plant requires without exporting any of the excess offsite. This system also lends itself very well to conversion when a plant expansion is being planned.

This system has a fair amount of flexibility in balancing the thermal energy demands with the turbine–generator output, since the extraction intermediate-pressure steam can be varied to suit the process requirements, because

E. Gas-Turbine Topping Cycle

Figure 28-3 is a simplified diagram of the basic gas-turbine topping unit cycle. Note that the generator, switching station, control, and connections to the utility grid and plant's electrical system are the same as for the steam-turbine system. This is also true for all of the cogeneration systems, varying only in the size and complexity of the generator, control, and switchgear to meet the actual plant and system requirements. In the system shown, there is a speed-reduction gear required to reduce the turbine speed to the more normal generator speed, since the gas-turbine speed is much

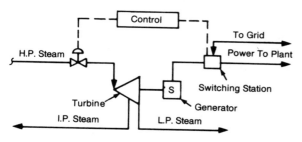

Fig. 28-2. Basic steam-turbine topping cycle.

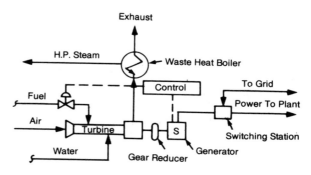

Fig. 28-3. Basic gas-turbine combined topping cycle.

higher than the generators are usually operated at.

In the basic system shown, the industrial-style gas-turbine unit pulls in and compresses several times as much air as is needed for the combustion process. The excess is bypassed around the combustor, and is injected into the exhaust stack, reducing its temperature to between 800 and 950°F. This temperature is sufficient to produce about 150 psig steam in a waste heat boiler, resulting in an exhaust to atmosphere of about 300°F.

The combined efficiency of this cycle varies between about 70% and 85%, depending on accessories, size, and output relative to the system's rated capacity.

The water injection shown is usually required for suppression of contaminants in the exhaust to meet emission standards.

The waste heat boiler may also be used to produce hot fluids, such as water, oil, water/glycol mixtures, or any of the other process fluids found in industry. The waste heat boiler is subject to all of the servicing and maintenance problems usually associated with fired boilers. Also, it is capable of being operated with economizers and superheaters, all of which are of the conventional design.

One additional possibility is a booster firing chamber between the turbine exhaust connection and the waste heat boiler that consumes the excess oxygen in the exhaust and raises the temperature of the gases entering the waste heat boiler, which increases the thermal output and the final temperature capability of the water or steam produced. Steam pressures of 600 psig are possible with this arrangement.

F. Internal-Combustion-Engine Cycle, Ebullient Cooling

More prevalent in the average to small-sized plant is the cogeneration package designed and installed based on the internal-combustion en-

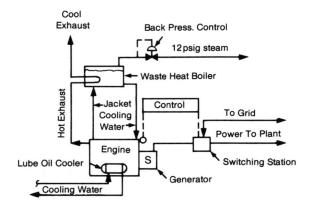

Fig. 28-4. Basic internal-combustion-engine cycle, ebullient cooling.

gine as the basic prime mover. Figure 28-4 shows the typical elementary cycle of such a system, in which the engine jacket cooling is attained by using the enging heat rejection as a source of heat for producing steam. By regulating the steam pressure with a back-pressure control valve, the engine jacket temperature is maintained at a constant temperature, which is the saturation, or boiling, temperature of the water; in the case shown it is 250°F. Note that all of the engine heat rejected and most of the heat in the exhaust are recovered in this system. The exhaust temperature leaving the engine is usually around 1000–1200°F, and it should not leave the steam separator at temperatures low enough to cause condensation in the discharge stack, usually about 300°F minimum.

Very seldom is an attempt made to use the lube-oil-cooler-water heat rejection as a source of heating the process water, since the two demands seldom match, and the supply of cooling water to the lube oil cooler must be maintained constant. The same is true for the cooling water for any turbocharger aftercooler installed.

The steam-generation equipment is subject to all of the problems found in the conventional boiler system of the same pressure rating. The water treatment is especially critical, since any scaling tendency will show up on the inside surfaces of the engine jacket, and is very difficult to detect and remove.

There are many variations of this system available in packages suitable for adaption to the plant requirements. Some are packaged on a skid with an absorption chiller for cooling or air-conditioning service.

G. Plant Hot Water from an Internal Combustion Engine

For those installations in which only a continual source of hot water is required, Fig. 28-5 shows the basic cycle for this situation. The engine and the heat-recovery muffler for the exhaust simply replace the conventional boiler in a hot-water circulating system, with a few refinements to satisfy the constant demands of the engine and balance the loads between the engine and the plant.

Note that it is a closed-loop system, in which the same water is constantly being recirculated through the system, and the plant's demands are being met from a heat exchanger. This minimizes the possibility of scaling of the engine jacket cooling surface.

There is a three-way valve in the hot water line to the heat exchangers that controls the flow of water into the jacket and maintains a constant outlet water temperature from the jacket.

The load balance heat exchanger "floats" on the system, to maintain a constant source of hot water to the plant and to absorb any excess heat rejected by the engine jacket. In short, the closed-loop system is designed to supply the plant load demands while at the same time regulating the engine jacket temperature.

The remainder of the system is self-explanatory, and similar to the requirements and restrictions given for Fig. 28-4 in regards to the engine and exhaust limitations.

The closed-loop hot-water cycle is similar to the normal hot-water systems already covered elsewhere in this book. Please refer to them for maintenance and operating problems, which are of a similar nature to Fig. 28-5.

H. Bottoming Power

In Section A of this chapter we briefly described the difference between "topping power" and "bottoming power." All four cycles described so far have been assumed to be topping power, if we make the assumption that the steam turbine in Fig. 28-2 is driven by high-pressure steam produced in a conventional-style boiler. However, should that high-pressure steam be produced in a waste heat boiler as found in many refineries, chemical plants, and other large users of thermal energy, then the steam-turbine cycle shown becomes bottoming power. Also, note that under certain conditions the steam-turbine cycle may be provided with high-pressure steam from the gas turbine cycles

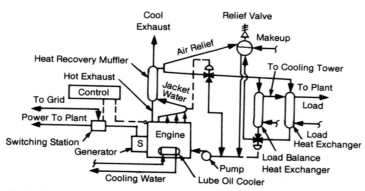

Fig. 28-5. Basic internal-combustion-engine cycle, hot-water supply.

shown, which makes the steam-turbine cycle a bottoming cycle.

There is yet another variation to be found in industry. Quite frequently in the past, the prime mover, usually a steam turbine, has been used to drive some major item in the plant, such as a centrifugal blower, rather than an electrical generator, as shown in Fig. 28-2. In this case, intermediate-pressure steam extracted from one of the middle stages of the turbine may be used to drive a generator in a bottoming cycle. By the same token, the cycles shown in Figs. 28-3 and 28-4 may also be applied in a similar manner. If the internal-combustion engine or the gas turbine drives a blower, with the resulting steam being used to drive an electrical generator, then the system becomes a bottoming power system. This arrangement will not be found very often, however.

The preceding functions are accomplished by signals between the control system and the switching station, which may be in the same cabinet or in a separate panel. The switching station usually consists of transfer switches, disconnect switches, circuit breakers, and any other related equipment required.

The boiler plant operator will probably have nothing to do with this part of the cogeneration plant. However, he should be aware that at any moment the cogeneration plant may be disconnected from the plant system or the utility grid, without any warning and with no indication as to the reason. The complexities of the modern electrical utility grid system are such that the electrical utility is not going to let the "tail wag the dog," so to speak. When the cogeneration system is proposed, the electrical utility will usually dictate in very strong terms the circuitry in the control cabinet.

I. Control Systems

One of the keys to the success of cogeneration has been the advent of precise control of the prime mover, the power output, and the intertie with the utility grid supplying outside power. Today's package systems usually include a sophisticated electronic control cabinet, which must be maintained properly by qualified servicemen. These cabinets contain sensors, meters, relays, control feedback, and signals. The functions that the control cabinet perform consist mainly of the following:

1. Maintain the electrical output voltage, frequency, and phase relationships compatible with the plant's demands and that of the electrical utility grid.

2. Govern the speed of the prime mover to attain the above goals.

3. Sense, meter, and control the flow of electrical power among the cogeneration unit, the plant, and the utility grid.

4. Protect the plant and the utility grid from faults in either system.

J. Upgrading Low-Pressure Steam

In the attempts to balance the steam output from the cogeneration packages with the plant demands, some designers have added steam compression to the systems. This concept was investigated and put to use when vapor-compression distillation plants were developed in the 1940–1960 era, with some success.

In the modern cogeneration package, the application consists of the addition of a compressor that boosts the low-pressure-steam output to any desired level required for the plant. Referring to Figs. 28-2 and 28-4, you will note that large quantities of low-pressure steam are produced, and in the case of the steam-turbine exhaust, the steam may be as low as 5 psig and the steam may be very wet. In the case of the ebullient cooling system, the 12-psig steam may have small amounts of moisture entrained in it, but it is usually drier than the turbine exhaust.

Compressing this low-pressure steam produces high-pressure steam that may have varying degrees of superheat. This may not be suitable for use in the plant processes, especially if the steam is used in heat-exchanger equipment, since superheated steam has a much lower heat-transfer coefficient than that for saturated steam. (We refer you to Chapter II for a review of the properties of superheated steam.) It then becomes necessary to reduce the superheat to close to saturation temperature, by one of the means described in Chapter II.

Starting with very wet low-pressure steam, the most efficient and cost effective method is with the helical screw compressor, and this is most likely the one to be found in today's cogeneration system. If the compression ratio is fairly low, then the old-style lobe-type blower may be used.

If the low-pressure steam is of high quality, and volumes are beyond the range of the helical screw blower, then centrifugal blowers may be used, often with intercoolers between the stages to reduce the output temperatures to close to saturation. The blades of the centrifugal compressor rotate at high tip speeds, and any entrained moisture in the steam has an eroding effect on the blades. This is usually solved by recirculating some of the superheated steam from the discharge into the suction line.

Another characteristic of the centrifugal blower limits its turndown ratio to about 50% of maximum capacity. Below that point, a condition known as "surging" may occur.

Reciprocating compressors have been used for this service, but not to the extent that the three mentioned previously have been used. Maintenance on the reciprocating compressor is usually higher than for the other two styles. Also, regardless of how much the designers attempt to make them oil-free, there is always a chance that reciprocating compressors will contaminate the steam with oil. As we mentioned earlier in this book, oil in the boiler causes much trouble. To remove it, it may be necessary to use what are known as "Loofsa Sponges" in the condensate return tank to absorb the oil, or to use one of the more sophisticated methods.

Regardless of which method is used to com-press the low-pressure steam, the result is advantageous, because nearly all of the work expended in the compression appears as increased heat energy in the final steam discharge, and thus may be recovered in the process where the steam is used.

K. Effect on the Boiler Plant Operator

A close, hard look at the cycles given here for the various types of cogeneration systems will readily disclose one basic fact: all of them require fairly complicated rotating mechanical prime movers as the heart of the system. Obviously, this is going to have an effect on the boiler plant operator, especially if the plant adopts the cogeneration concept as a conversion from an existing plant. The effect will not be as profound if the plant is new construction from the ground up, since the new plant operators will simply be hired as required to meet the new demands.

Another obvious factor is common to all mechanical devices; none of them are available 100% of the time, since they are subject to both planned and emergency shutdowns for servicing and repair. The plant managers are then faced with the necessity of how to handle this dilemma, and there are several alternatives, as follows;

1. Provide a complete standby boiler plant to supply the thermal energy demands when the cogeneration plant is down.

2. Provide for a standby cogeneration unit or prime mover when the decision is made to convert the plant.

3. Have contingency plans for bringing in a portable steam-generating plant on short notice, if such is available in the area.

4. Accept the possibility that the plant's production facilities will be down during the servicing period.

There is no problem with obtaining the full amount of plant electrical power during the shutdown, since the local utility will be supplying it as part of the contract for power.

If an existing plant converts to cogeneration, then the existing boiler plant will probably be retained as standby equipment. This means that the boiler and its accessories must be placed on either wet or dry layup, ready for quick lightoff when needed. It will then be the responsibility of the boiler plant operator to determine which is the best layup method to use for the particular plant situation. There is a possibility that the boiler plant will deteriorate faster than normal, unless it is activated regularly and a program is initiated to maintain it properly in operable condition.

If this situation exists, then the present boiler plant operator will probably be expected to expand his expertise to include other duties in the plant, either on the remainder of the steam system or on the new cogeneration plant.

Many of the cogeneration plants are being sold with contracts for maintenance included, which makes much sense, since the suppliers of the packages know the equipment, and it is to their advantage to see that the units perform at top efficiency and availability for public relations purposes.

However, should the plant decide to utilize its own personnel to maintain the new cogeneration equipment, then the boiler plant operator will no doubt be expected to learn the operation and servicing of the equipment as part of his regular duties. Maintenance of rotating machinery is more complex than operating a boiler, and the boiler plant operator will have to expand his expertise to meet the new challenge.

Of course, the new steam-generating portion of the cogeneration unit is subject to practically the same maintenance and servicing problems as the more conventional steam plant. Should the existing boiler operator be required to service and maintain that portion only of the new equipment, there should be little difficulty if the boiler plant operator is well versed in basics of steam generation.

If cogeneration expands as it appears to be doing, then we expect there will be changes in the future that will affect boiler operators. We

predict that there may arise a new type of boiler service contract firm, one which will offer a service of supplying qualified boiler plant operators on short notice, to go into a cogeneration plant and start up their boiler plant, and supply all operating personnel during the periods when the cogeneration equipment is down for servicing. There may already be firms offering this service.

If this prediction comes to pass, then the boiler operator of the future will be required to have all the necessary licenses, training, and experience to go into any boiler plant, quickly acquaint himself with the equipment, and get steam up and on the line with only a few hours notice. This will require a high degree of mobility and a willingness to leave home for periods of time from a few hours to several weeks. Also, he should take steps to acquaint himself with the different types and makes of boilers, controls, and burners, and accumulate all operating and servicing data that comes across his desk or through his hands.

As often happens when a new technology or combination of new concepts takes off in high fashion throughout an industry, a number of new firms spring up almost overnight with their version of the new approach. If the cogeneration installation in your plant is supplied by one of the well-known, established firms, then you probably have nothing to worry about. However, if your installation was supplied by a firm with only a short history of cogeneration, put together for the express purpose of getting in on the ground floor of a new industry, then we have some suggestions to make. Of course, some of these new firms do survive, but many do not, most often due to lack of business sense and accumen, even though their engineering expertise is excellent.

The only way that a plant operator can protect himself and his firm against such a supplier from going out of business is to be well prepared to take over complete charge of the installation. This means becoming an expert on the system while the supplier's installers and servicemen are around and still on the job, and accumulating all of the available data, instructions, specifications for the accessories as well as the basic equipment, flow diagrams, and control schematics. Also, it would be well to attempt to identify the parts that will be the major source of trouble in the system, and order sufficient quantities of spares to keep the plant in operation for several years. In short, be prepared to live with a cogeneration system that may be very well designed and installed, but is not backed up with any outside assistance.

C H A P T E R 2 9
HIGH ALTITUDE OPERATION

A. Scope

This chapter is included to answer the questions some boiler operators may have concerning the effects of altitude on the boiler and burner performances. The discussion is based on the very exhaustive analysis and calculations presented in the North American *Combustion Handbook*, Second Edition, as published by the North American Manufacturing Co., with some additional comments and analysis by the author.

With only a few exceptions, boilers are permanently mounted pieces of equipment. They are selected, designed, and furnished to be installed in one definite location, and this includes a known elevation above sea level. The few exceptions at this time consist mostly of truck-mounted portable boilers and rental boilers, which will be excluded from this discussion as being too few to be worthy of consideration.

Table 29-1 shows the approximate elevation above sea level of the major cities of the world that are above 1,000 ft, and from this it is obvious that very few boilers will be found above about 5,000 ft elevation. As it is safe to say that there are probably no boilers below sea level, we shall also not include that category of installation.

As we shall show in later sections of this chapter, there is only an insignificant difference in boiler and burner performance between sea level and about 1,500 ft elevation, and the boiler plant operator will not be aware of any effect of elevation if his station is between those limits. In fact, if the boiler and burner were properly designed for the location, he should not be aware of any appreciable adverse effect on his boiler performance at any elevation.

This reduces the discussion to those plants above 1,500 ft elevation, and as we shall see later, to operating pressures not higher than about 15 psig since it is in this range that the effects, if any, of elevation will become evident. Even then, it may not be an important factor in the over-all plant performance.

B. Basic Effects of Altitude

Nearly everyone is aware that water boils at temperatures below 212°F at elevations above sea level, and at atmospheric pressure. This is because atmospheric pressure, 0 psig, is on a floating base, and therefore moves up and down, not only with changes in elevation above sea level, but also to a lesser extent with atmospheric conditions. At sea level, 0 psig is based on a pressure of 14.7 psi above absolute zero pressure, or its equivalent, 29.92 in. Hg absolute.

Referring to page 390 in the Appendix, we see that this atmospheric pressure base re-

Table 29-1. Elevations (ft) of Some Cities more than 1,000 feet above Sea Level

Africa		North America (Continued)	
Algeria, Constantine	2170	Mexico, Monterey	1624
Ethiopia, Addis Ababa	9850	Mexico, Puebla	7150
Morocco, Marrakech	1600	U.S.A., Albuquerque, NM	4950
U.S. Africa, Johannesburg	5689	U.S.A., Atlanta, GA	1050
		U.S.A., Canton, OH	1030
Asia		U.S.A., Denver, CO	5280
Australia, Canberra	1875	U.S.A., El Paso, TX	3695
Iran, Tehran	4000	U.S.A., Oklahoma City, OK	1195
Saudi Arabia, Riyadh	1897	U.S.A., Omaha, NE	1040
Syria, Alep	1400	U.S.A., Phoenix, AZ	1090
Turkey, Ankara	2250	U.S.A., Salt Lake City, UT	4390
		U.S.A., Santa Fe, NM	6950
Central America		U.S.A., Spokane, WA	1890
Costa Rica, San Jose	3868	U.S.A., Tucson, AZ	2390
Guatemala, Guatemala	4850	U.S.A., Wichita, KS	1290
Honduras, Tegucigalpa	3500		
Salvador, San Salvador	2178	South America	
		Argentina, Tucuman	1500
Europe		Bolivia, La Paz	12200
Bulgaria, Sofia	1700	Brazil, Sao Paulo	2700
Germany, Munich	1700	Chile, Santiago	1800
Germany, Nurnberg	1150	Colombia, Bogota	8630
Spain, Madrid	2150	Colombia, Medellin	4880
Switzerland, Zurich	1360	Colombia, Palmira	3000
Yugoslavia, Belgrade	2270	Ecuador, Quito	9300
		Peru, Cuzco	11440
North America		Venezuela, Caracus	3164
Canada, Edmonton, Alberta	2200		
Mexico, Mexico City	7349		

duces as elevations above sea level increase. For instance, the city of Denver, at an elevation of about 5,300 ft, has a base atmospheric pressure of about 24.90 in. Hg absolute, which is about 5 in. below the base pressure at sea level. From Fig. 29-1, we see that the resultant boil-

ing point at atmospheric pressure is thus 203°F in the city of Denver. This is why food takes longer to cook when being boiled at elevated locations.

What are the effects of elevation on boiler plant operation or design?

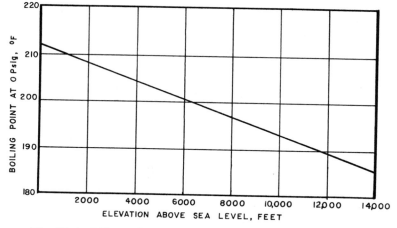

Fig. 29-1. Effect of altitude on the boiling point of water.

Probably the first one to be considered is the decreased density of air as elevations increase. The air quantities given in Chapters 4 and 5 for proper combustion of the various fuels are based on the air being supplied at sea level. Burners require, not volume, but weight of air; therefore, as the density decreases, the volume must increase to supply the proper weight of oxygen for the combustion process to occur. Figure 29-2 indicates the increase in volume for increase in elevation above sea level. Thus, for firing at 5,000 ft elevation, the burners will require 1.20 times (20% more) than the volume of air required at sea level.

In general, this means that the design of the following elements will be affected when the boiler–burner combination is assembled:

Draft fan—increased volume capacity required.

Burner air registers—larger openings required to keep the velocity and flame front within bounds.

Gas burner ports—gas volume will be increased at higher elevations.

Boiler breeching—larger cross-sectional areas needed.

Stack—larger cross-sectional area required.

For boilers operating on natural draft, the above changes are also required. The height should not be affected by the altitude, since, referring to Fig. 4-6, we see that the draft is the *difference* in weight between the two columns of air and flue gas, and this *difference* in weight should not be appreciably affected by the altitude.

We do not suggest that the boiler operator attempt to convert his boiler and burner combination to fire at a different altitude, should that ever become a problem. It is best to call in the supplier of the combination to redesign the changes required. There are several methods available, and there are many other things to take into consideration in making the change. For one thing, there may be a change in turndown ratio for the gas burner when converting it to high-altitude firing. If the fuel is one of the solid or heavy oil varieties, then other factors must be considered.

C. Effect on Steam Tables

There is some effect on the steam table values based on sea level conditions, which may conceivably involve some boiler plants using

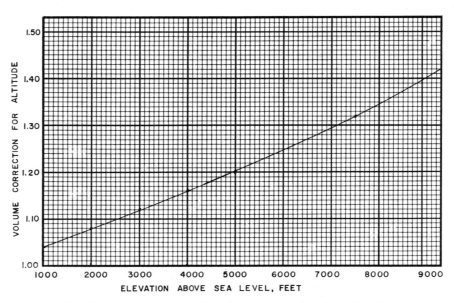

Fig. 29-2. Combustion air volume correction for altitude.

steam at low pressure for process applications.

The table on page 390 indicates that at 5,000 ft elevation, the atmospheric pressure is 24.90 in. Hg absolute, which is 5 in. below the atmospheric pressure at sea level. This 5 in. difference is the equivalent of 2.5 psi difference, since there are 2.04 in. Hg in 1 lb pressure. The net effect of this 2.5 psi difference is to raise the pressures listed in the left-hand column on page 394 upward in the column by 2.5 psig in every case. This places the horizontal 0 psig row at the horizontal 5-in.-Hg row, half way between the 4- and the 6-in. row. Note that this brings the 0 psig saturation temperature up to the 203°F position, half way between 201°F and 205°F. This checks with the value of 203°F found for 5,000 ft elevation in Figure 29-1.

We now have relocated the 0 psig base on which the steam tables are constructed. In doing this, every listing across the table must be moved with it, so that, for instance, the latent heat of evaporation for 0 psig is now 976 BTU/lb, instead of the previous value of 970 BTU/lb, a gain of 6 BTU/lb.

The major differences are evident at pressures below about 15 psig. At pressures above 15 psig, the differences become insignificant as the pressures rise. As an illustration of this, note Table 29-2.

Obviously, the increase in latent heat, which is the portion of the total heat in the steam being used in the plant process, is of no useful con-

Table 29-2. Altitude Effect on Steam Tables at 5,000 ft Elevation

Gauge Pressure (psig)	Saturation Temperature (°F)	Increase in Latent Heat of Evaporation (BTU/lb)
0	203	6
15	245	3
100	336	0.5

sequence at the higher pressures, and very little at lower pressures. The 6 BTU/lb increase at 0 psig is only 0.6% of the latent heat at sea level.

The net result of this discussion is that the boiler operator need not concern himself with the effects of any altitude that his plant is situated, with the possible exception of a change in turndown ratio for gas burners above about 1,500 ft elevation. There may be more of an effect at the higher elevation on the operator himself than on the equipment.

There is one possible exception to the last paragraph, however, and that is the case of purchasing a used boiler from a lower altitude and moving it to one of the higher areas in the country. This could also happen when renting a boiler as a replacement during repair or overhaul of the plant's regular boiler. Moving the boiler from 1000 ft elevation to the Denver area, for instance, would require the boiler to either be modified considerably or be derated by 15–20%.

C H A P T E R 3 0
SOME FINAL ADVICE

A. There's More to it than Skill

By now it should be quite obvious that there is a lot more to your job than just perfecting your skills, for which you are being paid the going wage in the industry. The firm paying your wages is made up of people, many of whom go about their own prescribed job functions every day, with their own particular skills.

The one guiding instrument for this entire force is what is known as "plant policy," usually formulated by unknown individuals in the higher echelons of the firm, and split into lesser documents for the guidance of the departments down through the hierarchy to the lowliest position in the plant. Each of those lesser plant policy statements are interpreted by individuals, with their own ideas and understanding of the terms of the policy. That, coupled with the variations in temperments of the plant personnel from top to bottom can, and does, often lead to conflicts throughout the plant. These little differences in personality and drives have to be understood and considered by any plant worker who is interested in advancement and fulfillment of his objectives before he can hope to reach his maximum potential in the plant.

What does it take for today's boiler plant worker to function within that jungle of personalities, other than perfecting his job skills?

B. Personal Factors

Probably the most important item that will shape the worker's success or failure on the job is his attitude toward the firm, his fellow employees, his supervisor, and his chosen career. A negative attitude in any of those areas will seriously hamper a worker's rise up the ladder.

So the first rule is: TRY AND MAINTAIN A POSITIVE, UPBEAT ATTITUDE TOWARD YOUR JOB AND THE PLANT PERSONNEL.

The next thing affecting your survivability is whether or not your supervisor can rely on you to perform consistently at your very best. When you are assigned to a particular task, will you perform it willingly and skillfully? Or will your performance vary considerably from day to day, depending on how you feel each day?

Can your supervisor rely on you being on the job on time, not extending your lunch break over the prescribed time, and working your coffee breaks into your daily schedule?

How about your personal behavior and grooming? Are they such as to inspire your fellow workers and your supervisor to have confidence in your performance? As stated in Chapter 11, a boiler consists of a large amount of concentrated power, and thus it must be treated with respect if you are to survive. This rules out any thought of "horseplay" while on the job or on company premises. Save it for when you are on your own time, if you are inclined to hilarity and playfulness.

Remember: BOILER OPERATION IS SERIOUS BUSINESS!

Also remember: ALCOHOL, DRUGS, AND BOILER WATER DO NOT MIX!

If you are addicted to any form of these, we strongly urge you to change your occupation to one less dangerous, then get treatment immediately for your habit, before you kill yourself and others on the job.

What this all reduces to is: any personal habit that reduces a worker's alertness and caution on the job places himself and others at high risk.

C. Plant-Related Factors

Keep in mind that your job depends on the firm remaining in operation. In order for the firm to remain in operation, it must make a profit. For both of these to happen, the employees must perform their tasks with a reasonable degree of skill, diligently applied. Among other things, this requires that the employees show a sense of loyalty to the firm or plant.

One of the reasons why West Germany and Japan have been able to rise to prominence since the destruction of World War II is the basic loyalty of the workers to their firm. And loyalty to the firm means a feeling of respect for those in authority, among other things.

A feeling of respect for those in authority may, at times, be difficult to muster by an employee. Regardless of what you may think of those in authority, and, specifically, your immediate supervisor, the fact remains that the firm has seen fit to appoint him to that position, and you are expected to take direction from him.

As an example, we cite the principle behind the military salute. The salute is a mark of respect to the uniform and the rank, and not the personality behind that uniform and rank.

In short, you do not have to like your supervisor, but you should show him the respect due his position and authority.

In most cases, the boiler plant operator works with tools and equipment supplied by the plant.

There are very good reasons why the plant workers should show a healthy respect for the firm's property, and especially for the tools they have supplied.

Poor employee attitude is displayed when he mistreats the tools and equipment supplied by the firm, since mistreatment is caused by either carelessness or disrespect for the property of others.

Remember: A WORKER WHO IS CARELESS WITH HIS TOOLS MAY BE EQUALLY CARELESS IN THE PERFORMANCE OF HIS DUTIES!

The subject of "teamwork" has been given much play in the business literature in recent years. A certain amount of this will be found in the plant's maintenance department, since most maintenance tasks are, or should be, carried out with at least two workers. This is usually done for reasons of safety, and for training an apprentice.

Our suggestion in this respect is to be flexible and cooperative. Do not expect that your team will always be compatible, but be willing to work with whomever is assigned to work with you. In fact, the more often the team composition is altered, the better chance for each worker to learn to cooperate with others, and thus become a better worker and a more valued employee.

Probably one of the best examples of cooperation would be the willingness to exchange shift assignments on a temporary or emergency basis with fellow employees. Another gesture of cooperation is the willingness to put in extra time and effort when a situation arises requiring it, such as when the production facilities are down due to problems in the steam plant.

D. Plant Politics and Gossip

In the beginning of this chapter we mentioned the diverse personalities of the plant personnel, and how this often leads to conflicts within the organization. The solution to these conflicts leads to political maneuvers, gossip, and power plays of all types.

There is a general pattern to these political movements and activities. Generally speaking, the higher up in the hierarchy, the more intense are the politics and the gossip that goes with it. This follows from the higher salary brackets, with more money and power at stake.

Fortunately, the boiler plant worker will probably not find these political fights too intense at this level. They will often exist, but to a much less extent than on the higher plant levels.

The best advice we can offer should you find the political atmosphere permeating your level is to keep your eyes, ears, and above all, your mind open at all times, and your mouth shut! Be aware of what is going on in the plant and the department, but do not take any position openly with one faction or another. Any maneuvering on your part should be quiet, cautious, and subtle, and aimed only at protecting your own position to prevent any reflection or fall-out on you.

In short: PLAY IT COOL AND SMART!

Much the same can be said about the gossip that often circulates in plants at all levels. This is often in connection with the politics being practiced in the plant, and should be regarded in the same light. When you are made a party to gossip, be completely noncommittal in your response, and do not encourage further embellishment. Absorb it into your mind, evaluate it as to source, the person being maligned, your own knowledge of all parties, and your sense of fairness and logical behavior, then store it away in your mind.

Above all: DO NOT REPEAT IT!

If you then find yourself passing judgment at a later date on the recipient of that gossip, discount it considerably, based on your own evaluation.

E. Conflicts

We have mentioned several times in this chapter the possibility of conflicts developing between individuals or groups in any plant or organization. With the primary goal of the firm being to assemble the many and various skills required to perform the plant operations, it is inevitable that there will at times be clashes between personalities. These are often a direct result of the political power plays.

Where do these clashes come from? They come mainly from the strive for power, but also the following differences play their part:

Personalities.

Approach to the job.

Interpretation of the job description.

Undefinable subconscious biases.

The one thing to keep in mind is that any conflict between two people or groups does not always reduce to one being right and the other being wrong. Once this fact is recognized, a compromise may be easier to find, and your position in the plant made more secure.

When confronted with a clash with someone else in the plant, attempt to ascertain which of the above differences is causing the confrontation. Be fair and objective, then govern yourself with that idea in mind.

F. Plant-Sponsored Activities

Many plants initiate a complete program of outside leisure activities among the employees. This is done to promote cooperation among the employees, provide relaxation and education, and, in general, improve the morale of the employees. These activities also provide a release of energy in a safe manner while accomplishing all of the preceding goals.

Generally, the larger the plant, the more plant-sponsored activities will be found. The degree of participation in those activities may be used as an approximate measure of the morale among the employees. They usually are not, and should not, be used as an evaluation of the merits of an employee, and the more progressive firms will take steps to see that is not done, other than a possible notation in the personnel file.

Whether or not you as a boiler plant operator take part in any of these activities is entirely at your own discretion. Do not feel that it is expected of you, and that therefore you should sign up for at least one of these activities. That is all we shall say concerning plant-sponsored activities, other than to point out that they are an excellent way to get to know your fellow workers and enlarge your circle of acquaintances. You work with these people during the day, and getting to know them in a relaxed atmosphere can help considerably in smoothing your work load on the job. Also, remember that many jobs are obtained through the help of acquaintances, either directly or indirectly through references.

G. Moving Up

By applying the principles and procedures established in this book, after months—and possibly, years—of hard work, the time will come when you may be offered a promotion. The new position, whether Watch Chief, Department Supervisor, or whatever is needed to fulfill the plant's requirement, will no doubt carry increased responsibilities, with an increase in monetary compensation.

Congratulations!

But before you accept the offer, stop and consider the full impact on your life and the lives of your "buddies" in the watch-standing fraternity. We shall present here a few things to be considered.

Can you leave them and become their leader?

Will you still feel at ease in their back yard parties, their outside activities, such as the firm's bowling team? Or does the firm expect you to change—"grow" will probably be the term used? You may be expected to change your circle of friends and acquaintenances, as the new position may put you in the category known as "lower management." The new po-

sition, if accepted, may find you drifting automatically into a new peer group, with different aspirations, interests, and outlook toward the plant and its employees.

Consider, also, how much you have learned about your fellow watch standers. You probably know all of their habits, attitudes, and other little quirks. For instance, "good ol' Joe." Good watch-tender, likable, outgoing, fun at a party, knows his job well. But he has one fault; he has a habit of walking off the property with the firm's small tools, and sometimes the plant materials, for use in his workshop or home. This is a very common problem in industry. Do you feel you can now become the plant watch-dog? How would you handle such a situation? Suggestion; call a closed meeting of the crew, give them a heart-to-heart talk on the subject, do not accuse anyone, just explain that you want the practice to stop, and give the consequences if it doesn't stop. This procedure can be used for other situations that need correcting.

Consider, also, the effect on your future in the firm should you not accept the offer of a promotion. What has been the past practice of the firm when other employees have rejected a promotion? Any reason to believe your situation is any different? You will probably be kept in your present position, someone else on the crew will be offered the promotion, or someone from another department may be brought in to fill it. You may be compensated partially with steady increases in wages, but that is only possible up to the lower level of the position you just rejected. This places a definite ceiling on your future income from this firm.

So, after all of this, you have decided to take the offer, and will do your best to fulfil your new responsibilities. Fine, you are now off your old plateau, and onto a new one, with increased opportunities.

Go for it! And give it your best.

H. Moving On

Our society is known for its mobility, including the movement of workers between plants. In the past, plant-sponsored pension plans have tended to keep employees tied to one job for life. Fortunately, that is changing slowly for the better, so that now workers are feeling free to move from job to job, and do so quite readily. This results in a steady rise in the individual's experience and worth to the employer, as well as an improvement in the worker's feelings and personal life.

Regardless of the real reason you are leaving a firm, keep in mind that the reasons given will be recorded on your personnel file and will forever be used when any new prospective employer checks on your reasons for leaving.

Some careful thought should be given when you leave your present firm for a new one. Therefore, we shall list here some of the more acceptable reasons, with the suggestion that you choose one, and be very careful how you use it, so that it will not come back to haunt you later. Pick one, and stay with it!

Your new job opens a new field with unlimited potential, not available in your present position.

You have been offered a promotion in another firm that opens a new opportunity to you.

You wish to broaden your experience, and you feel you have reached the limit of your possibilities in this firm.

Transportation is a problem and you have a job offer closer to home.

A friend or a relative needs your help and skills in a new venture.

You may feel frustrated, cheated, washed-out, stressed-out, and brow-beaten by your supervisor, but resist the urge to list any such reasons for leaving. They will leave the impression that you were unable to get along with others in the plant, and this is what the prospective employer will think when he is given those reasons on record in your personnel file. If it is obvious to all that you are leaving for one of those reasons listed in the first sentence of this paragraph, and it cannot be covered up, simply say that you do not appear to be as effective an employee as you should be, and you

have doubts about your ability to continue to function in the present conditions. You plan on taking a vacation, take further stock of yourself, and reevaluate your future before trying again at another firm. None of these reasons place the blame on either yourself or your supervisor. There simply was a conflict of personalities.

Remember, that supervisor that you would like to "tell off" may be asked by a prospective employer to give a reference for you at some future date.

Rule number 1: DO NOT BURN YOUR BRIDGES BEHIND YOU! Leave the firm on a smooth note if possible.

Rule number 2: Before leaving, contact as many of your fellow workers as possible, smoothing out any past difficulties or differences.

Rule number 3: Assemble as many promises of references as possible, with all names and addresses.

Rule number 4: In the exit interview, do not unburden your gripes. Be sure and give only positive and constructive comments about the firm and your opinion of it and the employees. Soft-pedal any known conflicts existing between you and others in the firm.

That is all the advice we can offer. The rest is up to you. Good Luck!

I. Conclusion

We have given you only a few of the more advisable and easily attained heat reclaiming and energy conservation ideas to be applied in the average boiler plant. There are many more, and we refer you again to the government publication mentioned at the beginning of Chapter 26. Also on page 405 in the Appendix you will find a list of many more ideas.

The energy shortage and the increasing cost of fuel will put a premium on those plant operators who are able to produce a constant stream of energy conservation ideas in the plant. We hope we have given you not only the concrete advice necessary, but also the incentive to go even further in your BTU-chasing phase of the job.

Remember to keep in mind always that

$$Heat = Energy = Money$$

Also, the ideal situation is achieved when any waste product leaving the plant premises, whatever the form may be, air, gas, vapor, solid, or liquid, leaves or is discharged at temperatures as close as practically possible to ambient temperature. Learn to watch for waste heat going out of the plant or into the air whenever you walk around the plant, and you will be on the way to becoming a first-class BTU Chaser.

APPENDICES

Contents

S E C T I O N   A
HOT WATER BOILERS

RECOGNITION

In addition to the Cleaver-Brooks Engineering and Service Departments, we wish to thank the following companies for their friendly help and cooperation in the preparation of this bulletin:

1. Bell and Gossett Company, Morton Grove, Illinois.

2. Johnson Service Company, Milwaukee, Wisconsin.

3. Minneapolis-Honeywell Regulator Company, Minneapolis, Minnesota.

4. The Powers Regulator Company, Skokie, Illinois.

PREFACE

Whenever a hot water system is laid out, all of the components (e.g. boiler, water temperature controls, pumps, circuits, valves, heat users, etc.) must be selected to work together to achieve the design intent. This harmony of operation of the components results in longer system life, better system control and operation, and greater customer satisfaction.

The operation of the boiler in relation to the components in the system is an important design consideration.

The purpose of this bulletin is to present various methods of boiler application to hot water systems and to recommend the best methods based on experience. While many of the comments will apply to any boiler, it should be understood that all of the comments are specifically directed to Cleaver-Brooks' products and their application.

The information in this bulletin is intended to be used as a general guide to system layout and to the use of Cleaver-Brooks' products in hot water systems.

Hot Water System, Components and Controls

Section 1. BOILER

The design of today's hot water boiler must provide for air removal from the boiler, high differentials between outlet and return water temperature, and proper tempering of return water before it comes in contact with the heating surface. *The selection of the boiler manufacturer is an important decision.*

The following boiler features help assure the engineer and the customer that he has a unit specifically designed for today's hot water systems:

A. Forced Internal Circulation — The Cleaver-Brooks hot water boiler design creates a flow pattern within the boiler which provides two distinct advantages:

1) The design assists the normal thermal circulation pattern existing within the boiler.

2) The design tempers return water before it comes in contact with the boiler heating surfaces. This results in more uniform water temperatures throughout the boiler.

B. Air Removal from the Boiler — All Cleaver-Brooks hot water boiler outlet connections include a dip tube which extends 2 to 3 inches into the boiler. This dip tube does not allow any air (which may be trapped at the top of the shell) to get back into the system.

Any oxygen or air which is released in the boiler will collect or be trapped at the top of the boiler shell.

The air vent tapping on the top center line of the boiler should be piped into the expansion or compression tank. Any air which is trapped at the top of the boiler will find its way out of the boiler through this tapping.

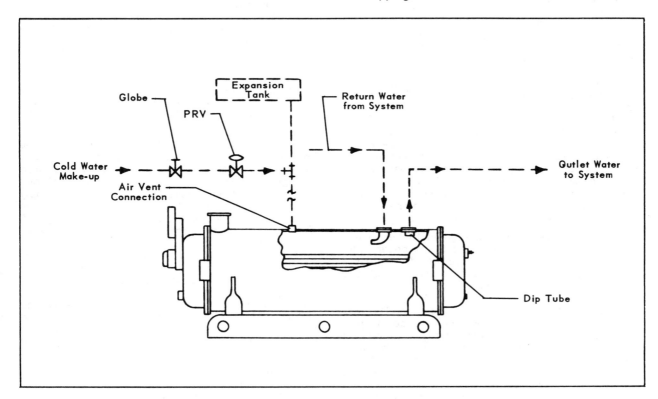

Cleaver-Brooks Hot Water Boiler Design

Section 2. BOILER OPERATION

Peculiar as it may seem, some hot water systems are installed with, what appears to be, very little regard for the proper operation of the boiler. The following points should be carefully considered during the layout of a hot water system:

A. Boiler Water Temperature — Boilers constructed in accordance with Section IV, Low Pressure Heating Boilers, of the ASME Boiler and Pressure Vessel Code can be operated with water temperatures up to 250 degrees F.

Boilers for operation over 250 degrees F must be constructed in accordance with Section I, Power Boilers, of the ASME Boiler and Pressure Vessel Code.

B. Minimum Boiler Water Temperature — The minimum recommended boiler water temperature is 170 degrees F. When water temperatures lower than 170 degrees F are used, the combustion gases are reduced in temperature to a point where the water vapor condenses. The net result is that corrosion occurs in the boiler and breeching.

This condensation problem is more severe on a unit which operates intermittently and which is greatly oversized for the actual load. This is not a matter which can be controlled by boiler design, since an efficient boiler extracts all the possible heat from the combustion gases. However, this problem can be minimized by maintaining boiler water temperatures above 170 degrees F.

Another reason for maintaining boiler water temperature *above* 170 degrees F is to provide a sufficient temperature "head" when No. 6 fuel oil is to be heated to the proper atomizing temperature by the boiler water in a safety type oil preheater. (The electric preheater on the boiler must provide *additional heat* to the oil if boiler water temperature is not maintained above 200 degrees F.)

Note: If the water temperature going to the system must be lower than 170 degrees F, the boiler water temperature should be a minimum of 170 degrees F (200 degrees F if used to preheat No. 6 oil) and mixing valves should be used.

C. Rapid Replacement of Boiler Water — The system layout and controls should be arranged to prevent the possibility of pumping large quantities of cold water into a hot boiler, thus causing shock, or thermal stresses. A formula, or "magic number" cannot be given, but it should be borne in mind that 200 degrees or 240 degrees water in a boiler cannot be completely replaced with 80 degrees water in a few minutes' time without causing thermal stress. This applies to periods of "normal operation" as well as during initial start-up.

This problem can be avoided in some systems by having the circulating pump interlocked with the burner so that the burner cannot operate unless the circulating pump is running.

When individual zone circulating pumps are used, it is recommended that they be kept running — even though the heat users do not require hot water. The relief device or by-pass valve will thus allow continuous circulation through the boiler and can help prevent rapid replacement of boiler water with "cold" zone water.

D. Continuous Flow Through the Boiler — The system should be piped and the controls so arranged that there will be water circulation through the boiler under all operating conditions. The operation of three way valves and system controls should be checked to make sure that the boiler will not be by-passed. Constant circulation through the boiler eliminates the possibility of stratification within the unit and results in more even water temperatures to the system.

A rule of thumb of 1/2 to 1 gpm per boiler horsepower can be used to determine the minimum continuous flow rate through the boiler under all operating conditions.

E. Multiple Boiler Installations — When multiple boilers of equal or unequal size are used, care must be taken to insure adequate or proportional flow through the boilers. This can best be accomplished by use of balancing cocks and gauges in the supply line from each boiler. If balancing cocks or orifice plates are used, a significant pressure drop (e.g. 3-5 psi) must be taken across the balancing device to accomplish this purpose.

If care is not taken to insure adequate or proportional flow through the boilers, this can result in wide variations in firing rates between the boilers.

In extreme cases, one boiler may be in the "high fire" position, and the other boiler or boilers may be loafing. The net result would be that the common header water temperature to the system would not be up to the desired point. This is an important consideration in multiple boiler installations.

F. Pressure Drop Through Boiler — There will be a pressure drop of less than three feet head (1 psi-2.31 ft. hd.) through all standardly equipped Cleaver-Brooks boilers operating in any system which has more than a 10 degree F temperature drop.

Section 3. CONTROL OF WATER TEMPERATURE TO SYSTEM

There are two main methods of controlling the water temperature to the system:
1) Fixed (constant) water temperature, 2) Variable water temperature. The following variations of each method will be described and evaluated:

    A. Fixed Water Temperature, Controls Located on Boiler, Single or Multiple Boilers.

    B. Fixed Water Temperature, Controls Located in Header, Multiple Boilers.
        1. Firing in Unison.
        2. Combination of Lead-Lag and Firing in Unison.

    C. Variable Water Temperature, Indoor-Outdoor Control System, Single or Multiple Boilers.
        1. Resetting the Controls Located on the Boiler.
          a. Electric — "on-off" boiler operation.
          b. Electronic — full modulation of boiler.
          c. Pneumatic — full modulation of boiler.
        2. Controlling a 3-way Mixing Valve (Fixed boiler water temperature).
        3. Combination of Resetting Controls Located on Boiler and Controlling a 3-way Mixing Valve.

A. Fixed Water Temperature (Controls Located On Boiler) (Single Or Multiple Boiler) (See Fig. 1) —

All Cleaver-Brooks boilers have an operating limit temperature control, which turns the boiler ON or OFF, depending on the selected setting of the cut-in and cut-out points. The difference between the cut-in and cut-out points is termed the differential.

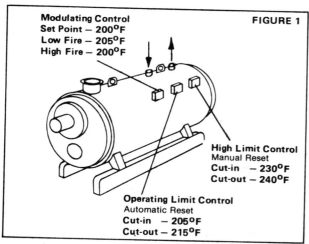

FIGURE 1

Modulating Control
Set Point — 200°F
Low Fire — 205°F
High Fire — 200°F

High Limit Control
Manual Reset
Cut-in — 230°F
Cut-out — 240°F

Operating Limit Control
Automatic Reset
Cut-in — 205°F
Cut-out — 215°F

In addition to the operating limit temperature control, each boiler is equipped with a high limit temperature control. This control is normally of the manual reset type, and is used as a backup control to the operating limit control.

Boilers which operate on the full modulation principle have another control called the modulating temperature control, which varies the firing rate in direct relation to the load demand (temperature of the boiler water).

The modulating control (normally set below the cut-in point of the operating limit control) controls the operation of the boiler from 25 - 100 percent of load, or rated boiler output.

As the temperature of the water in the boiler rises, the firing rate decreases. When the burner is in the low fire position and the temperature continues to climb because the low fire rate is more than required, the operating limit control will turn the unit off when the temperature reaches the cut-out point. The limit control will turn the unit on when the water temperature drops to the cut-in point.

Thus, fixed water temperature to the system is obtained by setting the limit control (and modulating control if used) on the boiler at the desired water temperature. This is the simplest method of controlling water temperature.

B. Fixed Water Temperature (Controls Located In Common Header) (Multiple Boilers)

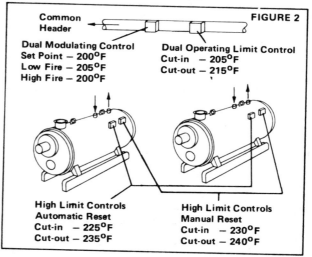

FIGURE 2

Common Header

Dual Modulating Control
Set Point — 200°F
Low Fire — 205°F
High Fire — 200°F

Dual Operating Limit Control
Cut-in — 205°F
Cut-out — 215°F

High Limit Controls
Automatic Reset
Cut-in — 225°F
Cut-out — 235°F

High Limit Controls
Manual Reset
Cut-in — 230°F
Cut-out — 240°F

No. 1 FIRING IN UNISON — On multiple boiler installations, where the load is comparatively constant, a system or method of control referred to as simultaneous operation with unison firing is frequently employed. (See Fig. 2.) When this system is used, the operating limit temperature control and the high limit temperature control are retained on the individual boilers. The modulating control is deleted.

A dual operating limit control and a dual modulating control are header mounted. These header mounted controls are so located to detect the blend water temperature as exists in the header, or at a location that is downstream of the boiler closest to the system circulating pump.

On a call for heat, both boilers are brought on the line simultaneously. If low fire is sufficient to handle the load, both boilers will remain in operation in low fire.

In the event that both boilers firing at low fire will not handle the load and there is a further decrease in header water temperature, both boilers will then modulate toward the high fire position and as required to meet the load demand.

On a decrease in load demand, which in turn results in an increase in header water temperature, the boilers will then modulate toward the low fire position.

Under condition of an extremely light load, both boilers firing in low fire will continue to increase the header water temperature until such time as the cut-off temperature is reached.

This system is preferred to system "A" (on multiple boiler installations), since both boilers share the load equally at all times. Both boilers come up to temperature together and remain in operation until the load drops to the point which shuts off both boilers.

No. 2. LEAD-LAG START WITH UNISON FIRING —

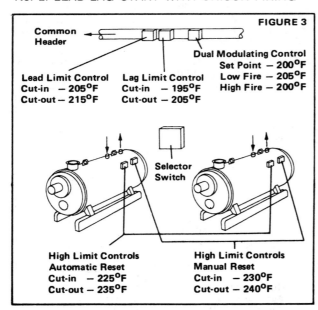

FIGURE 3

Common Header

Dual Modulating Control
Set Point — 200°F
Low Fire — 205°F
High Fire — 200°F

Lead Limit Control
Cut-in — 205°F
Cut-out — 215°F

Lag Limit Control
Cut-in — 195°F
Cut-out — 205°F

Selector Switch

High Limit Controls
Automatic Reset
Cut-in — 225°F
Cut-out — 235°F

High Limit Controls
Manual Reset
Cut-in — 230°F
Cut-out — 240°F

On occasion when systems have a large load swing, or there are periods during the year when one boiler will handle the load the greater portion of the time, it is advisable to employ a method or type of control system known as lead-lag start with unison firing. (See Fig. 3.) When this system of boiler control is used, the operating limit tempera-

ture control and the high limit temperature control are retained on the boilers. The modulating control is deleted.

Mounted in the header are two individual operating limit temperature controls and a dual modulating control. (See Figure 4 for control location)

Under condition of light load, only one boiler will operate and will modulate between low and high fire as the load demands. If there is a sudden load swing, creating a condition under which one boiler cannot handle the load, and there is a further drop in header water temperature, the second, or lag boiler, will be brought on the line and both boilers will share the load equally.

A selector switch is provided to enable the operator to select which of the boilers will be the lead boiler.

With this system each boiler shares the load and the wear and aging of the boiler is equalized by changing the lead boiler.

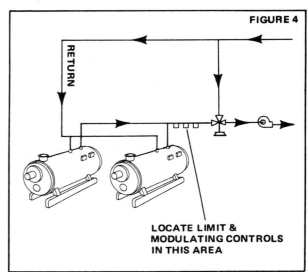

FIGURE 4

RETURN

LOCATE LIMIT & MODULATING CONTROLS IN THIS AREA

C. Variable Water Temperature (Indoor-Outdoor Control System) (Single or Multiple Boilers) —

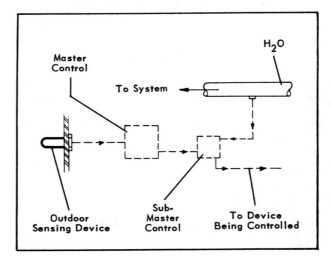

H₂O

Master Control

To System

Outdoor Sensing Device

Sub-Master Control

To Device Being Controlled

The indoor-outdoor system varies the temperature of the water going to the system in direct relation to the outdoor temperature. There are three types of systems commonly used:

Electric.
Electronic.
Pneumatic (Air Operated).

The electronic and pneumatic systems use an outdoor sensing device, a master control panel, and a submaster control panel.

The outdoor device transmits the outside temperature to the master panel which compares this temperature to a predetermined scale and then sends an adjusted message to the submaster control. The submaster compares the message from its master with the actual water temperature going to the system and then sends an adjusted message to the device which it is controlling.

There are three common methods in which the indoor-outdoor system is used to control water temperature to the system. The first of these methods is:

1. *Resetting the Controls Located on the Boiler (or Resetting the Boiler Water Temperature) (Minimum 170 degrees F)*

a) Electric — "on-off" boiler operation.

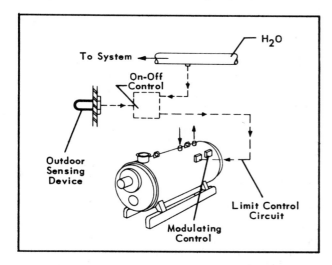

The electric indoor-outdoor system ties into the limit control circuit on the boiler and "cuts out" the boiler whenever the water temperature (going to the system) gets too high. In effect, this prevents the addition of heat to the boiler water and monitors the boiler water temperature to the system.

Similarly, it turns the unit on when water temperature to the system falls below the desired point.

The tie-in of this system in the limit circuit over-rides the standard limit control on the boiler.

This system is *not preferred* since it can "cut out" the boiler while it is in high fire position and thus "shorts out" the full modulation feature. It is *satisfactory* on boilers which have *"on-off"* burner operation.

b) Electronic — full modulation of boiler.

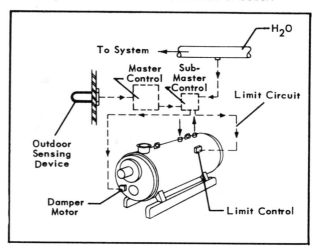

This is the preferred system on full modulation boilers since it allows the unit to operate as designed.

The modulating temperature control is removed from the boiler. The submaster control contains a damper motor which repositions controls that operate the modulating damper motor on the boiler. In effect, the submaster is thus directly regulating the firing rate in relation to the desired water temperature going to the system.

An auxiliary switch on the panel damper motor acts as the limit control to turn the boiler "on" or "off." The limit control which was left on the boiler serves as an extra high limit temperature control.

c) Pneumatic — full modulation of boiler.

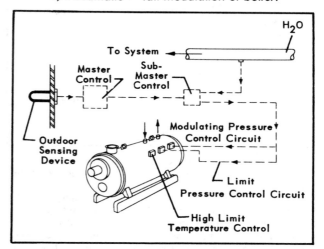

The pneumatic indoor-outdoor control system requires the use of a modulating and a limit *pressure* control on the boiler in addition to the limit temperature control. The modulating temperature control is deleted.

Monitored air pressure from the submaster control operates the modulating pressure control which regulates the firing rate. Thus, the increase or decrease in firing rate varies the water temperature to the system.

Monitored air pressure also operates the limit pressure control in conjunction with the modulating pressure control. This feature is desirable since the limit pressure control can then "cut out" the unit at the desired water temperature determined by the indoor-outdoor system.

The limit temperature control on the boiler acts as a high limit safety control.

This is a preferred pneumatic system since it does not cut out the boiler in the high fire position and thus allows full modulation to occur, as designed.

Note: With the proper controls, the electric, electronic, or pneumatic systems could also be used to operate the boiler under the "lead-lag" or "unison" methods described on Pages 5 and 6.

The second method in which the indoor-outdoor system is used to control water temperature to the system is:

2. *Controlling a 3-way mixing valve (Fixed Boiler Temperature)*

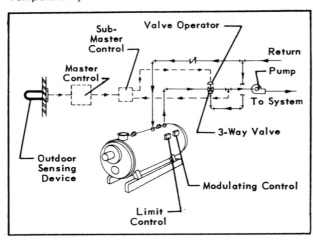

This is perhaps the most common application of an indoor-outdoor system for variable water temperature to the system.

The submaster control sends its message to the operator or controller of the 3-way mixing valve which has (2) inlets and (1) outlet. The operator or controller then positions or repositions the valve stem which changes the proportion of hot and cooler water entering each of the two inlets. The result is a change in the water temperature leaving the outlet and going to the system.

Whenever 3-way valves are used, they must be carefully sized so that the valve will not "hunt" or operate in an "almost closed" position. Valve manufacturers have indicated that there should be a 5-10 psi drop across the valve for proper operation. The

size of the valve *should not* be selected by making it the same size as the pipe in that part of the system.

During the sizing and selection of the valve, consideration must also be given to:

a. Possible rapid replacement of boiler water (Reference 2-C, Page 4).

b. Continuous flow through the boiler (Reference 2-D, Page 4).

The third method in which the indoor-outdoor system is used to control water temperature to the system is:

3. *A combination of resetting controls located on boiler and controlling a 3-way mixing valve*

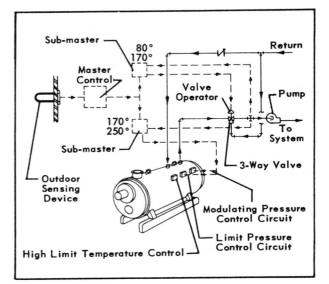

This system could be applied and is recommended rather than a single 3-way mixing valve (System C 2) when the water temperature to the system must be varied over a wide range (e.g. 80 degrees F to 250 degrees F). Two submaster controls are used.

When variable temperatures between 170 and 250 degrees F are desired, one of the submaster controls operates the controls on the boiler and resets the water temperature as desired (Operation similar to System C 1). The 3-way mixing valve operates as a Straight through 2-way valve under this condition (the inlet port from the return line remains closed).

During the time that variable temperatures between 80 and 170 degrees F are desired, one of the submaster controls operates the boiler controls to hold the boiler water temperature at 170 degrees F (minimum setting), and the other submaster operates the 3-way mixing valve. Flow proportions and size of 3-way valve are selected so that there is *always* a flow of water through the boiler. (Boiler is not bypassed.)

North American	**ATLAS 3-PASS GENERATORS** **DETAILED DESCRIPTION** **Series 4 and 5 — Specifications**	**150 to 700 hp** **Low Pressure Water** **11-75**

W-1   The following description is for standard low pressure hot water units. Equipment or accessories not shown below are extras--additions for such extras must be included in quotations.

W-2   All welded construction, designed in accordance with the ASME Code for Heating Boilers for 30 psi water design pressure and applicable state codes. Furnace is of the plain cylindrical type. Tubes are 2½" OD with 0.105" wall thickness, rolled and beaded into tube sheets. All units are of full wet-back construction.

W-3   **BOILER SHELL ACCESSORIES:** Structural steel skids, insulated flush front reversing chamber and rear flue collector, hinged and bolted front access doors, bolted access door on rear flue collector, 16" diameter rear furnace access door, rear observation port, 3" x 4" handholes with fittings, 11" x 15" manhole, 2" fiberglass insulation, green hammertone enamel steel jacket, lifting lugs and flue brush.

W-4   **HOT WATER TRIM:**

   A.   McDonnell & Miller 193D low water cutoff with switches for low water cutoff and alarm mounted on right hand side of unit. A second low water cutoff (Warrick probe type) is available at extra cost.

   B.   Marshalltown or equal pressure gage (6" diameter Fig. 41 with 0 - 100 psig & 0 - 231 ft range on units thru 300 hp--8½" diameter Fig. 29C with 0 - 60 psig & 0 -138 ft range on units 350 hp and larger).

   C.   American 5-6360AH or equal thermometer, (5" diameter, 50 - 300° F range on units thru 300 hp--Marshalltown Fig. 17 or equal thermometer, 8½" diameter, 30 - 300° F range on units 350 hp and larger).

   D.   Watts Fig. 740 or equal water relief valves per ASME Code.

W-5   **COMBUSTION EQUIPMENT:** Designed with 100% of air supplied by centrifugal blower. No secondary air or induced draft fans required to overcome boiler friction. Accessories include large burner observation port, 3450 rpm blower motor (1750 rpm on units 350 hp thru 600 hp), air intake silencer with acoustic lining and hinged burner construction for inspection and cleaning of burner internals.

W-6   **FUEL ARRANGEMENTS:**

   A.   NATURAL GAS: Gas equipment includes main gas regulator, one motor-operated main gas shutoff valve, pilot gas regulator, pilot gas solenoid and motor-operated butterfly valve. Lubricated plug valve for main gas line and pilot gas cock are furnished with unit, unpiped.

   B.   LIGHT OIL: Burner is a medium pressure air atomizing type (11 to 50 psig). Equipment includes oil nozzle, oil solenoid, motor-operated oil flow control valve, air compressor with motor (furnished for remote mounting on 700 hp unit), oil pump set with motor, single basket suction strainer and pressure relief valve. Ignition: Electric with gas pilot.

   BA. COMBINATION GAS AND LIGHT OIL: All equipment in A and B. Air compressor and oil pump set does not run during gas operation.

   C.   NO. 5 OIL: Burner is a medium pressure air atomizing type (17 to 50 psig). Maximum oil viscosity: 1000 SSU at 100° F. (This includes some light No. 6 oils.) Equipment includes oil nozzle, oil solenoid, motor-operated oil flow control valve, air compressor with motor (furnished for remote mounting on 500 hp thru 700 hp units), burner electric oil heater, electric heater oil relief valve, oil thermometer, (return line electric oil heater optional), oil pressure regulator, oil pressure gage and basket strainer. Electric oil heaters have low watt density (12 watts per sq in., maximum). On units 250 hp and larger, a hot water safety type oil heater with control aquastat, oil thermometer, oil pressure gage and hot water circulating pump is included, piped on boiler. An oil pump set is furnished for remote mounting with unassembled accessories consisting of pump safety relief valve, single basket suction strainer, vacuum-pressure gage, oil thermometer, pressure gage and circulating loop relief valve (on single unit installations circulating relief valve installed at burner). Ignition: Electric with gas pilot.

C_A. NO. 5 OIL AND GAS: All equipment in A and C. Air compressor does not run during gas operation.

D. NO. 6 OIL: Burner is a medium pressure air atomizing type (17 to 50 psig). Maximum oil viscosity: 5000 SSU at 100° F. Equipment includes oil nozzle, oil solenoid, motor-operated oil flow control valve, air compressor with motor (furnished for remote mounting on 500 thru 700 hp units), burner electric oil heater, electric heater oil relief valve, oil thermometer, (return line electric oil heater optional), oil pressure regulator, oil pressure gage and basket strainer. Electric oil heaters have low watt density (12 watts per sq in, maximum). A hot water safety type oil heater with control aquastat, oil thermometer, oil pressure gage and hot water circulating pump is included, piped on boiler. An oil pump set is furnished for remote mounting with unassembled accessories consisting of pump safety relief valve, single basket suction strainer, vacuum-pressure gage, oil thermometer, pressure gage and circulating loop relief valve (on single unit installations, circulating relief valve installed at burner). Ignition: Electric with gas pilot.

D_A. NO. 6 OIL AND GAS: All equipment in A and D. Air compressor does not run during gas operation.

W-7 **CONTROL EQUIPMENT:**

A. TYPE OF CONTROL: Full modulating with turndown ratio of four to one, which is equivalent to a firing range of 25% through 100% of maximum capacity.

B. SENSING ELEMENT: One Honeywell L4008A operating aquastat, 8" insertion with separable well and one Honeywell T991A modulating aquastat. (One Honeywell L4008E safety high limit aquastat optional.)

C. COMBUSTION SAFEGUARD: Honeywell maximum safety programmer, flame rectification type using flame rod to prove gas pilot and main gas flame, and photo-electric cell to prove light oil flame. Honeywell C7012A ultra-violet flame sensor to prove gas pilot and main heavy oil flame (supplied on heavy oil and combination heavy oil and gas units and all units 500 hp and larger).

D. SEQUENCE OF OPERATION: (1) 60-second pre-purge, (2) 10-second pilot proving, (3) 10-second main flame proving with pilot still on, (4) main flame operation with pilot off, (5) 15-second post-purge.

E. SAFETY INTERLOCKS: Honeywell C645A blower air switch, Honeywell microswitch hinge interlock and Honeywell enforced low fire start switch. An atomizing air interlock is provided on oil units and a low oil temperature interlock is provided on heavy oil units.

F. CONTROL LIGHTS: Six lights are furnished: one each for low water cutoff, flame failure, limit, pilot, oil operation and gas operation.

G. CONTROL SWITCHES: Start-stop switch, manual purge switch and fuel selector switch on dual-fuel units.

H. CONTROL PANEL: NEMA I enclosure is 14 gage with internal surfaces painted white. All control wiring is type TWC, 105 C rating, continuously coded by color and number. Customer terminal strip provided for remote operating controls, remote low water alarm, system circulating pump interlock and remote flame failure alarm. (Remote controls and alarms not furnished.) Snap-cover duct used for control wiring in panel.

I. RATE OF FIRE POTENTIOMETER: A potentiometer, equipped with manual lock nut, is provided to manually limit the maximum rate of fire at any point throughout the firing range.

W-8 **STANDARD ELECTRICAL CHARACTERISTICS:**

Blower motor:	208 or 230-460 volt/3 phase/60 cycle - on all units
Oil pump motor:	208 or 230-460 volt/3 phase/60 cycle - on light and heavy oil units
Air compressor:	208 or 230-460 volt/3 phase/60 cycle - on light and heavy oil units
Control circuit:	250 va, 115 volt/1 phase/60 cycle - on gas and light oil units
	750 va, 115 volt/1 phase/60 cycle - on heavy oil units

A step-down transformer for control circuit is provided at no extra cost. Magnetic starters are furnished for blower motors, light oil pump motors and air compressors. Manual starters for heavy oil pump motors are furnished for field mounting. No fused disconnect switches are furnished with units.

SECTION B
STEAM BOILER SPECIFICATIONS

	Detailed Description Series 4 and 5 — Specifications	40 to 125 hp Low Pressure Steam
		11-75

H-1 **GENERAL:** The following description is for standard low pressure steam units. Equipment or accessories not shown below are extras--additions for such extras must be included in quotations.

H-2 **BOILER SHELL:** All welded construction, designed in accordance with the ASME Code for Heating Boilers for 15 psi steam design pressure and applicable state codes. Furnace is of the plain cylindrical type. Tubes are 2½" OD with 0.105" wall thickness, rolled and beaded into tube sheets. All units are of full wet-back construction.

H-3 **BOILER SHELL ACCESSORIES:** Structural steel skids, insulated flush front reversing chamber and rear flue collector, hinged and bolted front access doors, bolted access door on rear flue collector, 16" diameter rear furnace access door, rear observation port, 3" x 4" handholes with fittings, 11" x 15" manhole (except 50 and 60 hp Series 4 units and 40 and 50 hp Series 5 units), 2" fiberglass insulation, green hammertone enamel steel jacket, lifting lugs and flue brush.

H-4 **STEAM TRIM:**

A. McDonnell & Miller 193D combination water column with switches for low water cutoff, low water alarm and pump control mounted on right hand side of unit; ½" Penberthy 36A or equal manual operated water gage set with glass and guards. A second low water cutoff (Warrick probe type) is available at extra cost.

B. Marshalltown Fig. 24 or equal steam gage, 6" diameter, 0 - 30 psig range with syphon and cock.

C. Kunkle Fig. 930 or equal safety valves per ASME Code.

H-5 **COMBUSTION EQUIPMENT:** Designed with 100% of air supplied by centrifugal blower. No secondary air or induced draft fans required to overcome boiler friction. Accessories include large burner observation port, 3450 rpm blower motor, air intake silencer with acoustic lining and hinged burner construction for inspection and cleaning of burner internals.

H-6 **FUEL ARRANGEMENTS:**

A. NATURAL GAS: Gas equipment includes main gas regulator, one motor-operated main gas shutoff valve, pilot gas regulator, pilot gas solenoid and motor-operated butterfly valve. Lubricated plug valve for main gas line and pilot gas cock are furnished with unit, unpiped.

B. LIGHT OIL: Light oil equipment is of the mechanical atomizing type (air atomizing type available on units 60 hp and larger at extra cost) and includes oil nozzles, motor-operated oil flow control valve, oil pump set with motor, internal suction filter and pressure regulator. Ignition: Electric with gas pilot.

BA. COMBINATION GAS AND LIGHT OIL: All equipment in A and B. Oil pump does not run during gas operation.

C. NO. 5 OIL: Burner is a medium pressure air atomizing type (17 to 50 psig). Maximum oil viscosity: 1000 SSU at 100° F. (This includes some light No. 6 oils.) Equipment includes oil nozzle, oil solenoid, motor-operated oil flow control valve, air compressor with motor, burner electric oil heater, electric heater oil relief valve, oil thermometer, (return line electric oil heater optional), oil pressure regulator, oil pressure gage and basket strainer. Electric oil heaters have low watt density (12 watts per sq in., maximum). An oil pump set is furnished for remote mounting with unassembled accessories consisting of pump safety relief valve, single basket suction strainer, vacuum-pressure gage, oil thermometer, pressure gage and circulating loop relief valve (on single unit installations circulating relief valve installed at burner). Ignition: Electric with gas pilot. (No. 5 oil burner not available on 40 & 50 hp units.)

CA. NO. 5 OIL AND GAS: All equipment in A and C. Air compressor does not run during gas operation.

Courtesy of E. J. Walsh & Co., Ltd.

D. NO. 6 OIL: Burner is a medium pressure air atomizing type (17 to 50 psig). Maximum oil viscosity: 5000 SSU at 100° F. Equipment includes oil nozzle, oil solenoid, motor-operated oil flow control valve, air compressor with motor, burner electric oil heater, electric heater oil relief valve, oil thermometer, (return line electric oil heater optional), oil pressure regulator, oil pressure gage and basket strainer. Electric oil heaters have low watt density (12 watts per sq in, maximum). A steam oil heater with temperature regulator, oil thermometer, oil pressure gage, strainer, check valve and bucket trap is included, piped on boiler. An oil pump set is furnished for remote mounting with unassembled accessories consisting of pump safety relief valve, and circulating loop relief valve (on single unit installations circulating relief valve installed at burner). Ignition: Electric with gas pilot. (No. 6 oil burner not available on 40 and 50 hp units.)

D_A. NO. 6 OIL AND GAS: All equipment in A and D. Air compressor does not run during gas operation.

H-7 CONTROL EQUIPMENT:

A. TYPE OF CONTROL: Full modulating on natural gas, No. 5 oil, No. 6 oil, combination No. 5 oil and gas, and combination No. 6 oil and gas with a turndown ratio of three to one, which is equivalent to a firing range of 33% through 100% of maximum capacity. Low-fire-start-high-fire-run on light oil. (Full modulating on light oil available on 60 hp and larger units at extra cost.)

B. SENSING ELEMENT: One Honeywell P455A combination on-off and modulating pressuretrol (except on straight light oil units, one L404A operating pressuretrol is furnished). (One Honeywell L404C safety high limit pressuretrol optional.)

C. COMBUSTION SAFEGUARD: Honeywell maximum safety programmer, flame rectification type, using flame rod to prove gas pilot and main gas flame, and photo-electric cell to prove light oil flame. Honeywell C7012A ultra-violet flame sensor to prove gas pilot and main heavy oil flame (supplied on heavy oil and combination heavy oil and gas units).

D. SEQUENCE OF OPERATION: (1) 60-second prepurge, (2) 10-second pilot proving, (3) 10-second main flame proving with pilot on, (4) main flame operation with pilot off, (5) 15-second postpurge.

E. SAFETY INTERLOCKS: Honeywell C645A blower air switch, Honeywell microswitch hinge interlock and Honeywell enforced low fire start switch. Low temperature and atomizing air interlocks furnished on heavy oil units.

F. CONTROL LIGHTS: Six lights are furnished: one each for low water cutoff, flame failure, limit, pilot, oil operation and gas operation.

G. CONTROL SWITCHES: Start-stop switch, manual purge switch and fuel selector switch on dual-fuel units.

H. CONTROL PANEL: NEMA 1 enclosure is 14 gage with internal surfaces painted white. All control wiring is Type TWC, 105 C rating, continuously coded by color and number. Customer terminal strip provided for remote operating controls, remote low water alarm, boiler feed pump and remote flame failure alarm. (Remote controls and alarms not furnished.) Snap-cover duct used for control wiring in panel.

I. RATE OF FIRE POTENTIOMETER: A potentiometer, equipped with manual lock nut, is provided on all but straight light oil units to manually limit the maximum rate of fire at any point throughout the firing range.

H-8 STANDARD ELECTRICAL CHARACTERISTICS:

Blower motor:	115-230 volt/1 phase/60 cycle	- all except 125 hp & heavy oil
	208 or 230-460 volt/3 phase/60 cycle	- 125 hp and heavy oil units
Oil pump motor:	115 volt/1 phase/60 cycle	- all light oil units except 125 hp
	208 or 230-460 volt/3 phase/60 cycle	- 125 hp and heavy oil units
Air compressor:	208 or 230-460 volt/3 phase/60 cycle	- all air atomizing units
Control circuit:	250 va, 115 volt/1 phase/60 cycle	- on gas and light oil units
	750 va, 115 volt/1 phase/60 cycle	- on heavy oil units

A step-down transformer for control circuit is provided at no extra cost. Magnetic starters are furnished for blower motors, light oil pump motors and air compressors. Manual starters for heavy oil pump motors are furnished for field mounting. No fused disconnect switches are furnished with units.

| ATLAS 3-PASS GENERATORS DETAILED DESCRIPTION Series 4 and 5 — Specifications | 40 to 125 hp High Pressure Steam 11-75 |

P-1   GENERAL: The following description is for standard high pressure steam units. Equipment or accessories not shown below are extras--additions for such extras must be included in quotations.

P-2   BOILER SHELL: All welded construction designed in accordance with the ASME Code for Power Boilers for 150 psi steam design pressure and applicable state codes. Furnace is of the plain cylindrical type (Morrison corrugated type furnace on the Series 5-125 hp unit only). Tubes are 2½" OD with 0.105" wall thickness, rolled and beaded into tube sheets. All units are of full wet-back construction.

P-3   BOILER SHELL ACCESSORIES: Structural steel skids, insulated flush front reversing chamber and rear flue collector, hinged and bolted front access doors, bolted access door on rear flue collector, 16" diameter rear furnace access door, rear observation port, 3" x 4" hand-holes with fittings, 11" x 15" manhole (except 50 & 60 hp Series 4 units and 40 & 50 hp Series 5 units), 2" fiberglass insulation, green hammertone enamel steel jacket, lifting lugs and flue brush.

P-4   STEAM TRIM:

A.   McDonnell & Miller 193D combination water column with switches for low water cutoff, low water alarm and pump control mounted on right hand side of unit; ½" Penberthy 36A or equal manual operated water gage set with glass, guards and three Penberthy SB-2 or equal manual operated try cocks. A second low water cutoff (Warrick probe type) is available at extra cost.

B.   Marshalltown Fig. 24 or equal steam gage, 6" diameter, 0 - 300 psig range with syphon and cock.

C.   Consolidated Fig. 1543 or equal safety valves per ASME Code.

P-5   COMBUSTION EQUIPMENT: Designed with 100% of air supplied by centrifugal blower. No secondary air or induced draft fans required to overcome boiler friction. Accessories include large burner observation port, 3450 rpm blower motor, air intake silencer with acoustic lining and hinged burner construction for inspection and cleaning of burner internals.

P-6   FUEL ARRANGEMENTS:

A.   NATURAL GAS: Gas equipment includes main gas regulator, one motor-operated main gas shutoff valve, pilot gas regulator, pilot gas solenoid and motor-operated butterfly valve. Lubricated plug valve for main gas line and pilot gas cock are furnished with unit, unpiped.

B.   LIGHT OIL: Light oil equipment is of the mechanical atomizing type (air atomizing type available on units 60 hp and larger at extra cost) and includes oil nozzle, motor-operated oil flow control valve, oil pump set with motor, internal suction filter and pressure regulator. Ignition: Electric with gas pilot.

BA.   COMBINATION GAS AND LIGHT OIL: All equipment in A and B. Oil pump does not run during gas operation.

C.   NO. 5 OIL: Burner is a medium pressure air atomizing type (17 to 50 psig). Maximum oil viscosity: 1000 SSU at 100° F. (This includes some light No. 6 oils.) Equipment includes oil nozzle, oil solenoid, motor-operated oil flow control valve, air compressor with motor, burner electric oil heater, electric heater oil relief valve, oil thermometer, (return line electric oil heater optional), oil pressure regulator, oil pressure gage and basket strainer. Electric oil heaters have low watt density (12 watts per sq in., maximum). An oil pump set is furnished for remote mounting with unassembled accessories consisting of pump safety relief valve, single basket suction strainer, vacuum-pressure gage, oil thermometer, pressure gage and circulating loop relief valve (on single unit installations circulating relief valve installed at burner). Ignition: Electric with gas pilot. (No. 5 oil burner not available on 40 & 50 hp units.)

CA.   No. 5 OIL AND GAS: All equipment in A and C. Air compressor does not run during gas operation.

D. NO. 6 OIL: Burner is a medium pressure air atomizing type (17 to 50 psig). Maximum oil viscosity: 5000 SSU at 100° F. Equipment includes oil nozzle, oil solenoid, motor-operated oil flow control valve, air compressor with motor, burner electric oil heater, electric heater oil relief valve, oil thermometer, (return line electric oil heater optional), oil pressure regulator, oil pressure gage and basket strainer. Electric oil heaters have low watt density (12 watts per sq in, maximum). A steam oil heater with temperature regulator, oil thermometer, oil pressure gage, steam pressure regulator, strainer, check valve and bucket trap is included, piped on boiler. An oil pump set is furnished for remote mounting with unassembled accessories consisting of pump safety relief valve, single basket suction strainer, vacuum-pressure gage, oil thermometer, oil-pressure gage and circulating loop relief valve (on single unit installations, circulating relief valve installed at burner). Ignition: Electric with gas pilot. (No. 6 oil burner not available on 40 & 50 hp units.)

D_A. NO. 6 OIL AND GAS: All equipment in A and D. Air compressor does not run during gas operation.

P-7 CONTROL EQUIPMENT:

A. TYPE OF CONTROL: Full modulating on natural gas, No. 5 oil, No. 6 oil, combination No. 5 oil and gas, and combination No. 6 oil and gas with a turndown ratio of three to one, which is equivalent to a firing range of 33% through 100% of maximum capacity. Low-fire-start-high-fire-run on light oil. (Full modulating on light oil available on 60 hp and larger units at extra cost.)

B. SENSING ELEMENT: One Honeywell P455A combination on-off and modulating pressuretrol (except on straight light oil units, one L404A operating pressuretrol is furnished). (One Honeywell L404C safety high limit pressuretrol optional.)

C. COMBUSTION SAFEGUARD: Honeywell maximum safety programmer, flame rectification type, using flame rod to prove gas pilot and main gas flame, and photo-electric cell to prove light oil flame. Honeywell C7012A ultra-violet flame sensor to prove gas pilot and main heavy oil flame (supplied on heavy oil and combination heavy oil and gas units).

D. SEQUENCE OF OPERATION: (1) 60-second pre-purge, (2) 10-second pilot proving, (3) 10 to 15-sec. main flame proving with pilot on, (4) main flame operation with pilot off,(5) 15-sec. post-purge.

E. SAFETY INTERLOCKS: Honeywell C645A blower air switch, Honeywell microswitch hinge interlock and Honeywell enforced low fire start switch. Low temperature and atomizing air interlocks furnished on heavy oil units.

F. CONTROL LIGHTS: Six lights are furnished: one each for low water cutoff, flame failure, limit, pilot, oil operation and gas operation.

G. CONTROL SWITCHES: Start-stop switch, manual purge switch and fuel selector switch on dual-fuel units.

H. CONTROL PANEL: NEMA 1 enclosure is 14 gage with internal surfaces painted white. All control wiring is Type TWC, 105 C rating, continuously coded by color and number. Customer terminal strip provided for remote operating controls, remote low water alarm, boiler feed pump and remote flame failure alarm. (Remote controls and alarms are not furnished.) Snap-cover duct used for control wiring in panel.

I. RATE OF FIRE POTENTIOMETER: A potentiometer, equipped with manual lock nut, is provided on all but straight light oil units to manually limit the maximum rate of fire at any point throughout the firing range.

P-8 STANDARD ELECTRICAL CHARACTERISTICS:

Blower motor:	115-230 volt/1 phase/60 cycle	- all except 125 hp & heavy oil
	208 or 230-460 volt/3 phase/60 cycle	- 125 hp and heavy oil units
Oil pump motor:	115 volt/1 phase/60 cycle	- all light oil units except 125 hp
	208 or 230-460 volt/3 phase/60 cycle	- 125 hp and heavy oil units
Air compressor:	208 or 230-460 volt/3 phase/60 cycle	- all air atomizing units
Control circuit:	250 va, 115 volt/1 phase/60 cycle	- on gas and light oil units
	750 va, 115 volt/1 phase/60 cycle	- on heavy oil units

A step-down transformer for control circuit is provided at no extra cost. Magnetic starters are furnished for blower motors, light oil pump motors and air compressors. Manual starters for heavy oil pump motors are furnished for field mounting. No fused disconnect switches are furnished with units.

Electrical Grounding of the Boiler

The combination of electrical equipment and controls, moving water and fuel, together with high combustion rates, all under high agitation, can cause stray currents around and in the boiler and its accessories on the boiler skid. Some of the things that could happen are electrical shocks, sparking, and galvanic action inside the water chambers of the boiler.

Boiler water is an electrolyte, and this means that local "cells" similar to those which produce electric power from batteries may result in pitting of the boiler heat-transfer surfaces or pressure-retaining parts.

To prevent any of these problems from occurring, we strongly suggest that all boilers be grounded electrically, to feed any such stray currents into the ground before they have a chance to do any harm. Any good boiler service firm will be able to assist in this, and Section 250 of the National Electrical Code should be followed.

If the boiler was supplied as a skid-mounted package, and was UL approved, then originally all attached accessories were no doubt grounded to the skid. Then all that is required is to ground the boiler skid or base, by one ground at a corner for small boilers, or two grounds at opposite corners for the larger boilers.

If the boiler was assembled on the site from parts purchased separately, then we suggest that a competent electrician be engaged to check the entire boiler package for proper grounding, as well as the proper method of grounding the boiler skid.

The accompanying illustration shows a common grounding method used on heavy industrial machinery. The grounding conductor size is given in the National Electrical Code and is based on the amount of amperage fed to the boiler package or by actual testing of the package for stray currents.

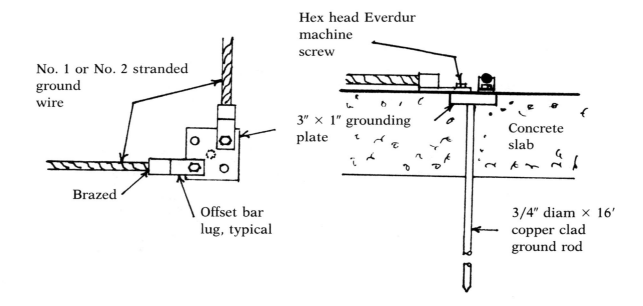

Section 3-1A
Page 15

2 Boilers, 2 Pumps—Automatic Standby

120 volt proportioning valves with end switches in branch lines to each boiler

Elementary Piping Diagram—1DPD08-B

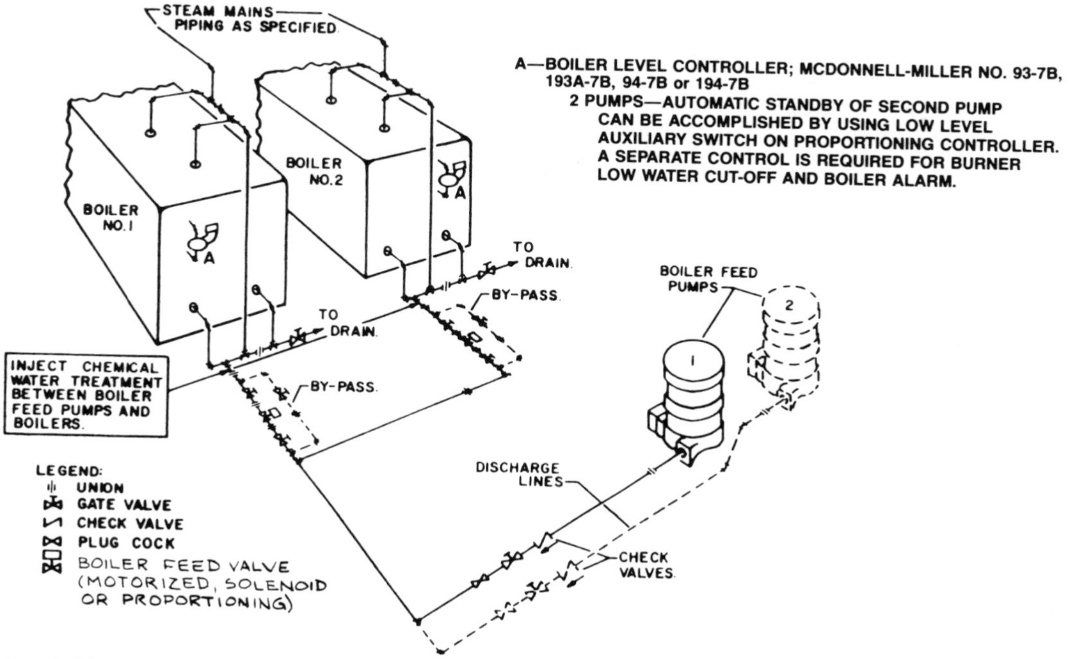
STEAM MAINS PIPING AS SPECIFIED
A—BOILER LEVEL CONTROLLER; MCDONNELL-MILLER NO. 93-7B, 193A-7B, 94-7B or 194-7B
2 PUMPS—AUTOMATIC STANDBY OF SECOND PUMP CAN BE ACCOMPLISHED BY USING LOW LEVEL AUXILIARY SWITCH ON PROPORTIONING CONTROLLER. A SEPARATE CONTROL IS REQUIRED FOR BURNER LOW WATER CUT-OFF AND BOILER ALARM.
BOILER NO. 2
BOILER NO. 1
A
A
TO DRAIN
TO DRAIN
BY-PASS
BY-PASS
BOILER FEED PUMPS
1
2
INJECT CHEMICAL WATER TREATMENT BETWEEN BOILER FEED PUMPS AND BOILERS.
DISCHARGE LINES
CHECK VALVES
LEGEND:
UNION
GATE VALVE
CHECK VALVE
PLUG COCK
BOILER FEED VALVE (MOTORIZED, SOLENOID OR PROPORTIONING)

Suggested Control Specifications (To be added to Unit Guide Specification)

The unit manufacturer shall furnish, mount on the pump unit and wire a NEMA 2 control cabinet with piano hinged door, enclosing the following:

2 Combination magnetic starters (having 3 overload relays) with fused disconnects and cover interlocks.
2 "Off-Continuous-Lead-Lag" pump selector switches.
2 Pump running pilot lights.
1 Control circuit disconnect switch with cover interlock.
1 Numbered terminal block.
1 Fused control circuit transformer when the motor voltage exceeds 130 volts.

Control cabinet shall contain U.L. listed or recognized components. Control cabinet shall be listed by Underwriter's Laboratory.

Control components shall be provided by the unit manufacturer, for operation as follows: as the level in the boiler recedes, the pump controller will start to open the proportioning valve; the end switch on the valve will start "lead" pump. The controller and valve will seek an equilibrium point of optimum feeding. If the level in the boiler continues to recede, the lower contacts on the controller will close, starting the "lag" pump. As the level is restored, the switch will open, and stop the "lag" pump. The feed valve will close as the level is restored to normal. "Lead" pump operation shall be nearly continuous. The pump selector switches shall provide "Off-Continuous-Lead-Lag" positions.

The unit manufacturer shall furnish (1) ITT McDonnell-Miller pump control,

a. No. 93-7B rated for 150 psi for boilers with separate water columns, *or*
b. No. 193A-7B rated for 150 psi with a water column type body, *or*
c. No. 94-7B rated for 250 psi for boilers with separate water columns, *or*
d. No. 194-7B rated for 250 psi with a water column type body

for mounting on each boiler and (1) 120 volt proportioning electric motorized feed valve with end switch to be installed in each boiler feed line.

The installing contractor, in addition to the above noted pump control, shall provide and install a low water burner cut-off switch, a low water boiler alarm switch and associated circuits in accordance with local codes.

The unit shall be factory tested as a complete unit and a certified test report of pump characteristics shall be submitted prior to shipment. The unit manufacturer shall furnish complete elementary and connection wiring diagrams, (2DW079), piping diagrams, (1DPD08-B), installation and operation instructions.

Manufacturer shall be ITT Domestic Pump, Morton Grove, IL.

Courtesy Bell and Gossett Division, ITT.

SECTION C

FUELS

GUARANTEED FUEL-TO-STEAM EFFICIENCIES ON MODEL CB® FIRETUBE BOILERS AT 50% AND 100% OF RATING

BOILER SIZE BHp	*Gas Fuel				*No. 2 Oil				*No. 5 and No. 6 Oils			
	15# Design 10 PSIG Operation		150# Design 125 PSIG Operation		15# Design 10 PSIG Operation		150# Design 125 PSIG Operation		15# Design 10 PSIG Operation		150# Design 125 PSIG Operation	
	% of Load		% of Load		% of Load		% of Load		% of Load		% of Load	
	50%	100%	50%	100%	50%	100%	50%	100%	50%	100%	50%	100%
100	83.5	83.5	80.5	81.0	86.5	86.5	83.5	84.0	87.0	87.0	84.0	84.5
125	82.0	82.0	79.0	79.5	85.0	85.5	82.0	83.0	85.5	86.0	82.5	83.5
150	83.0	83.0	80.0	80.5	86.0	86.5	83.0	84.0	86.5	87.0	83.5	84.5
200	83.5	83.5	80.5	81.5	87.0	87.0	84.0	84.5	87.5	87.5	84.5	85.0
250	82.0	82.0	79.5	80.0	85.5	85.5	83.0	83.0	86.0	86.0	83.5	83.5
300	82.5	82.5	80.0	80.5	86.0	86.0	83.5	83.5	86.0	86.0	83.5	83.5
350	83.0	83.0	80.5	81.0	86.5	86.5	84.0	84.5	86.5	86.5	84.0	84.0
400	83.0	83.0	80.0	80.5	86.0	86.5	83.5	84.0	87.0	87.0	84.5	85.0
500	84.0	84.0	81.0	81.5	87.0	87.5	84.0	85.0	86.5	87.0	84.0	84.5
600	84.0	84.5	81.0	82.0	87.5	87.5	84.0	85.0	87.5	88.0	84.5	85.5
700	84.5	84.5	81.5	82.0	87.5	87.5	84.5	85.5	88.0	88.0	85.0	86.0
800	84.5	84.5	81.5	82.0	88.0	88.0	85.0	85.5	88.5	88.5	85.5	86.0

*Fuel Heating Values: No. 2 Oil – 140,000 BTU/Gal.
Natural Gas – 1,000 BTU/Cu. Ft.   No. 5 & 6 Oil – 150,000 BTU/Gal.

NOTE: CB FTSE includes Radiation and Convection boiler heat losses to the boiler room.

NATURAL GAS FUEL BURNING RATES (THERMS/HR.) AT VARIOUS EFFICIENCIES

Boiler Size BHp	Fuel-To-Steam Efficiency																
	60.0	62.5	65.0	67.5	70.0	72.5	75	76	77	78	79	80	81	82	83	84	85
100	55.8	53.6	51.5	49.6	47.9	46.2	44.7	44.1	43.5	43.0	42.4	41.9	41.4	40.9	40.4	39.9	39.4
125	69.8	67.0	64.4	62.0	59.8	57.7	55.8	55.1	54.4	53.7	53.0	52.3	51.7	51.1	50.4	49.8	49.3
150	83.7	80.4	77.3	74.4	71.8	69.3	67.0	66.1	65.2	64.4	63.6	62.8	62.0	61.3	60.5	59.8	59.1
200	111.6	107.2	102.0	99.2	95.7	92.4	89.3	88.1	87.0	85.9	84.8	83.7	82.7	81.8	80.7	79.7	78.8
250	139.5	133.9	128.8	124.0	119.6	115.5	111.6	110.1	108.7	107.3	106.0	104.6	103.4	102.1	100.9	99.7	98.5
300	167.4	160.7	154.5	148.8	143.5	138.6	133.9	132.2	130.5	128.8	127.2	125.5	124.0	122.5	121.0	119.6	118.2
350	195.3	187.5	180.3	173.6	167.4	161.6	156.2	154.2	152.2	150.2	148.3	146.5	144.7	142.9	141.2	139.5	137.9
400	223.2	214.3	206.0	198.4	191.3	184.7	178.6	176.2	173.9	171.7	169.5	167.5	165.3	163.3	161.4	159.4	157.6
500	279.0	267.8	257.5	248.0	239.1	230.9	223.2	220.3	217.4	214.6	211.9	209.3	206.7	204.2	201.7	199.3	197.0
600	334.8	321.4	309.0	297.6	287.0	277.1	267.8	264.3	260.9	257.5	254.3	251.0	248.0	245.0	242.0	239.1	236.3
700	390.6	374.9	360.5	347.2	334.8	323.2	312.5	308.3	304.3	300.4	296.6	293.0	289.3	285.8	282.3	279.0	275.7
800	446.4	428.5	412.0	396.8	382.6	369.4	357.1	352.4	347.8	343.4	339.0	335.0	330.6	326.6	322.7	318.8	315.1

Gas = 1,000 BTU/Cu. Ft.

$$\frac{\text{Output (BTU/Hr)}}{\text{FTSE}} = \text{Input (BTU/Hr)}$$

FTSE = Fuel-to-Steam Efficiency

$$\frac{\text{Input (BTU/Hr)}}{100,000 \text{ BTU/Therm}} = \text{Therms/Hr}$$

Courtesy of Cleaver-Brooks Co.

NO. 2 OIL FUEL BURNING RATES (GPH) AT VARIOUS EFFICIENCIES

Boiler Size BHp	Fuel-To-Steam Efficiency																			
	60.0	62.5	65.0	67.5	70.0	72.5	75	76	77	78	79	80	81	82	83	84	85	86	87	88
100	40.0	38.5	37.0	35.5	34.0	33.0	32.0	31.5	31.0	30.5	30.5	30.0	29.5	29.0	29.0	28.5	28.0	28.0	27.5	27.0
125	50.0	48.0	46.0	44.5	42.5	41.0	40.0	39.5	39.0	38.0	38.0	37.5	37.0	36.5	36.0	35.5	35.0	35.0	34.5	34.0
150	60.0	57.5	55.0	53.0	51.0	49.5	48.0	47.0	46.5	46.0	45.5	45.0	44.5	43.5	43.0	42.5	42.0	41.5	41.0	41.0
200	79.5	76.5	73.5	71.0	68.5	66.0	64.0	63.0	62.0	61.5	60.5	60.0	59.0	58.5	57.5	57.0	56.5	55.5	55.0	54.5
250	99.5	95.5	92.0	88.5	85.5	82.5	79.5	78.5	77.5	76.5	75.5	74.5	74.0	73.0	72.0	71.0	70.5	69.5	68.5	68.0
300	119.5	115.0	110.5	106.5	102.5	99.0	95.5	94.5	93.0	92.0	91.0	89.5	88.5	87.5	86.5	85.5	84.5	83.5	82.5	81.5
350	139.5	134.0	129.0	124.0	119.5	115.5	111.5	110.0	108.5	107.5	106.0	104.5	103.5	102.0	101.0	99.5	98.5	97.5	96.0	95.0
400	159.5	153.0	147.0	141.5	136.5	132.0	127.5	126.0	124.0	122.5	121.0	119.5	118.0	116.5	115.0	114.0	112.5	111.0	110.0	108.5
500	199.5	191.5	184.0	177.0	171.0	165.0	159.5	157.5	155.5	153.5	151.5	149.5	147.5	146.0	144.0	142.5	140.5	139.0	137.5	136.0
600	239.0	229.5	220.5	212.5	205.0	198.0	191.5	189.0	186.5	184.0	181.5	179.5	177.0	175.0	173.0	171.0	169.0	167.0	165.0	163.0
700	279.0	268.0	257.5	248.0	239.0	231.0	223.0	220.0	217.5	214.5	212.0	209.0	206.5	204.0	201.5	199.0	197.0	194.5	192.5	190.0
800	319.0	306.0	294.5	283.5	273.5	264.0	255.0	251.5	248.5	245.0	242.0	239.0	236.0	233.5	230.5	227.5	225.0	222.5	220.0	217.5

No. 2 Oil = 140,000 BTU/Gal.

$$\frac{\text{Output (BTU/Hr)}}{\text{FTSE}} = \text{Input (BTU/Hr)}$$

FTSE = Fuel-to-Steam Efficiency

$$\frac{\text{Input (BTU/Hr)}}{140,000 \text{ BTU/Gal.}} = \text{GPH}$$

NO. 6 OIL FUEL BURNING RATES (GPH) AT VARIOUS EFFICIENCIES

Boiler Size BHp	Fuel-To-Steam Efficiency																			
	60.0	62.5	65.0	67.5	70.0	72.5	75	76	77	78	79	80	81	82	83	84	85	86	87	88
100	37.0	35.5	34.5	33.0	32.0	31.0	30.0	29.5	29.0	28.5	28.5	28.0	27.5	27.0	27.0	26.5	26.0	26.0	25.5	25.5
125	46.5	44.5	43.0	41.5	40.0	38.5	37.0	36.5	36.0	36.0	35.5	35.0	34.5	34.0	33.5	33.2	33.0	32.5	32.0	31.5
150	56.0	53.5	51.5	49.5	48.0	46.0	44.5	44.0	43.5	43.0	42.5	42.0	41.5	41.0	40.5	40.0	39.5	39.0	38.5	38.0
200	74.5	71.5	68.5	66.0	64.0	61.5	59.5	58.5	58.0	57.0	56.5	56.0	55.0	54.5	54.0	53.0	52.5	52.0	51.5	50.5
250	93.0	89.5	86.0	82.5	79.5	77.0	74.5	73.5	72.5	71.5	70.5	69.5	69.0	68.0	67.0	66.5	65.5	65.0	64.0	63.5
300	111.5	107.0	103.0	99.0	95.5	92.5	89.5	88.0	87.0	86.0	85.0	83.5	82.5	81.5	80.5	79.5	79.0	78.0	77.0	76.0
350	130.0	125.0	120.0	115.5	111.5	107.5	104.0	103.0	101.5	100.0	99.0	97.5	96.5	95.5	94.0	93.0	92.0	91.0	90.0	89.0
400	149.0	143.0	137.5	132.5	127.5	123.0	119.0	117.5	116.0	114.5	113.0	111.5	110.0	109.0	107.5	106.5	105.0	104.0	102.5	101.5
500	186.0	178.5	171.5	165.5	159.5	154.0	149.0	147.0	145.0	143.0	141.0	139.5	138.0	136.0	134.5	133.0	131.5	130.0	128.5	127.0
600	223.0	214.0	206.0	198.5	191.5	184.5	178.5	176.0	174.0	171.5	169.5	167.5	165.5	163.5	161.5	159.5	157.5	155.5	154.0	152.0
700	260.5	250.0	240.5	231.5	223.0	215.5	208.5	205.5	203.0	200.5	198.0	195.5	193.0	190.5	188.0	186.0	184.0	181.5	179.5	177.5
800	297.5	285.5	274.5	264.5	255.0	246.5	238.0	235.0	232.0	229.0	226.0	223.0	220.5	217.5	215.0	212.5	210.0	207.5	205.0	203.0

No. 6 Oil = 150,000 BTU/Gal.

$$\frac{\text{Output (BTU/Hr)}}{\text{FTSE}} = \text{Input (BTU/Hr)}$$

FTSE = Fuel-To-Steam Efficiency

$$\frac{\text{Input (BTU/Hr)}}{150,000 \text{ BTU/Gal.}} = \text{GPH}$$

Courtesy of Cleaver-Brooks Co.

THESE GAS PRESSURE REQUIREMENTS EFFECTIVE
ON ALL BOILERS SHIPPED AFTER OCTOBER 1, 1974

CHART NO. 4

Minimum Required Gas Pressures

at the Entrance to Standard and Factory Mutual Gas Trains
(Based on 1000 btu/cu. ft. Natural Gas and Elevation up to 700 Feet)

BHP	Model CB		Model CBH		Models Progress and Monitor		
	Connection Size (Inches)	*Net Regulated Pressure Required (Inches W.C.)	Connection Size (Inches)	*Net Regulated Pressure Required (Inches W.C.)	Connection Size (Inches)	Minimum Pressure Required (Inches W.C.)	Maximum Permissible Pressure
15	1¼	4.0	–	–	1	4.0	8 oz./sq. in.
20	1¼	4.0	–	–	1	4.0	8
25	–	–	2	4.5	1½	4.0	8
30	2	4.0	2	4.5	1½	4.0	8
40	2	4.5	2	4.5	2	4.0	8
50A	2	6.5	2	7.0	2	6.5	8
50	2	4.5	2	4.5	2	4.0	8
60	2	5.5	2	4.5	2	4.0	8
70	2	8.0	2	5.0	–	–	–
80	2	9.5	2	5.5	–	–	–
100A	2	9.5	2	9.5	–	–	–
100S	2	11.0	2	9.5	–	–	–
100	2	9.5	2	8.5	–	–	–
125A	2½	11.5	–	–	–	–	–
125	2½	7.5	–	–	–	–	–
125S	2½	8.5	–	–	–	–	–
150	2½	8.5	–	–	–	–	–
150S	2½	11.5	–	–	–	–	–
175A	2½	11.5	–	–	–	–	–
175S	2½	16.0	–	–	–	–	–
200	2½	15.5	–	–	–	–	–
250	3	16.0	–	–	–	–	–
300	3	20.5	–	–	–	–	–
350	3	29.0	–	–	–	–	–
400	4	18.5	–	–	–	–	–
500	4	29.0	–	–	–	–	–
600	4	41.0	–	–	–	–	–
700	4	57.0	–	–	–	–	–
800	4	73.0	–	–	–	–	–

* Net regulated pressure means the pressure at the gas train entrance for the Maximum Gas Consumption Rate. To accomplish this, a gas pressure regulator is required.

To obtain minimum required gas pressures at altitudes above 700 feet, multiply the pressures by the following factors:

Altitude (Ft)	Correction Factor	Altitude (Ft)	Correction Factor
1000	1.04	6000	1.57
2000	1.13	7000	1.70
3000	1.22	8000	1.84
4000	1.33	9000	2.01
5000	1.44		

Inches W.C. x 0.577 = Oz/Sq In.	Oz/Sq In. x 1.732 = Inches W.C.
Inches W.C. x 0.0361 = psig	Oz/Sq In. x 0.0625 = psig
psig x 27.71 = Inches W.C.	psig x 16.0 = Oz/Sq In.

Courtesy of Cleaver-Brooks Co.

THESE GAS PRESSURE REQUIREMENTS EFFECTIVE
ON ALL BOILERS SHIPPED AFTER OCTOBER 1, 1974

CHART NO. 5

Minimum Required Gas Pressures

at the Entrance to Gas Trains Equipped with Industrial Risk Insurers (IRI) Requirements

BHP	Model CB		Model CBH		Models Progress and Monitor	
	Connection Size (Inches)	Minimum Pressure Required (Inches W.C.)	Connection Size (Inches)	Minimum Pressure Required (Inches W.C.)	Connection Size (Inches)	Minimum Pressure Required (Inches W.C.)
15	1¼	5.0	–	–	1	–
20	1¼	5.0	–	–	1	–
25	–	–	2	5.0	1½	–
30	2	5.0	2	5.0	1½	–
40	2	5.0	2	5.0	2	–
50A	2	7.5	2	7.5	–	–
50	2	5.5	2	5.0	2	–
60	2	6.5	2	5.5	2	–
70	2	9.0	2	6.0	–	–
80	2	10.5	2	7.5	–	–
100A	2	11.0	2	13.0	–	–
100S	2	12.0	2	13.0	–	–
100	2	11.0	2	11.5	–	–
125A	2½	11.5	–	–	–	–
125	2½	7.5	–	–	–	–
125S	2½	8.5	–	–	–	–
150	2½	9.5	–	–	–	–
150S	2½	11.5	–	–	–	–
175A	2½	12.5	–	–	–	–
175S	2½	16.0	–	–	–	–
200	2½	15.5	–	–	–	–
250	3	16.0	–	–	–	–
300	3	20.5	–	–	–	–
350	3	29.0	–	–	–	–
400	4	18.5	–	–	–	–
500	4	29.0	–	–	–	–
600	4	41.0	–	–	–	–
700	4	57.0	–	–	–	–
800	4	73.0	–	–	–	–

Courtesy of Cleaver-Brooks Co.

Gas Pilot Data
for No. 5 and No. 6 Oil Fired Boilers
Model CB Boilers

BHP	Connection Size (Inches) (NPT)	Minimum Required Gas Pressure (Up to 700 Feet) Inches WC	Maximum Permissible Gas Pressure (psig)	Length of Time That Pilot is On (Sec)	Ignition Transformer Voltage
50-100	1/2	4	3	25	6000 V
125-800	1/2	2	3	25	6000 V

BHP	Average Gas Pilot Consumption — (CFH) Rate				
	Manufactured 500 Btu/Cu. Ft	Mixed 800 Btu/Cu. Ft	Natural 1000 Btu/Cu. Ft	Propane 2500 Btu/Cu. Ft	Butane 3200 Btu/Cu. Ft
50-100	120	75	60	25	20
125-200	200	125	100	40	30
250-350	300	180	150	60	45
400-800	400	250	200	80	60

BHP	Minimum Required Gas Pilot Pressure at Different Altitudes (Based on 1000 Btu/Cu. Ft Natural Gas)									
	700 Feet (Inches WC)	1000 Feet (Inches WC)	2000 Feet (Inches WC)	3000 Feet (Inches WC)	4000 Feet (Inches WC)	5000 Feet (Inches WC)	6000 Feet (Inches WC)	7000 Feet (Inches WC)	8000 Feet (Inches WC)	9000 Feet (Inches WC)
50-100	4	4.5	4.5	5.0	5.5	6.0	6.5	7.0	7.5	8.5
125-800	2	2.5	2.5	2.5	3.0	3.0	3.5	3.5	4.0	4.0

Approximate Gas Usage:

a. Multiply the CFH rate by 0.007 to obtain the number of cu. ft of gas used in 25 sec (Length of (1) light-off)

b. Multiply the number of cu. ft/light (Item a) by the estimated number of lights/hour or per day to obtain the approximate usage in cu. ft/hour or cu. ft/day.

Note: If the customer does not have gas service in his plant, or if the installed gas system is not adequate to serve the gas pilot, a bottled gas system may be desirable. This system has proven very satisfactory for pilot service since the regulation is positive and the gas consumption is relatively low.

Courtesy of Cleaver-Brooks Co.

COMMERCIAL STANDARD CS12-48
Detailed Requirements for Fuel Oils (A)

No.	Grade of Fuel Oil (B)	Flash Point Degrees F Min.	Pour Point Degrees F Max.	Water and Sediment Per Cent Max.	Carbon Residue On 10 Per Cent Residuum, Per Cent Max.	Ash Per Cent Max.	Distillation Temperatures, Degrees F 10 Per Cent Point Max.	90 Per Cent Point Max.	End Point Max.
1.	Distillate oil intended for vaporizing pot-type burners and other burners requiring this grade. (D)	100 or legal	0	trace	0.15		420		625
2.	A distillate oil for general purpose domestic heating for use in burners not requiring No. 1	100 or legal	20	0.10	0.35		(E)	675	
4.	An oil for burner installations not equipped with preheating facilities	130 or legal	20	0.50		0.10			
5.	A residual type oil for burner installations equipped with preheating facilities	130 or legal		1.00		0.10			
6.	An oil for use in burners equipped with preheaters permitting a high-viscosity fuel	150 or legal		2.00(F)					

(A) Recognizing the necessity for low sulphur fuel oils used in connection with heat-treatment, non-ferrous metal, glass and ceramic furnaces and other special uses, a sulphur requirement may be specified in accordance with the following table.

Grade of Fuel Oil	Sulphur, Max., Per Cent
No. 1	0.5
No. 2	1.0
Nos. 4, 5, and 6	no limit

Other sulphur limits may be specified only by mutual agreement between buyer and seller.

(B) It is the intent of these classifications that failure to meet any requirement of a given grade does not automatically place an oil in the next lower grade unless in fact it meets all requirements of the lower grade.

(C) Lower or higher pour points may be specified whenever required by conditions of storage or use. However, these specifications shall not require a pour point lower than 0 degrees F under any conditions.

(D) No. 1 oil shall be tested for corrosion in accordance with par. 15 for three hours at 122 degrees F. The exposed copper strip shall show no gray or black deposit.

(E) The 10 per cent point may be specified at 440 degrees F maximum for use in other than atomizing burners.

(F) The amount of water by distillation plus the sediment by extraction shall not exceed 2.00 per cent. The amount of sediment by extraction shall not exceed 0.50 per cent. A deduction in quantity shall be made for all water and sediment in excess of 1.0 per cent.

Courtesy of Cleaver-Brooks Co.

COMMERCIAL STANDARD CS12-48

Detailed Requirements for Fuel Oils (A)

Grade of Fuel Oil (B)	VISCOSITY								Gravity Degrees API Min.
	Sayb. Univ. at 100 Degrees F		Sayb. Furol at 122 Degrees F		Kin. C.S. at 100 Degrees F		Kin. C.S. at 122 Degrees F		
No.	Max.	Min.	Max.	Min.	Max.	Min.	Max.	Min.	
1. A distillate oil intended for vaporizing pot-type burners and other burners requiring this grade. (D)									35
2. A distillate oil for general purpose domestic heating for use in burners not requiring No. 1.	40				2.2	1.4			26
4. An oil for burner installations not equipped with preheating facilities	125	45			(26.4)	(5.8)			
5. A residual type oil for burner installations equipped with preheating facilities		150				(32.1)	(81)		
6. An oil for use in burners equipped with preheaters, permitting a high-viscosity fuel			300	45			(638)	(92)	

(A) Recognizing the necessity for low sulphur fuel oils used in connection with heat-treatment, non-ferrous metal, glass and ceramic furnaces and other special uses, a sulphur requirement may be specified in accordance with the following table.

Grade of Fuel Oil	Sulphur, Max., Per Cent
No. 1	0.5
No. 2	1.0
Nos. 4, 5, and 6	no limit

Other sulphur limits may be specified only by mutual agreement between buyer and seller.

(B) It is the intent of these classifications that failure to meet any requirement of a given grade does not automatically place an oil in the next lower grade unless in fact it meets all requirements of the lower grade.

(C) Lower or higher pour points may be specified whenever required by conditions of storage or use. However, these specifications shall not require a pour point lower than 0 degrees F under any conditions.

(D) No. 1 oil shall be tested for corrosion in accordance with par. 15 for three hours at 122 degrees F. The exposed copper strip shall show no gray or black deposit.

(E) The 10 per cent point may be specified at 440 degrees F maximum for use in other than atomizing burners.

(F) The amount of water by distillation plus the sediment by extraction shall not exceed 2.00 per cent. The amount of sediment by extraction shall not exceed 0.50 per cent. A deduction in quantity shall be made for all water and sediment in excess of 1.0 per cent.

Courtesy of Cleaver-Brooks Co.

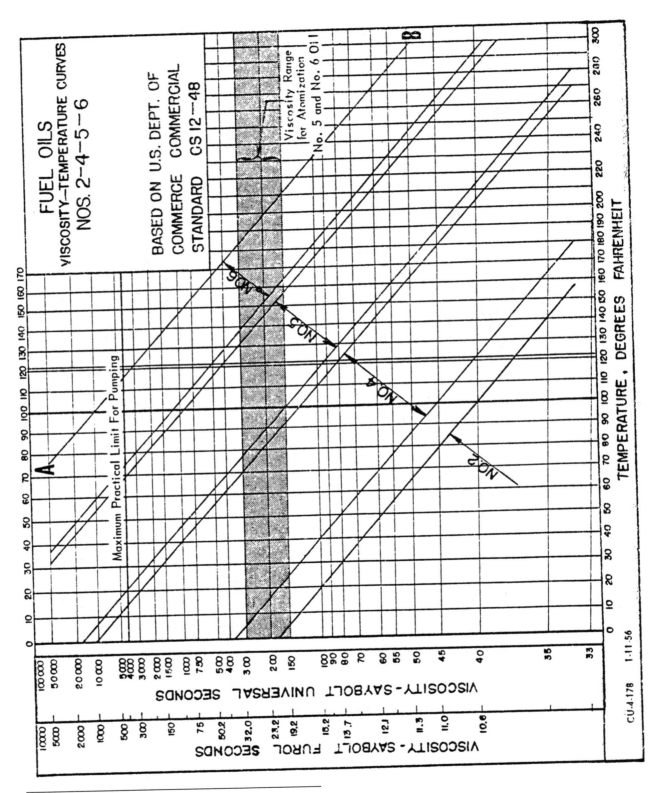

FUEL OILS
VISCOSITY—TEMPERATURE CURVES
NOS. 2—4—5—6

BASED ON U.S. DEPT. OF
COMMERCE COMMERCIAL
STANDARD CS 12—48

Courtesy of Cleaver-Brooks Co.

NORTH AMERICAN Mfg. Co., Cleveland, Ohio 44105

PHYSICAL PROPERTIES OF LIQUIFIED PETROLEUM GASES

	Commercial Propane	Commercial Butane
Formula	C_3H_8	C_4H_{10}
Normal State, 60 F, 30" Mercury	Gas	Gas
Freezing Point (or Melting), F	-310	-210
Boiling Point F at 30" Mercury	-44	32
Specific Gravity of Gas (Air = 1) at 60 F, 30" Hg	1.52	2.00
Specific Gravity of Liquid (Water = 1) 60 F/60 F	0.51	0.58
Weight per Gallon of Liquid at 60 F, lb	4.24	4.85
Gallons per Pound of Liquid at 60 F	0.236	0.206
Btu per Gallon (Vaporized)	91,500	102,600
Btu per Pound (Vaporized)	21,560	21,180
Btu per Cubic Foot (Vaporized)	2,520	3,260
Latent Heat of Vaporization at Boiling Point, Btu/lb	185	166
Latent Heat of Vaporization at Boiling Point, Btu/gal	785	807
Specific Heat of Gas, Btu/lb/F at 60 F	0.405	0.385
Specific Heat of Liquid, Btu/lb/F at 60 F	0.590	0.550
Density of Gas, Pounds per cubic foot	0.116	0.153
Density of Liquid, Pounds per cubic foot	31.7	36.0
cu ft of gas/gal liquid at 60 F, 30" Hg	36.5	31.0
cu ft of gas/lb liquid at 60 F, 30" Hg	8.55	6.51
Vapor Pressure, psi gauge, at -10 F temperature	20	--
0 F	28	--
10 F	45	--
32 F	60	--
60 F	100	12
80 F	130	22.5
100 F	190	37
120 F	240	55

Combustion Data

	Commercial Propane	Commercial Butane
cu ft Air required to burn 1 cu ft gas	23.5	30.0
cu ft Oxygen required to burn 1 cu ft gas	5	6.25
cu ft Air required to burn 1 lb of gas	200	195
Ignition Temperature	920-1020	900-1000
Maximum Flame Temperature	3600	3625
% Gas in Air for Maximum Flame Temperature	4.4	3.5
Maximum Rate of Flame Propagation in 1" tube, inches/sec	32	33
% Gas in Air-Gas Mixture at maximum flame propagation rate	4.8	3.8
% Gas in Air-Gas Mixture at lower limit of flammability	2.4	1.9
% Gas in Air-Gas Mixture at upper limit of flammability	9.5	8.5
Octane Number (Iso-Octane = 100)	125	91

Products of Complete Combustion

	Commercial Propane	Commercial Butane
cu ft CO_2 per cu ft gas	3.0	3.9
cu ft water vapor per cu ft gas	4.0	5.0
cu ft N_2 per cu ft gas	18.5	23.6
pounds CO_2 per pound gas	3.0	3.1
pounds H_2O per pound gas	1.6	1.5
pounds N_2 per pound gas	12.0	11.8
Ultimate % CO_2 by volume	13.9	14.1

Courtesy of North American Mfg. Co.

SECTION D

PIPING SYSTEMS

North American	ENGINEERING DATA – RECOMMENDED PIPING FOR LOW PRESSURE STEAM GENERATORS	ATLAS BOILER ROOM & STACK DATA
		3-65

NOTES:

1. IF LOW PRESSURE STEAM OR GRAVITY SYSTEM HAS LONG RETURNS (DOES NOT HAVE VACUUM OR PUMPED RETURNS). OVERSIZE OR DOUBLE THE CONDENSATE RECEIVER SIZE IN ORDER TO HAVE 15 MINUTES STORAGE CAPACITY. PROPER SIZE CONDENSATE RECEIVER PREVENTS FLOODING OF SYSTEM AND EXCESSIVE RAW WATER MAKE-UP DURING COLD STARTS.

2. INSTALL STOP CHECK VALVE ON TOP OF EACH STEAM GENERATOR. SHOULD GENERATOR OPERATION BE SEQUENCED. BE CAREFUL CORRECT SIZE AND TYPE OF STOP CHECK VALVE IS SELECTED FOR PROPER OPERATION WITH MINIMUM STEAM PRESSURE DROP ACROSS IT.

3. TO INSURE EQUAL OUTPUT OF EACH STEAM GENERATOR (ONE STEAM GENERATOR DOES NOT HOG OR MONOPOLIZE THE LOAD). IT IS RECOMMENDED TO HAVE SIMULTANEOUS OPERATION OF THE GENERATORS FROM A COMMON MASTER CONTROL.

4. FOR MAXIMUM STEAM GENERATOR PROTECTION A SECOND SEPARATE LOW WATER FUEL CUTOFF IS RECOMMENDED (PIPED AND WIRED ON EACH STEAM GENERATOR).

5. OFTEN, IN A LOW PRESSURE STEAM SYSTEM AN INDUCED VACUUM IN THE SUPPLY HEADER WILL LOWER THE WATER LINE IN THE STEAM GENERATOR AND SIPHON FEED WATER INTO THE MAINS AND FLOOD THEM. TO PREVENT THE ABOVE, IT IS RECOMMENDED A MOTORIZED SHUT-OFF (ON-OFF OR MODULATING TYPE) VALVE BE USED IN THE FEEDWATER LINE TO EACH GENERATOR WHICH IS CONTROLLED BY THE INDIVIDUAL FEEDWATER CONTROLLER.

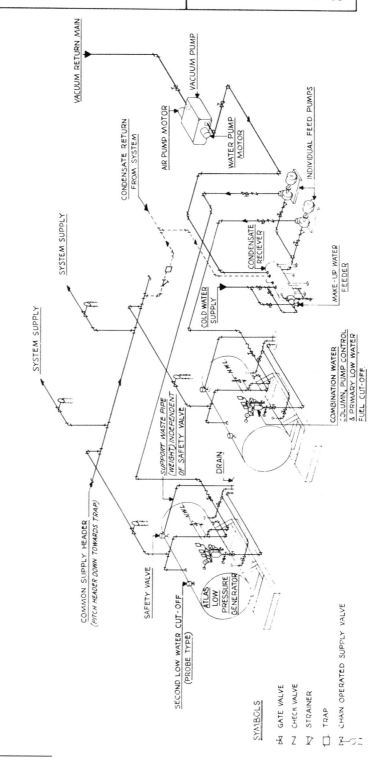

SYMBOLS

GATE VALVE
CHECK VALVE
STRAINER
TRAP
CHAIN OPERATED SUPPLY VALVE

Courtesy of North American Mfg. Co.

NORTH AMERICAN Mfg. Co.
Cleveland, OH 44105 USA

ATLAS BLR. RM. & STACK DATA
Section Q, Page 4　　　2-76

RECOMMENDED PIPING FOR HIGH PRESSURE STEAM GENERATORS

NOTES:

1. SECOND SEPARATE MEANS OF FEEDING WATER INTO STEAM GENERATOR SHOULD BE PROVIDED IN ADDITION TO BOILER FEED PUMP

2. SECOND SEPARATE LOW WATER CUTOFF IS RECOMMENDED FOR MAXIMUM SAFETY

3. TO REMOVE SUSPENDED SOLIDS, A SURFACE BLOWOFF VALVE OR A CONTINUOUS BLOWOFF SYSTEM IS RECOMMENDED

4. FOR MAXIMUM REMOVAL OF BOTTOM SLUDGE OR MUD A FRONT BOTTOM BLOWDOWN IS RECOMMENDED PIPED IN ADDITION TO REAR BLOWDOWN

5. VENTS FROM SAFETY VALVES, BLOWOFF TANK SHOULD EXTEND THROUGH BOILER ROOM ROOF AND TERMINATE 6 FT. ABOVE ROOF. VENT LINES SHOULD BE INDEPENDENTLY SUPPORTED FROM ROOF.

6. ON MULTIPLE BOILER INSTALLATIONS, IT IS RECOMMENDED TO INSTALL A MOTORIZED VALVE (NOT SPRING LOADED TYPE) IN EACH INDIVIDUAL FEED WATER LINE. MOTORIZED FEED VALVE SHOULD BE ENERGIZED BY PUMP CONTROL AND END SWITCH ON MOTORIZED VALVE SHOULD, IN TURN, ENERGIZE BOILER FEED PUMP.

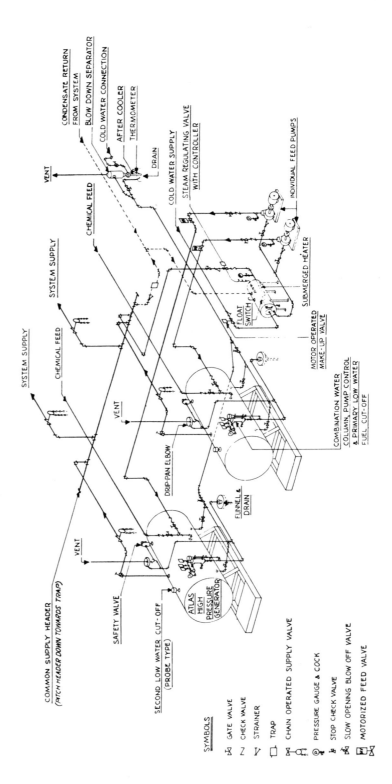

SYMBOLS

⋈ GATE VALVE
∠ CHECK VALVE
ν STRAINER
☐ TRAP
CHAIN OPERATED SUPPLY VALVE
Ⓟ PRESSURE GAUGE & COCK
STOP CHECK VALVE
SLOW OPENING BLOW OFF VALVE
MOTORIZED FEED VALVE

STEAM PIPE CAPACITIES
POUNDS PER HOUR OF SATURATED STEAM AT GAUGE PRESSURES SHOWN, AND IN PRESSURE DROPS PER 100 FEET OF PIPE

Pipe Size	5 PSIG		15 PSIG			25 PSIG		
	1/4	1/2	1/4	1/2	1	1/2	1	1.5
3/4	18	27	24	35	47	40	60	75
1	35	55	47	70	95	80	120	150
1-1/4	80	120	100	140	200	160	240	300
1-1/2	120	170	150	210	300	240	350	450
2	225	340	280	425	600	475	700	900
2-1/2	350	550	450	650	950	750	1100	1400
3	650	950	800	1200	1700	1300	2000	2500
4	1300	1950	1600	2400	3500	2700	4000	5000
6	4000	6000	4800	7000	10,000	8000	12,000	15,000
8	8000	12,000	10,000	14,000	20,000	16,000	25,000	30,000
10	14,000	21,000	18,000	25,000	36,000	30,000	44,000	55,000

Pipe Size	50 PSIG			75 PSIG		
	1/2	1	1.5	1/2	1	1.5
3/4	38	55	65	60	85	105
1	77	110	130	120	170	210
1-1/4	170	240	290	250	350	450
1-1/2	260	370	450	360	500	650
2	530	750	900	700	1000	1300
2-1/2	870	1240	1500	1100	1600	2000
3	1600	2270	2700	2000	3000	3600
4	3400	4800	6200	4000	6000	7500
6	10,200	14,500	19,000	12,000	18,000	22,000
8	21,400	30,300	38,000	25,000	36,000	45,000
10	40,000	56,000	70,000	45,000	65,000	80,000

Pipe Size	100 PSIG			125 PSIG		
	1/2	1	1.5	1	2	3
3/4	65	100	125	105	150	200
1	130	200	250	210	300	400
1-1/4	290	400	500	450	680	820
1-1/2	400	600	750	700	1000	1250
2	800	1200	1500	1300	2000	2500
2-1/2	1300	2000	2400	2100	3000	3800
3	2300	3400	4000	3600	5200	7000
4	4600	7000	8400	7500	11,000	14,000
6	13,000	21,000	25,000	22,000	33,000	42,000
8	27,000	42,000	50,000	45,000	68,000	84,000
10	50,000	75,000	90,000	81,000	120,000	150,000

STEAM PIPE CAPACITIES, (Cont.)

150 PSIG / 200 PSIG

Pipe Size	1	2	3	2	3	4
3/4	120	175	220	200	240	295
1	240	350	440	400	490	590
1-1/4	500	700	900	800	1000	1200
1-1/2	720	1000	1300	1200	1500	1800
2	1400	2000	2600	2400	3000	3500
2-1/2	2300	3300	4000	3700	4900	5500
3	4000	6000	7200	6500	8200	10,000
4	8000	12,000	15,000	14,000	17,000	20,000
6	25,000	36,000	45,000	42,000	51,000	60,000
8	50,000	72,000	90,000	84,000	102,000	120,000
10	90,000	130,000	160,000	150,000	190,000	220,000

250 PSIG / 300 PSIG

Pipe Size	2	4	6	2	4	6
3/4	215	325	400	240	350	450
1	430	650	800	475	700	900
1-1/4	900	1350	1650	1000	1450	1950
1-1/2	1400	2000	2500	1500	2300	2800
2	2600	3800	4800	2800	4100	5400
2-1/2	4000	6000	7500	4500	7000	8400
3	7500	11,000	14,000	8400	12,000	15,000
4	15,000	22,000	28,000	16,800	24,000	30,000
6	45,000	66,000	84,000	50,000	72,000	90,000
8	90,000	132,000	168,000	100,000	150,000	180,000
10	160,000	250,000	300,000	180,000	270,000	350,000

NOTES ON THE USE OF THE STEAM PIPE CAPACITY CHARTS

1. The column headings, ¼, ½, 1, etc., refer to the pressure drop, in pounds per square inch for 100 feet of pipe or equivalent piping.

2. When three columns are given from which to choose, the highest may be used for branch run-outs, the middle one for average main runs, and the lowest one for special cases.

3. A safe rule to follow is that the total pressure drop throughout a main run of 5% to 10% of the available boiler pressure may usually be tolerated.

RETURN PIPE CAPACITIES

In Pounds Per Hour

Pipe Size In.	Low Pressure		High Pressure Return, Steam Pressure in Psig					
	Gravity Return	Vacuum Return	25	50	100	150	200	250
3/4			236	312	419	560	682	1,074
1	200	350	474	617	823	1,120	1,138	2,150
1-1/4	400	600	989	1,306	1,755	2,330	2,880	4,450
1-1/2	700	950	1,610	2,126	2,850	3,800	4,710	7,350
2	1,200	2,000	3,280	4,325	5,785	7,700	9,550	14,875
2-1/2	1,650	3,350	5,400	7,160	9,640	12,800	15,850	24,600
3	2,600	5,350	9,890	13,070	17,550	23,300	28,850	34.750
4	6,500	11,000	20,800	27,360	36,550	49,200	60,900	73,350
5	10,400	19,400	38,850	52,925	70,000	91,500	114,500	127,600
6	18,000	31,000	61,200	83,700	112,700	150,000	185,500	223,100

NOTES ON USE OF ABOVE TABLE

1. The "Steam Pressure in PSIG" refers to the pressure ahead of the traps which discharge into the return line being sized.

2. If the condensate is being collected in a receiver, and then pumped back to a deaerator or hotwell, the return line should be sized with the use of tables giving pressure drop vs. flow in pipe.

3. For the pumped return system mentioned in 2, the pump must be selected to pump against the total of the following resistances:
    a) Return pipe friction from Note 2 above.
    b) Vertical difference in elevation from the level in the condensate receiver to the level in the hotwell or deaerator.
    c) The gage pressure on the surface of the water in the hotwell or deaerator.

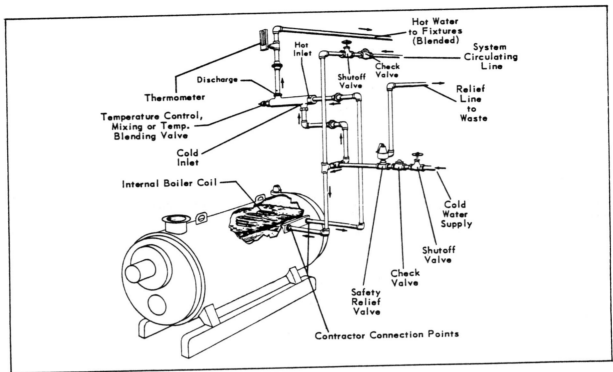

Figure No. 1 — Typical Internal Boiler Coil Piping Layout

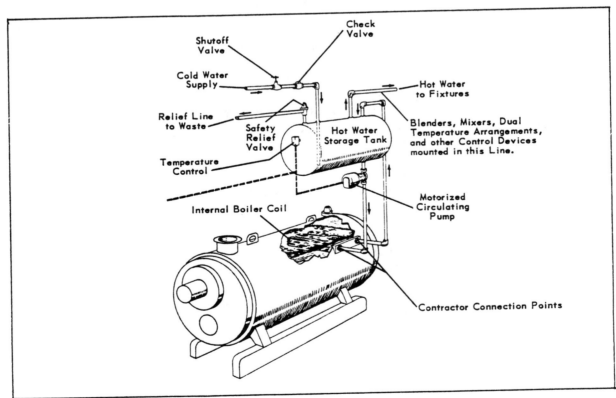

Figure No. 2 — Typical Internal Boiler Coil Piping Layout Using a Hot Water Storage Tank

North American
Mfg. Co.
Cleveland, OH 44105 USA

Light Oil System Piping
showing auxiliary transfer pump
and circulating loop

Drawing LO9
7-63

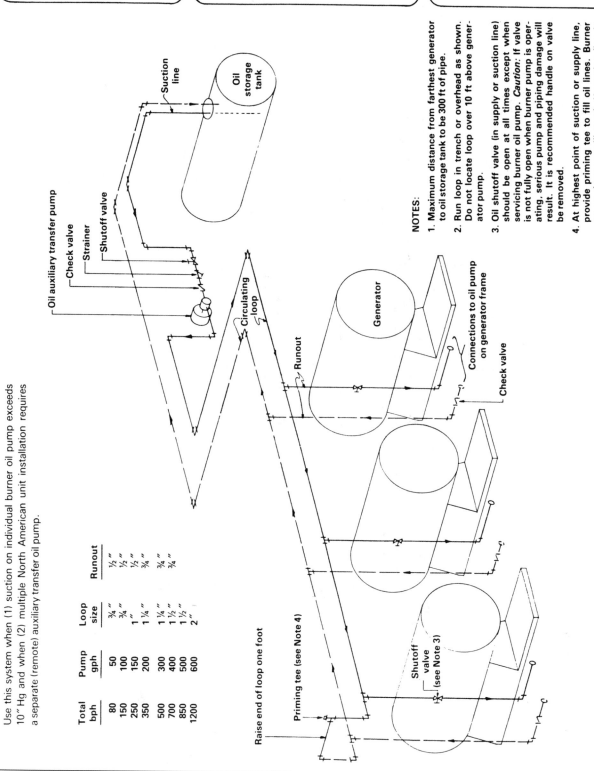

NOTES:

1. Maximum distance from farthest generator to oil storage tank to be 300 ft of pipe.

2. Run loop in trench or overhead as shown. Do not locate loop over 10 ft above generator pump.

3. Oil shutoff valve (in supply or suction line) should be open at all times except when servicing burner oil pump. *Caution:* If valve is not fully open when burner pump is operating, serious pump and piping damage will result. It is recommended handle on valve be removed.

4. At highest point of suction or supply line, provide priming tee to fill oil lines. Burner pump damage will result if suction line is not filled with oil.

Use this system when (1) suction on individual burner oil pump exceeds 10" Hg and when (2) multiple North American unit installation requires a separate (remote) auxiliary transfer oil pump.

Total bph	Pump gph	Loop size	Runout
80	50	¾ "	½ "
150	100	¾ "	½ "
250	150	1 "	½ "
350	200	1¼ "	¾ "
500	300	1¼ "	¾ "
700	400	1½ "	¾ "
850	500	1½ "	
1200	600	2 "	

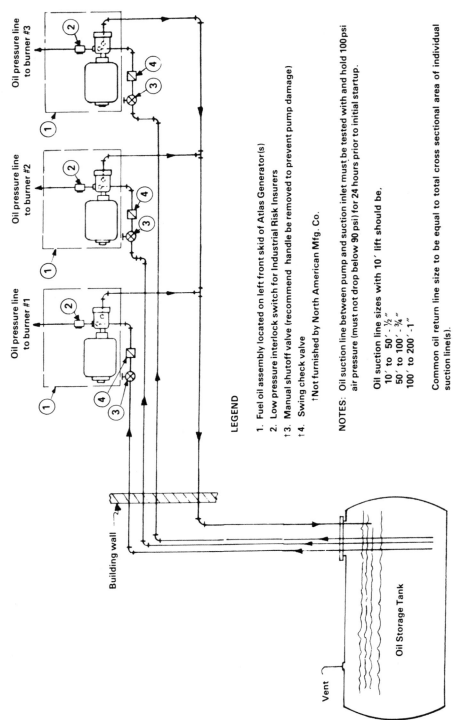

North American
Mfg. Co.
Cleveland, OH 44105 USA

Light Oil Piping Layout
for burner(s) on generator
sizes 80 hp to 150 hp

Drawing LO4
9-78

Oil pressure line
to burner #3

Oil pressure line
to burner #2

Oil pressure line
to burner #1

Building wall

Oil Storage Tank

Vent

Individual suction line(s) and
common return line for Industrial
Risk Insurers installations.

LEGEND

1. Fuel oil assembly located on left front skid of Atlas Generator(s)
2. Low pressure interlock switch for Industrial Risk Insurers
†3. Manual shutoff valve (recommend handle be removed to prevent pump damage)
†4. Swing check valve
  †Not furnished by North American Mfg. Co.

NOTES: Oil suction line between pump and suction inlet must be tested with and hold 100 psi
air pressure (must not drop below 90 psi) for 24 hours prior to initial startup.

Oil suction line sizes with 10′ lift should be.
  10′ to 50′ - ½″
  50′ to 100′ - ¾″
  100′ to 200′ -1″

Common oil return line size to be equal to total cross sectional area of individual
suction line(s).

Install priming tee at highest point of individual suction line(s).

No. 4 Oil Piping
Multiple Boiler Installation
Remote Oil Pumps

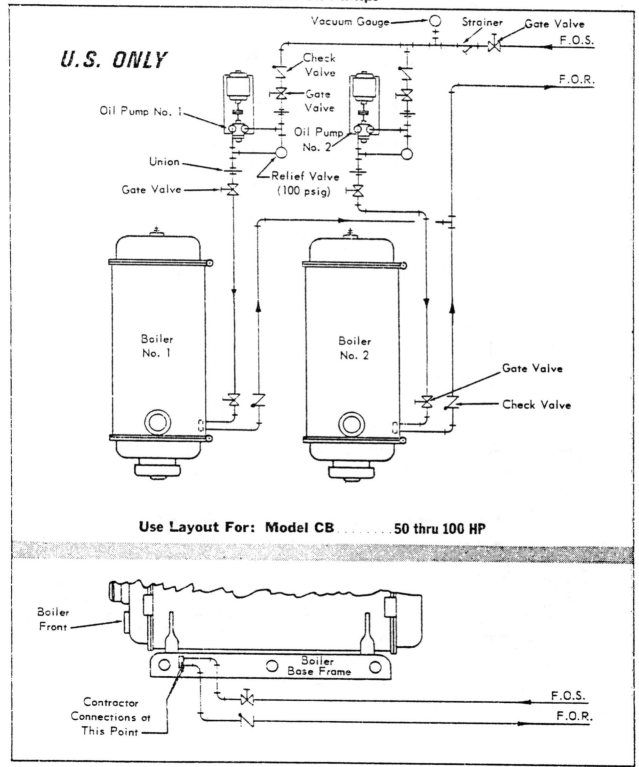

Use Layout For: Model CB........50 thru 100 HP

No. 5 and No. 6 Oil Piping
Multiple Boiler Installation

Remote Oil Pump (Stand-by pump Optional)   **Pressurized Loop System**

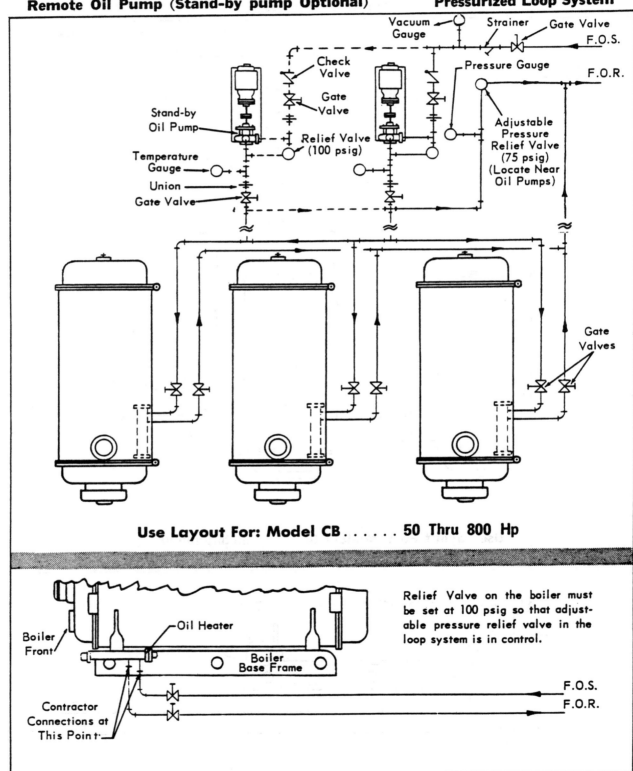

Use Layout For: Model CB......50 Thru 800 Hp

Relief Valve on the boiler must be set at 100 psig so that adjustable pressure relief valve in the loop system is in control.

SPECIFICATIONS FOR PIPING, VALVES AND FITTINGS

The following specifications are recommendations for use on steam and condensate lines for use in those plants desiring to cut maintenance costs to a reasonable limit. If these items are not readily available from your regular suppliers, then we suggest you be very careful in accepting any offered substitutes, as these specifications are used in the larger and better engineered construction projects for the power, refinery and chemical fields.

All joints should preferably be welded, except where piping joins flanged equipment.

Cast iron flanges, flat faced, should be used only on condensate systems, where the line joins equipment, and then only with the use of full faced gaskets.

Do not attempt to use galvanized piping when joints are to be welded, as welding galvanized pipe produces toxic gases.

All steam piping should be black steel.

Condensate lines may be galvanized, threaded.

For Condensate Lines and Steam, To 150# @ 366°F

PIPE

2" and smaller	Carbon Steel, Seamless Sch. 80	ASME SA-106 Gr. B
2-1/2" and up	Carbon Steel, Seamless Socket-Welding	ASME SA-106 Gr. B

FITTINGS

2" and smaller	3000 lb. Carbon Steel,	ASME SA-234 Gr. WPB
2-1/2" and up	Carbon Steel, Seamless Butt-Welding (Same Schedule as Pipe)	ASME SA-234 Gr. WPB

FLANGES

2" and smaller	150 lb. Carbon Steel Raised Face, Socket-Welding	ASME SA-105 or ASME SA-181 Gr. II
2-1/2 and up	150 lb. Carbon Steel, Seamless Butt Welding or Slip-on (Same Schedule as Pipe)	ASME SA-105 or ASME SA-181 Gr.II

GASKETS

	Compressed asbestos, Rubber Bonded Flat Ring 1/16" Thickness	ANSI B16.21

BOLTING

Continuous Thread Stud Bolts Hex Nuts, Semi-finished	ASME SA-193 Gr. B7 ASME SA-194 Gr. 2H

VALVES, see attached list
Carbon Steel, Ductile Iron and Bronze

		Valve Numbers		
		GATE	GLOBE	CHECK
2" & smaller	600# SW	#1	#3	#5
$2\frac{1}{2}$ & up	150# BW	#2	#4	#6
2" & smaller	150# scrd	Scrd. equivalent to 150 #BW		

For Steam To 500# @ 470°F

PIPE

3" and smaller	Carbon Steel, Seamless Sch. 80	ASME SA-106 Gr. B
4" and up	Carbon Steel, Seamless Sch. 40	ASME SA-106 Gr. B

FITTINGS

2" and smaller	3000 lb. Carbon Steel, Socket Welding	ASME SA-234 Gr. WPB
2-1/2" and up	Carbon Steel, Seamless Butt-Welding (Same Schedule as Pipe)	ASME SA-234 Gr. WPB

FLANGES

2" and smaller	300 lb. Carbon Steel Raised Face, Socket-Welding	ASME SA-105 or ASME SA-181 Gr. II
2-1/2 and up	300 lb. Carbon Steel, Raised Face Welding Neck (Bore to Match Pipe)	ASME SA-105 or ASME SA-181 Gr.II

GASKETS

All Sizes	Flexitallic Type CG-CA-304SS and C.S. Outer Ring Nominal original thickness .175"

BOLTING

Continuous Thread Stud Bolts	ASME SA-193 Gr. B7
Hex Nuts, Semi-finished	ASME SA-194 Gr. 2H

VALVES, see attached list
Carbon Steel

		Valve Numbers		
		GATE	GLOBE	CHECK
2" & smaller	600# SW	#1	#3	#5
$2\frac{1}{2}$" & 3"	300# BW	#7	#10	#12
4"	300# BW	#8	#11	#13
6" & up	300# BW	#9	#11	

VALVES

#1 600# Gate, 2" and smaller carbon steel, cast ASME SA216 Gr, WCB or forged ASME SA105, socket weld ends, bolted bonnet, hard faced solid wedge, hard faced integral or pressed in seats.

#2 Equivalent to #1 in 150# BW.

#3 600# Globe, 2" and smaller, carbon steel, forged ASME SA105, socket weld ends, bolted or union bonnet, hard faced seat and hard faced disc.

#4 Equivlanet to #3 in 150# BW.

#5 600# Check, 2" and smaller, carbon steel, cast ASME SA216 Gr, WCB or forged ASME SA105, socket weld ends, bolted or screwed cap, integral or welded in seat ring, piston type disc. Manufacturer's standard trim.

#6 Equilivalent to #5 in 150# BW.

#7 300# Gate, carbon steel, cast ASME SA216 Gr. WBC or forged ASME SA105, butt weld ends for schedule 80 pipe, bolted bonnet, O.S. & Y., hard faced solid wedge, integral or seal welded, hard faced seat rings.

#8 Same as Valve #187 except butt weld ends for schedule 40 pipe.

#9 300# Gate, carbon steel, cast ASME SA216 Gr, WCB or forged ASME SA105, butt weld ends for schedule 40 pipe, bolted bonnet, O.S. & Y., hard faced flexible wedge, integral or seal welded, hard faced seat rings.

#10 300# Globe, carbon steel, cast ASME SA216 Gr. WCB or forged ASME SA105, butt weld ends for schedule 80 pipe, bolted bonnet, O.S. & Y., hard faced fully seal welded seat rings, hard faced disc.

#11 300# Globe, carbon steel, cast ASME SA216 Gr. WCB or forged ASME SA105, butt weld ends for schedule 40 pipe, bolted bonnet, O.S. & Y., seal welded hard faced seat wring and hard faced disc.

#12 300# Check, carbon steel, cast ASME SA216 Gr. WCB or forged ASME SA105, butt weld ends for schedule 80 pipe, renewable seat ring, swing type disc, bolted cap. Manufacturer's standard trim.

#13 Equivalent to #12 except BW ends for schedule 40 pipe.

SECTION E
BOILER TRIM AND CONTROLS

TRIM LIST

NO.	SIZE	ITEM	SPECIFICATION	SHOP ASSEMBLED	FIELD INST. BY CO.	FIELD INST. BY PUR.
One		Water Column	Reliance W350X	x		
One		Level Alarm W/C.O.	Reliance EA-17	x		
One		Low Water C.O.	Reliance EA-100	x		
One		Level Gauge	Reliance prismatic			x
One		Gauge Valve	Reliance SG-854	x		
Three		Try-Cocks	Reliance 302	x		
One		Illuminator	Reliance WP-58			x
One	3/4	Column Drain Valve	R, P & C F-800	x		
One	3/4	Aux. L.W. Drain	R, P & C F-800	x		
One	3/8	Gauge Drain Valve	R, P & C F-800	x		
One	6	Steam Pressure Gauge	Ashcroft 1377RS	x		
One	1/2	Steam Gauge Siphon	Ashcroft 1377	x		
Two	1/2	Steam Shutoff and Test Valve	R, P & C F-560	x		
		Pipe and Fittings				
One	2½	S.B. Main Head Shutoff				
One	3/4	S.B. Drain Valve				
		Pipe and Fittings				

The piping for each boiler shall consist of connections between the blowoff outlet and the blowoff valve placed immediately outside of the setting, the necessary steam and water connections for the water column and the steam gauge piping. This piping shall be furnished by _____ The Company _____.

Trade or manufacturers' names and figure numbers listed above are to indicate quality furnished, and the COMPANY reserves the right to substitute equipment of equal quality without notice.

Courtesy of Zurn Industries Inc., Energy Division.

TRIM LIST (Cont'd.)

NO.	SIZE	ITEM	SPECIFICATION	SHOP ASSEMBLED	FIELD INST. BY CO.	FIELD INST. BY PUR.
		FEEDWATER SYSTEM				
One	3	F.W. Stop Valve	Powell 3003			x
One	3	F.W. Check Valve	Powell 3061A			x
One	–	F.W. Reg. Cont. Valve	Fisher			x
Two	2½	F.W. Reg. Isolation Valves	Powell 3003			x
One	3	F.W. Reg. By-pass Valve	Powell 3031			x
One	3/4	F.W. Reg. Drain Valve	R, P & C F-800			x
One	3/4	F.W. Reg. Shutoff	R, P & C F-56D			x
		Pipe and Fittings				
Two	4/3	Boiler Safety Valves	Consolidated 1811 NA			x
One	1	Steam Drum Vent	R, P & C F-80D			x
One	1½	Drum Blowoff Valves	Edward 1441/1443			x
		Header Blowoff Valves				
One	3/4	Chemical Feed Valve	R, P & C F-80D			x
		Chemical Feed Shutoff				
One	3/4	Cont. B.D. Shutoff Valve	R, P & C F-80D			x
		Flow Control Valve				
		Super. Safety Valve				
		Super. Drain Valve				
		Super. Vent Valve				
		Super. Therm.				
		Econ. Drain Valve				
		Econ. Vent Valve				

Courtesy of Zurn Industries, Inc.

North American

ENGINEERING DATA
STANDARD TRIM

GENERAL DATA
Section R, Page 1

2-76

STANDARD TRIM	350	360	380	390	3100	3100X	3125	3125X	3150	3200	3250	3300	3350	3400	3500	3600	3700
193D McDonnell-Miller Water Column, Pump Control, LWCO and Alarm—150# WP	H P	H P	H P	H P	H P	H P	H P	H P	H P	H P	H P	H P	H P	H P	H P	H P	H P
Penberthy Water Gauge, Fig. 36A, with glass ⅝" Dia. x 9⅝" Lg. — 150# WP	H P	H P	H P	H P	H P	H P	H P	H P	H P	H P							
Penberthy Try Cocks, Fig. SB2, ½" IPS — 150# WP (3 req'd)	P	P	P	P	P	P	P	P	P	P	P						
Ernst DCP Try Cocks—½" Chain Operated — Fig. 15 (3 req'd)											P	P	P	H P	H P	H P	
Ernst 6 — Chain Operated Water Gauge Set ½" IPS											P	P	P	H P	H P	H P	
Marshalltown Pressure Gauge Fig. 24, Range 0-30#	H 6"	H 6"	H 6"	H 6"	H 6"	H 6"	H 6"	H 6"	H 6"	H 6"	H 8½"	H 8½"	H 8½"	H 8½"	H 8½"	H 8½"	
Marshalltown Pressure Gauge Fig. 24, Range 0-300#	P 6"	P 6"	P 6"	P 6"	P 6"	P 6"	P 6"	P 6"	P 6"	P 6"	P 8½"	P 8½"	P 8½"	P 8½"	P 8½"	P 8½"	
Marshalltown Altitude Pressure Gauge, Fig. 41, 0 to 100 lbs., 0 to 231 ft.	W 4½"	W 4½"	W 6"	W 6"	W 6"	W 6"	W 6	W 6"	W 6"	W 6"	W 6"	W 6"					
American Dial Thermometer, 5" Dia., 50 to 300 F, 8" Probe, Fig. 6360AH-8	W	W	W	W	W	W	W	W	W	W	W	W					
64B McDonnell-Miller LWCO and Alarm 50# WP	W	W	W	W	W	W	W	W	W	W	W	W	W	W	W	W	
Penberthy Gauge Cock, ¼" IPS	H P	H P	H P	H P	H P	H P	H P	H P	H P	H P	H P	H P	H P	H P	H P	H P	
Marshalltown Pressure Altitude Gauge #29C, 0 to 60#, 0-138 ft., 8½" dia.													W	W	W	W	
Marshalltown Thermometer #17, 30-300 F, 8½" dial, 8" bulb, 8 ft. cap.													W	W	W	W	
OPTIONAL TRIM																	
Feed Water Globe Valve 200# WP	P 1¼"	P 1¼"	P 1¼"	P 1¼"	P 1¼"	P 1¼"	P 1¼"	P 1¼"	P 1½"	P 1½"	P 1½"	P 2"	P 2"	P 2"	P 2½"	P 2½"	
Feed Water Horizontal Swing Check Valve 200# WP	P 1¼"	P 1¼"	P 1¼"	P 1¼"	P 1¼"	P 1¼"	P 1¼"	P 1¼"	P 1½"	P 1½"	P 1½"	P 2"	P 2"	P 2"	P 2½"	P 2½"	
Injector Globe Valve 200# WP (2 req'd)	P ¾"	P 1"	P 1"	P 1¼"	P 1¼"	P 1¼"	P 1¼"	P 1¼"	P 1½"	P 1½"	P 1½"	P 2"	P 2"	P 2"	P 2½"	P 2½"	
Injector Horizontal Swing Check Valve 200# WP	P ¾"	P 1"	P 1"	P 1¼"	P 1¼"	P 1¼"	P 1¼"	P 1¼"	P 1½"	P 1½"	P 1½"	P 2"	P 2"	P 2"	P 2½"	P 2½"	
Penberthy Injector — 250# Steam Supply Pressure	P ¾"	P 1"	P 1"	P 1¼"	P 1¼"	P 1¼"	P 1¼"	P 1¼"	P 1½"	P 1½"	P 1½"	P 2"	P 2"	P 2"	P 2½"	P 2½"	
Drain Angle Valve 125# WP	H 1½"	H 1½"	H 1½"	H 1½"	H 1½"	H 1½"	H 2"	H 2"	H 2"	H 2"	H 2"	H 2½"	H 2½"	H 2½"	H 2½"	H 2½"	
Drain Gate Valve 125# WP	H 1½"	H 1½"	H 1½"	H 1½"	H 1½"	H 1½"	H 2"	H 2"	H 2"	H 2"	H 2"	H 2½"	H 2½"	H 2½"	H 2½"	H 2½"	
Y Type Blow Down Valve Jenkins #134 — 250# WP	P 1¼"	P 1¼"	P 1¼"	P 1¼"	P 1¼"	P 1¼"	P 1¼"	P 1¼"	P 1½"	P 1½"	P 1½"	P 1½"	P 2"	P 2"	P 2"	P 2"	
Lubricated Plug Valve Nordstrom #214 — 250# WP	P 1¼"	P 1¼"	P 1¼"	P 1¼"	P 1¼"	P 1¼"	P 1¼"	P 1¼"	P 1½"	P 1½"	P 1½"	P 1½"	P 2"	P 2"	P 2"	P 2"	
Stack Thermometer — American, 3" dial ½" N.P.S., #6360BH-S6, Range 200 to 700 F	H P W	H P W	H P W	H P W	H P W	H P W	H P W	H P W	H P W	H P W	H P W	H P W	H P W	H P W	H P W	H P W	

LEGEND: H — 15# Steam P — 150# Steam W — 30# Hot Water

Courtesy of E. J. Walsh & Co., Ltd.

LOGIC FLOW DIAGRAM 70D10

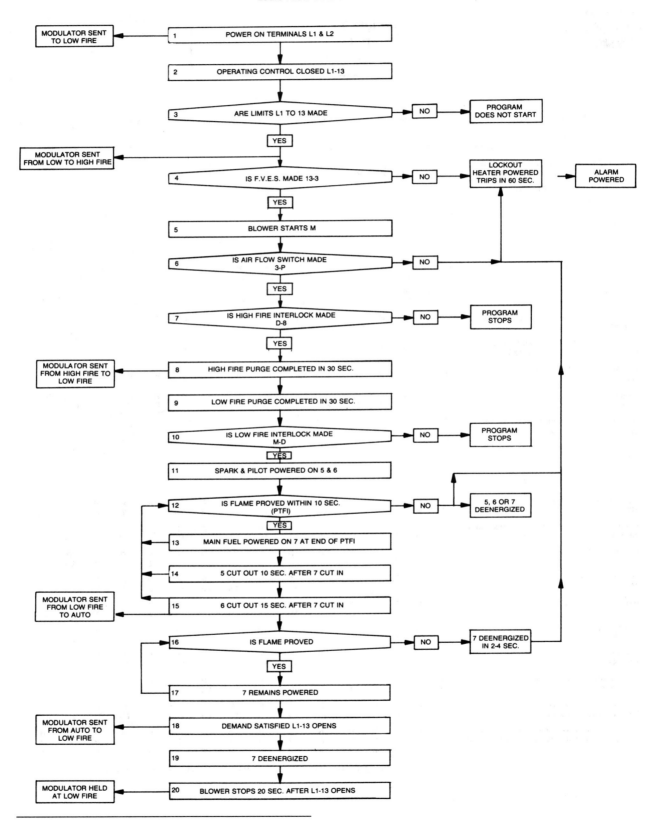

THE FLAME-MONITOR™ FLAME SAFEGUARD SYSTEM

The FIREYE FLAME-MONITOR is the most advanced single burner flame safeguard control available. Modular design lets you tailor the FIREYE FLAME-MONITOR to your specific burner cycle and flame amplifier characteristics. Proven quality and performance in thousands of installations makes it your best choice.

These unique features separate the FIREYE FLAME-MONITOR from all other flame safeguard controls:
- A vocabulary of 42 status messages are scrolled on the LED display in clear English language words and phrases to describe burner status and present historical information. No codes to learn, no confusion over message content.

A non-volatile memory which allows the control to remember its status and history when power is interrupted.
- Remote display capability.
- Built-in flame signal display.
- Multiple flame scanning amplifiers including ultra-violet, autocheck infrared and ultra-violet self-check.
- 9 interchangeable programmers for proper selection of timing functions, purge, flame failure response, display language and operation.

Service technicians like the FIREYE FLAME-MONITOR because of its ability to quickly lead them to the cause of a burner shutdown. When such a shutdown occurs, the display indicates the reason and where in the burner cycle it occurred. Without the FIREYE FLAME-MONITOR, it is often difficult for the technician to pinpoint the cause of the shutdown.

Check bulletin E1001 and E2001 for details on the operational specifications and additional features of the FIREYE FLAME-MONITOR.

E300 EXPANSION MODULE ADDS MULTIPLE INTERLOCK/LIMIT SWITCH READOUTS TO THE FIREYE FLAME-MONITOR™

The E300 EXPANSION MODULE adds additional monitoring and first-out annunciator capability for up to 16 specific interlock functions.

When a number of safety interlocks are wired in series, the FIREYE FLAME-MONITOR can only identify an open circuit in the string. By using the EXPANSION MODULE, it is possible for the FIREYE FLAME-MONITOR to display by name the exact interlock which caused the open circuit:

HIGH WATER	LOW WATER
HIGH GAS PRESSURE	LOW OIL PRESSURE
LOW GAS PRESSURE	LOW OIL TEMPERATURE
HIGH PRESSURE	LOW ATOMIZING MEDIA
HIGH TEMPERATURE	AIR FLOW
AUXILIARY LIMIT #1	AUXILIARY LIMIT #2
AUXILIARY LIMIT #3	AUXILIARY LIMIT #4
AUXILIARY LIMIT #5	AUXILIARY LIMIT #6

Servicemen can tell not only that a limit circuit has caused the boiler shutdown, but which specific limit is at fault.

The E300 EXPANSION MODULE will identify the first limit switch which opens to cause the shutdown and the FIREYE FLAME-MONITOR will display it (even if a power failure occurs) until the control is reset.

Fast, easy wiring from the FIREYE FLAME-MONITOR to the EXPANSION MODULE is accomplished via a convenient ribbon cable that has plug-in connectors at both ends.

Courtesy of Fireye Div., Allen-Bradley Co.

THE INDUSTRY'S MOST COMPREHENSIVE BURNER STATUS COMMUNICATIONS SYSTEM

The FIREYE FLAME-MONITOR System includes a unique communications capability for local and/or remote supervision of multiple burner installations. Current status for each burner plus historical data relating to hours of operation, number and cause of lockouts are readily available on both portable and workstation display monitors.

Maintenance and security personnel can monitor burner status in multiple-building complexes from a centralized supervisory station. The E500 COMMUNI-CATIONS MODULE brings real-time burner status data to any desired local or remote supervisory site.

THE E500 COMMUNICATION MODULE PROVIDES LOCAL OR REMOTE SUPERVISION OF EVERY FIREYE FLAME-MONITOR™ INSTALLATION

Communication of information in the commercial/industrial burner market has never been as easy as it is now. The FIREYE E500 COMMUNICATION MODULE allows you to be in constant contact with the FIREYE FLAME-MONITOR flame safeguard control system from virtually anywhere.

The E500 COMMUNICATION MODULE is mounted in the vicinity of the burner and monitors up to four FIREYE FLAME-MONITORS through a common communication output.

A built in telephone modem provides long distance data transmission via telephone lines. Using a desktop or portable data terminal, together with a modem, you can find out what is happening at the job site at any time. The E500 COMMUNICATION MODULE also allows you to interrogate the memory of the FIREYE FLAME-MONITOR to obtain such information as the number of lockouts experienced, burner cycle history, and the cause of the last six shutdowns.

An "AUTO CALL" switch on the COMMUNICATION MODULE allows the control to initiate a phone call if a lockout occurs. The user can program a phone number for the unit to call in the event of a failure.

Each unit can be individually serialized from a remote terminal by typing in a user selected 32 character description. It is also possible for the terminal operator to leave a 32 character message (non-volatile) on the E500 COMMUNICATION MODULE for the next caller.

When wiring direct to a local terminal, an easily accessible RS232c port on the E500 facilitates data transmission hook-up.

A clock keeps track of the current time and day of the week so that if an "AUTO CALL" is initiated the terminal operator will know when the lockout occurred. The clock can be reset from a terminal.

The menu list of 21 commands to the E500 COMMUNICATION MODULE provide the user with a flexibility never before offered in this market.

The ability to know what is happening in a boiler room located thousands of miles away is now an exciting reality. Even more exciting is the ability of the control to call you automatically when a shutdown occurs.

E500 System Connected to Four FLAME-MONITORS

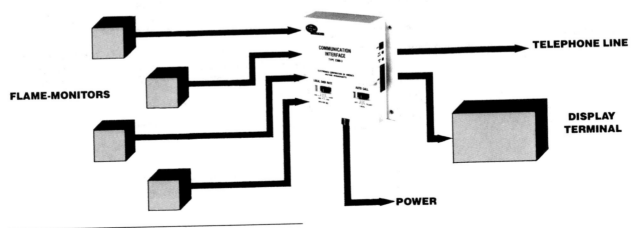

Courtesy of Fireye Div., Allen-Bradley Co.

Fusible Plugs

Most of the older designs of firetube, and some watertube, boilers have a "Last Ditch" safety device known as a fusible plug. This is a bronze plug with a tapered hole through the center, which has been filled with tin having a melting point of about 450°F. This plug is generally placed in a tube sheet, crown sheet, or similar surface, with one side exposed to the hot flue gases and the other side cooled by the boiler water. If the boiler water level drops below the plug, the increased temperature resulting from loss of cooling water melts the tin filling, and steam blows through it into the flue gas side. The sound is usually loud and distinctive enough to alert the operators to the dangers. It is placed at a level in the boiler that is just above the top row of tubes, clearing them by 1–2 in.

Some of the earlier designs located them in position to smother the fire in the furnace, on coal- or wood-fired units.

Fusible plugs must be kept free of all scale or soot, since the resulting insulating effect would cause them to overheat and blow prematurely. They must be replaced at every shutdown for inspection, since the tin filling varies from its set melting point with time and heat. It is possible to refill the plugs, but it is better to replace them with new ones.

Present practice is to supply them only in coal-fired boilers, since the ASME code only requires them for this service. The code also stipulates the location in the various design of boilers.

These plugs are not used in the higher-pressure boilers, only in the low- to medium-pressure units, up to about 125 psig. Above that pressure, the saturation temperature of the water is too close to the melting point of the tin filling to provide troublefree protection.

SECTION F
OPERATION AND MAINTENANCE

BOILER SERVICE MANUAL

NORTH AMERICAN ATLAS GENERATORS
Startup & Maintenance Instructions

November, 1975

Welcome to the growing family of North American Atlas Boiler owners. We have designed and built what we consider to be the most advanced generator on the market today. We have invested thousands of man hours in developing the most economical, maintenance-free and safe boiler that modern technology can provide.

This Service Manual is for your information. Study it, follow its instructions, and familiarize yourself with all of the equipment on your Atlas Generator for years of trouble-free service.

The Atlas nationwide service organization is now available to you. This organization is made up of local factory-trained personnel, backed by North American Service Engineers. These specialists are as near as your telephone. If you have questions, ask us. Only in this way can we be assured that your requirements are taken care of.

North American Mfg. Co. **Cleveland, Ohio, 44105**

340

SECTION F

North American Atlas Startup & Maint. Inst. page 3 of 8.

I. PREPARING BOILER FOR INITIAL STARTUP.
On first startup, all mill scale, oil and
foreign matter must be cleaned from the
boiler by "boiling out" the boiler as
described in Section II below. (The
boiler comes from the factory with a protective coating of oil on the tubes.
Until this oil is removed, there is always the possibility of the tubes becoming overheated.)

1. Startup should be planned for early
morning to allow enough time to properly
prepare the boiler for service.

2. The owner should have received his
Manufacturer's Data Sheets (approved by
the insurance company). Instruction
manuals for the burner and all auxil-
liaries should be on hand.

3. The owner should have on hand the
necessary chemical compounds (2 pounds
caustic soda and 2 pounds soda ash for
each 1000 pounds of water for boilout, plus
feedwater treatment compound as described
in Section VII).

4. The owner should see that all of his
operating personnel are present for instruction and to actually perform the
start-up (under the agent's supervision)
because they will only learn by doing.

5. The agent and the heating contractor's job foreman should be present.

a. If a unit is installed so that an
unsafe condition exists, something is
piped wrong, or piping leaks show up,
they should be settled while all parties
are present--not weeks later.

II. PREPARING BOILER FOR SERVICE.

1. Remove all handhole and manhole plates,
coat asbestos gaskets with a minimum of
graphite paste to prevent them from sticking to the plate or the boiler shell, then
replace. Tighten the handhole plates.

2. Leave top manhole or handhole plate
off so that air can escape when filling.

3. Open bottom drain or blowoff valves.

4. Check that all other valves are closed.

5. Fill boiler with city water pressure
or with feed pumps.

6. Allow water to flow into boiler and
out bottom drain for a couple of minutes.
Close bottom drain and fill unit to within 5" of top.

7. Remove safety valves, if possible, and
plug openings.

8. Mix 2 pounds of caustic soda and 2
pounds of soda ash for each 1000 pounds
of water in the boiler. The general
arrangement drawing gives the pounds of
water with the boiler full. If chemicals
are in solid form, dissolve in warm water
before pouring into boiler. Caution: warn
all personnel that the solution is extremely
caustic. Do not spill it on the boiler as
it will remove the paint. Pour the solution
into the manhole or handhole opening on top
of boiler.

9. Replace the manhole or handhole cover.
Leave safety pop valves blocked open or
pipe plug open. Close main steam valve
and return valves.

10. Start the burner, following the burner
operating instructions and check all limits
for proper operation. Slowly raise temperature of solution to boiling. Boil for
20 minutes and then close safety valve or
replace plug. Above procedure is necessary
to remove air in water.

11. Raise the boiler pressure to 5 psi
(pounds per square inch). Set the pressuretrols to maintain this pressure.

12. If the safety valves have been removed, the boiler should not be left unattended, under any circumstances. Check
the steam gauge continuously. If an insurance inspector is present, he will not, in
all probability, allow the safety valves
to be removed. Do not argue with him.
Do as he says.

13. After the boiler has boiled for at
least 5 hours, turn off the burner. Open
the front doors, rear doors and access
door. Inspect all tubes for tightness.
Some tubes may need hand rolling because

of shipment vibrations. Do not mistake moisture condensation on the tube sheet for a weeping tube. (The cleaning solution will show as a dirty white substance.)

14. Replace all doors and tighten lugs and nuts.

15. Open safety valves and release the steam pressure. When the pressure is completely released open the drain valve and lower the water level to the center of the gauge glass. If a hot water boiler, drain until the low water light comes on.

16. Remove manhole or top handhole plates. Don't worry if you drop the manhole plate. You can get it after you've drained the boiler. Don't drop the handhole plate. Tie a wire to the bolt before removing it.

17. Open the drain valve at the bottom of the boiler and run <u>hot</u> water into top of boiler until the shell of the boiler is cool enough to touch. If hot water is not available, open by-pass valve and slowly fill boiler to within 5" of top. Run cold water into the top and crack open the drain valve. Once the boiler has cooled, open bottom drain full open. Hose off interior of boiler with high pressure hose as the water line drops.

18. When the boiler is empty, remove handhole plates and inspect. If not clean, repeat cleaning. Don't hurry. You're only going to do this once in the life of the boiler. Hurrying will shock the boiler and loosen the tubes.

19. Coat asbestos gaskets with a minimum of graphite paste to prevent them from sticking, and replace the manhole and handhole plates.

20. Replace safety valves. Block safety pop off valves open and fill boiler to operating level.

21. When filling boiler for service, feedwater treatment compound should be added, both for hot water and steam boilers. Compound should have oxygen scavenger. See Section VII.

22. Start burner on Low Fire following Operating Instructions. Raise boiler water temperature slowly.

23. Steam should be coming out of the open safety valve. After all air is exhausted, close safety valve. Raise steam pressure slowly to 5 psi. Tighten all manhole and handhole plates. Raise the pressure until all safety valves pop. Check the pressure at which the safety valves relieved against stamping on valves.

24. Reset pressuretrols for operating pressure.

25. Again check all limits for proper operation under pressure.

26. Open main steam valve and let automatic controls operate the boiler normally.

III. STANDARD CONTROLS FOR LOW WATER, FEED WATER AND SAFETY RELIEF PRESSURE. Refer to Section IV for start-up procedure.

1. High Pressure Steam Boilers have:

   a. McDonnell-Miller 193D combination water column, pump control, low water cutoff and alarm. (150 psi W.P.)

   b. Warrick probe type IGIDO low water cutoff and alarm. (Optional--recommended as a secondary safety control.)

   c. Lonergan No. 11W safety valves set at 150 psi working pressure.

2. Low Pressure Steam Boilers have:

   a. McDonnell-Miller 193D combination pump control, low water cutoff and alarm. (150 psi W.P.)

   b. McDonnell-Miller 51-B2 water feeder, low water cutoff, and alarm. (35 psi W.P.) (optional)

   c. Warrick probe type IGIDO low water cutoff and alarm. (Optional--recommended as a secondary safety control.)

   d. Kunkle Figure 930 safety valves set at 15 psi W.P.

3. Low Pressure Hot Water Boilers have:

   a. McDonnell-Miller 64B low water cutoff and alarm. (50 psi W.P.)

   b. Warrick probe type IGIDO low water cutoff and alarm. (Optional--recommended as a secondary safety control.)

c. Watts Figure 740 relief valves set at 30 or 36 psi W.P.

All of the above water controllers are the automatic reset type. The McDonnell-Miller 193D and Warrick IGIDO relay may be specified with manual reset.

IV. PREPARING AUTOMATIC PUMP, LOW WATER, AND SAFETY VALVE CONTROLS FOR INITIAL STARTUP.

1. McDonnell-Miller controllers.

a. Remove the wooden plug that has been inserted in the float chamber to prevent damage during shipment. Replace it with the metal plug provided in a small white bag.

b. Check the operational levels--set at the factory. They may be changed if necessary to meet local conditions.

c. Check the low water cutoff by dropping the waterline in the boiler until the cutoff interrupts the fuel. Continue to drop the waterline until the second cutoff operates. Reset second cutoff with Blower Purge switch if probe type.

2. Warrick Probe Type IGIDO Low Water Cutoff (optional).

a. The probe is cut to length & shipped loose in the packing box or taped to the blower motor cable. Tighten the rod into the coupling with wrench, _tight_. Screw this rod & coupling assembly into the boiler coupling, _tight_.

b. Check its operation by blowing down boiler with drain valve or blow-down valves. Open control panel to watch relay drop out. Close drain valve and check that relay resets. You cannot check by watching low water alarm light.

3. Safety Valves. See Section II, Item 23. Check the capacity stamping on the safety valve against the pounds per hour stamped on the data plate on the boiler front. The total safety valve capacity should exceed the boiler capacity.

4. Water Column Gauge Glass. Open the pet cock and allow the water line to drop. Close the pet cock. The water level should return quickly.

5. Float Type Controls. On hot water boilers, loosen the pipe plug on top of

the equalizing line to bleed air from controller (64B).

6. Stack Thermometer (optional). American #6360AH-8 -3" dial thermometers, 200 F to 700 F range, are furnished. When new, their pointers normally rest at some point beyond the 200 F mark. Once heated, they return to 200 F.

7. Penberthy 36A Water Gauge and SB Try Cocks. Before boiler has pressure, tighten nut around glass. After boiler is under pressure, gently tighten packing nut on try cocks.

8. Casing. Occasionally the casing may be dented in shipment. If the dent is fairly large, a plumber's suction cup can be used to pull it out. If a small dent, drill 1/8" diameter hole, insert screw part way and pull dent out with pliers.

9. Safety Valves. If the safety pop-off valves leak, run the boiler pressure to within 5 psi of the set pressure. Manually hold valve open for 1 minute; then let it slam shut. Repeat 2 or 3 times. This will usually set the valve tightly in its seat.

10. Handhole Gasket. All handhole gaskets should be tightened under 5 psi pressure. If a leak still persists, check if gasket is even on handhole plate. If a leak still persists, drain boiler, remove handhole plate and check edge for roughness. If rough, clean with a file. Coat gasket with Permatex on both sides before tightening. If leak still persists, notify factory representative.

11. Flue Gas Condensation. Boilers operating on low fire often produce excessive moisture in the stack. This runs down over the back of the boiler. If stub stack, pry up 1½" and coat boiler vent with boiler putty. Lower stack into vent and smooth putty into all cracks. The best way to seal the stack to the vent is to weld it.

V. HYDROSTATIC BOILER TESTING IN FIELD.
If it is necessary to hydrostatically test a boiler in the field, use the following procedure:

1. Remove safety valves & plug openings.

North American Atlas Startup & Maint. Inst. page 6 of 8.

2. Subject the boiler to a test pressure 1½ times its design pressure:

Type Boiler	Rated pressure	Hydrostatic test pressure
low pressure steam	15 psi	45 psi
hot water	30 psi	45 psi
high pressure steam	150 psi	225 psi

3. Control the pressure properly as not to exceed the required test pressure by more than 6%.

4. Test with water at the temperature of the surrounding atmosphere--in no case less than 70 F.

5. Inspect for leaks while the boiler is under test pressure, release the pressure, and replace the safety valves.

VI. LAYING UP THE BOILER FOR SUMMER.

1. Run the boiler up to steaming temperature, then shut off.

2. Remove boiler manhole and open boiler drain. While boiler is draining, thoroughly flush down the waterside surfaces with a high pressure hose, using hot water.

3. Remove all handhole plates and leave all doors open so that the boiler will dry quickly. If the boiler room is not damp, the waterside may then be left as is for the summer.

4. If the boiler room is damp, replace all gaskets & refill the boiler with water. Add feedwater compounds, block the safety valve open, & boil the boiler for 2 hours to expel all oxygen from the water. Shut off burner, close all valves & openings. Leave the waterside surfaces in this condition for the summer.

5. Open all doors to the fireside surfaces. Clean the tubes thoroughly with a flue brush and wire-brush all other surfaces. Remove all soot with a vacuum cleaner and blow out all scale with an air hose.

6. Coat all fireside surfaces with SAE20 lubricating oil and leave them in this condition for the summer.

7. Close all fuel valves, swing open the burner, clean all strainers and the burner nozzle.

8. Clean all dust, soot and dirt from the safety valves.

9. Leave the disconnect to burner closed so that the protectorelay will remain dry.

10. Clean feedwater and condensate pumps. Replace packings if necessary.

11. Check all motor couplings.

12. Flush out all float chambers and check pivot points.

13. Blow out all tubing.

14. Check the burner instructions for summer shutdown procedure.

VII. BOILER FEEDWATER TREATMENT.
The feedwater for all modern high capacity boilers must be treated. This is a highly specialized field. Contact a reputable feedwater treatment organization prior to startup so that the boiler may be started at once. Read B&W's Bulletin TR537. "Eleven Ways to Avoid Boiler Tube Corrosion" This may be had by writing to: The Babcock & Wilcox Company, Tubular Products Division, Beaver Falls, Pennsylvania.

VIII. REGULAR BOILER MAINTENANCE.
(See attached log and explanation on back side.)

MAINTENANCE LOG Boiler Number _____ Date _____ to _____	Date	Mon	Tue	Wed	Thu	Fri	Sat	Sun	Mon	Tue	Wed	Thu	Fri	Sat	Sun	Mon	Tue	Wed	Thu	Fri	Sat	Sun	Mon	Tue	Wed	Thu	Fri	Sat	Sun
DAILY Records																													
1 Water Level																													
2 Steam Pressure																													
3 Feed Pump Pressure																													
4 Condensate Temperature																													
5 Flue Gas Temperature																													
6 Gas Pressure																													
7 Oil Pressure																													
8 Water Treatment																													
9 Expansion Tank Level-hot water																													
DAILY Checks																													
10 Low Water Cutoff																													
11 Water Level Cont.																													
12 Drain Gauge Glass																													
13 Comb. Safeguard																													
14 Main Fuel Valves																													
WEEKLY Checks																													
15 Clean Blower Wheel																													
16 Clean Perf. Plate																													
17 Clean Obs. Glasses																													
18 Clean Oil Nozzles																													
19 Check Air Comp. Filter																													
20 Check Air Comp. Oil Level (if req'd.)																													
21 Check Pres. Controls																													
MONTHLY Checks																													
22 Low Water Cutoff																													
23 Manually Pop Safety Valve																													
24 Clean Probe 2nd Low Water Cutoff																													
25 Clean Fireside Boiler																													
26 Clean Outside Boiler																													
27 Clean Oil Filters																													
YEARLY Checks																													
28 Replace manhole & handhole gaskets																													

North American Atlas Startup & Maint. Inst. page 8 of 8.

This log has been prepared for your convenience. Never rely on your memory or someone's word. Keep a record. Do not compromise with safety. Pay particular attention to the low water cutoff--75% of all boiler accidents result from low water.

A.   Explanation of Boiler Log Daily Checks:

Item 10.   Low Water Cutoff. Open blow-down valve below float chamber until burner circuit is interrupted and low water alarm lights.

Item 11.   Water Level Control. When blowing down column, observe that feed pump starts and stops.

Item 12.   Water Column Gauge Glass. Open pet cock so that water drops in glass. Close pet cock so that water level returns quickly. If level returns slowly, shut off burner, drop pressure on boiler, drain water in boiler to below equalizer connections. Remove water gauge, examine and rod through equalizer lines.

Item 13.   Combustion Safeguard. Consult Burner Instructions for procedure.

Item 14.   Main Fuel Valves. Consult Burner Instructions for procedure.

B.   Weekly Checks:

Items 15 to 20 consult Burner Instructions for procedures.

Item 21.   Pressuretrols (safety high limit). Remove cover and manually tilt mercury tube to interrupt burner circuit.

C.   Monthly Checks:

Item 22.   Low Water Cutoff. Simulate low water condition by lowering boiler water level with bottom drain valve, not with blow-down valve below float chamber. Dismantle and clean the low water cutoff every 6 months.

Item 23.   Safety Valve. Check safety valve by pulling the "try" lever to open position with steam at operating pressure. Release to allow valve to snap closed.

Item 24.   Clean Warrick Low Water Cutoff Probe. Relieve steam pressure and remove rod. Clean any corrosion present.

Item 25.   Clean Fireside. If the boiler is fired with gas only, and if the flue gas temperature is normal, it is only necessary to examine the fireside once every three months. Oil fired units should be examined every month. Warning: 1/32" of soot on boiler tubes causes 9.5% loss of efficiency; 1/16" of soot causes 26% loss of efficiency.

Item 26.   Clean outside of the Boiler.

   a.   Use an air hose or brush to clean the exterior surface of the boiler.

   b.   Take particular care to clean outside of safety valves by blowing compressed air. Examine for leaking valves as corrosion and failure is more likely to take place. If valve cannot be stopped leaking, replace with new valve. Have the valve repaired and save it as a spare.

   c.   Check all valves for leakage.

   d.   Perform burner maintenance.

Item 27.   Refer to Burner Section for Procedure.

D.   Yearly Checks: (see also Item 22 regarding semi-annual cleaning of low water cutoff.)

Item 28.   See Section II, Item 1.

SECTION III
INITIAL FIRING

3.1 BEFORE FIRING. Every Steam Generator is thoroughly tested and all adjustments are correctly made under actual operating conditions before shipment from the factory. Rough handling in shipment may cause loosening of plumbing connections or change some of the adjustments. It is recommended that the following procedure be carried out before initial starting to ensure satisfactory operation.

NOTE

At time of initial installation, remove plug from Feedwater Strainer and cover threads with Teflon tape or other nonseizing material then reinstall plug.

a. Visually inspect and tighten any loose plumbing connections and make sure Plant has been installed in accordance with Installation Manual R-7420, regarding fuel system, exhaust system, etc.

b. Check oil level in Water Pump Crankcase.

c. Check Flexible Coupling located between Water Pump, Blower and Drive Motor. To avoid excessive vibration and wear of Coupling, Motor and Pump Shafts must be within .005 inch of true axial alignment. A steel straightedge may be used across the Coupling Halves to check for correct alignment. A gap approximately 1/8 inch should be maintained between the metal Coupling Halves. Metal-to-metal engagement should be avoided.

NOTE

If Water Pump, Blower, or Drive Motor have been removed for any reason, replace any shims to original position. If necessary, remove or add shims under motor as required to maintain Coupling alignment to tolerance specified.

d. Check Drive Motor rotation before placing Plant in service.

e. When firing with oil, be sure Fuel Lines are open and fuel is circulating.

CAUTION

Check Fuel Pump Coupling clearance (1/32 inch) between metal-to-metal contact with Water Pump Shaft to compensate for end play of Water Pump during operation. Damage to Water Pump Shaft could result if clearance is too close.

f. Check operation of Water Softener. Suitable water-treating equipment should be installed before Steam Generator is placed in service. Adequate water treatment MUST be used from the time the Generator is first operated.

3.1.1 It is IMPORTANT that the Thermostat Control be tested after the Generator has been initially started and brought up to operating steam pressure. The Thermostat should be tested periodically to ensure continuous protection (see paragraph 7.6.7).

3.2 CONDITIONING OF NEW INSTALLATIONS.

3.2.1 To remove all contamination in the Water and Steam System, caused by new piping, the System should be cleaned after initial starting of the Unit.

a. Thoroughly dissolve 10 pounds of commercial trisodium phosphate (TSP) in water, using a separate container and add this solution to the Hotwell or Feedwater Make-up Tank. Operate the Plant for about one hour, then blowdown the Unit as instructed in paragraph 7.4.4.

b. To complete the cleaning operation, mix another solution of trisodium phosphate, this time using 5 pounds of the phosphate compound. Add this solution to the Hotwell and allow Plant to operate normally for 10 to 12 hours. Blowdown the Plant again as instructed in paragraph 7.4.4. This should completely clean the system of all residue.

Section IV
Gas-fired Operation

MODEL E-200

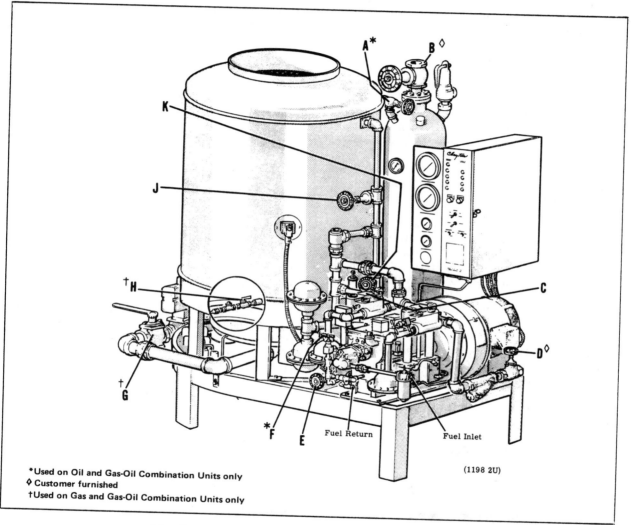

Fuel Return

Fuel Inlet

(1198 2U)

*Used on Oil and Gas-Oil Combination Units only
◊ Customer furnished
†Used on Gas and Gas-Oil Combination Units only

*A.	Soot Blower Valve		*F.	Burner Control Valve
◊ B.	Steam Discharge Valve		†G.	Main Gas Cock
C.	Feedwater Pump Housings		†H.	Pilot Gas Cock
◊ D.	Feedwater Intake Valve		J.	Coil Feed Valve
E.	Separator Drain Valve		K.	Coil Drain Valve

SECTION IV
GAS-FIRED OPERATION
(SEE SECTION V FOR OIL-FIRED OPERATION)

4.1 GENERAL. Instructions in this section are intended to acquaint the operator with each phase of operation. Cautions and notes are inserted to stress the importance on a particular instruction.

4.2 BEFORE STARTING. (See figure 4-1.)

a. Open water supply to Hotwell.

b. Close Steam Discharge Valve (B), Separator Drain Valve (E) and Coil Drain Valve (K). Also, close all Drain Cocks. On Gas-Oil Combination Units, close Soot Blower Valve (A).

c. Open Feedwater Intake Valve (D), Coil Feed Valve (J) and Steam Trap Valve (18, figure 7-2).

d. Close Main Gas Cock (G). Before initial start, it will be necessary to purge air from gas line to prevent automatic fuel shut-off after the Unit is started. To purge air, remove tube located between Pilot Gas Cock (H) and Pilot Solenoid Valve beneath Burner Manifold. Open Pilot Gas Cock (H) until air is purged. Hold flame at Pilot Cock outlet to check for positive presence of gas. Replace Tube when air has been expelled.

e. Place manual Low-Fire Switch (3, figure 7-2) in the "AUTO HI-LO" position. On Gas-Oil Combination Units, place "GAS-OIL" Switches in the "GAS" position.

f. Place "RUN-FILL" Switch (RFS) in the "FILL" position.

4.3 STARTING PLANT. (See figure 4-1.)

Be sure Main Gas Cock is closed before starting Plant.

a. Press "START" Button on Electrical Controls Box and hold in for 3 to 5 seconds until Air Pressure Annunciator Light is de-energized and circuit is maintained. Prime Feedwater Pump Housings (C) by opening Bleeder Cocks on front and rear of Housings, until air is expelled. If Pump fails to prime, loosen Intake Valve Caps two turns (wrench furnished) to eliminate air; then retighten. When Plant is started initially, or if it has been idle for a long time,

remove Intake and Discharge Valves from Feedwater Pump Housings and wipe the Discs and Seats with a clean cloth to ensure proper seating. Be sure Check Valves are installed into the same port from which they were removed.

NOTE

Check oil level in Water Pump Crankcase. Oil level should be at least half way in the sight glass when Pump is operating. Low oil level will reduce Pump capacity and cause Thermostat interruption. If necessary, add hydraulic oil with viscosity of 640 to 740 SSU at 100°F as required (see paragraph 7.7.4).

b. If Pump fails to prime upon initial start, or at any time the Unit is completely dry when started, stop Plant and close Feedwater Intake Valve (D). Remove Intake Valves from Feedwater Pump; then open Feedwater Intake Valve just enough to allow Pump columns to fill completely. When Pump is full, replace Check Valves. Reopen Feedwater Intake Valve and restart Plant.

c. Check Water Pump prime by throttling Coil Feed Valve (J) until feed pressure rises to 100 psi. If feed pressure does not rise to 100 psi, prime Water Pump in similar manner as explained in step a. Reopen Coil Feed Valve (J) after check.

d. If Plant is completely dry, allow 5 to 6 minutes for Pump to fill Heating Coil. If Plant is being restarted (not completely dry), allow 3 to 4 minutes for the Pump to fill Coil. Rising, or an appreciable pressure registered on the Steam Pressure Gauge, indicates that water has reached the Steam Separator and is being returned to the Hotwell through an orifice in the Steam Trap.

4.4 STARTING BURNER. (See figure 4-1.)

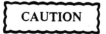

Do not start Burner until you are sure Feedwater Pump is fully primed and Plant is full (steps 4.3c and 4.3d).

a. With Plant fully charged with water, open Pilot Gas Cock (H) and Main Gas Cock (G). Place "RUN-FILL" Switch in the "RUN" position. After a prepurge period of 7 seconds, the "GAS" Indicator Pilot will light, indicating that a Gas Pilot has been established in the Burner.

NOTE

If Pilot Cock is not opened or if Pilot Flame is not established within approximately 12 seconds after the purge period, the Electronic Safety Control (ESC) will lock out the Burner Controls. In this case, wait 2 or 3 minutes for the Thermal Element to cool; then actuate the "RESET" Level located on the front of the Electronic Safety Control.

b. Every time the Burner cycles off due to excessive steam pressure, a purge period of seven seconds occurs before the Pilot Flame is ignited. After that, the Main Gas Valve (MGV) is energized. During Burner operation the Pilot Flame is extinguished.

NOTE

In the event of Burner flame failure, the Burner Control will shut off fuel immediately and try to light the Pilot. If Pilot and Burner do not ignite within 12 seconds, the Electronic Safety Burner Control (ESC) will lock out the Burner Control and require a "RESET" after Thermal Element cools.

c. After Burner ignites allow Plant to come up to operating steam pressure. The Burner will automatically cut back to "low fire" when steam pressure rises to about 10 psi below the maximum Steam Pressure Switch cut-off point. When steam pressure approaches maximum, SLOW-LY open Steam Discharge Valve (B) to gradually bring steam line up to pressure.

NOTE

In case of Burner shutdown before maximum steam pressure is reached, or in case of automatic Plant shutdown, refer to paragraph 6.1 for use of the Annunciator System in determining the cause of trouble.

d. If fire appears excessively blue or yellow, regulate the air supply to the Burner. (See paragraph 8.12.)

e. When Plant reaches full temperature (after initial start), check and record feed pressure for use in determining degree of Coil restriction (see paragraph 7.6.1). Also check adjustment of Thermostat Control (see paragraph 7.6.7).

4.5 AUTOMATIC OPERATION.

NOTE

Pressures given in the following text are typical for a unit adjusted for a maximum steam operating pressure of 195 psi. For Units adjusted for other pressures, the pressure ranges (in parentheses) given below the set maximum will apply.

4.5.1 When steam pressure rises to about 185 psi (or about 10 psi below maximum), the Modulating Pressure Switch will actuate the controls to cut Burner capacity to "low-fire" operation. On light steam demands the Steam Pressure Switch will cut-out and stop the Burner when pressure rises to 195 psi (or maximum). It will automatically cut-in and restart the Burner when pressure drops about 10 psi below maximum. The Burner will thus cycle on and off during light steam demands and the Motor and pump will operate continuously to maintain a balanced system.

NOTE

The 7-second purge period will be indicated by the "BURNER" Annunciator Light which is energized during this time period. The Annunciator will still function in its normal capacity while the Plant is operating automatically.

4.5.2 If demand for steam increases, full (high-fire) Burner operation will automatically be re-established by the Modulating Pressure Switch when steam pressure drops to about 175 psi (or about 20 psi below maximum). With moderate to heavy steam loads, the Burner will cycle between "low-fire" and "high-fire" operation, or will operate continuously at "high-fire" depending upon steam demand.

4.5.3 Low water or water failure will actuate the Thermostat Switch to stop the Burner. If the Burner shuts down due to thermostatic action, shut down Plant, find and correct the cause of insufficient water before resuming operation. Reset the Thermostat Switch manually to resume operation.

CAUTION

Be sure an adequate supply of water is available to the system at all times and thus eliminate any possibility of damage.

4.5.4 Overload or low voltage will cause the Thermal Overload Relay (OL located in the Electrical Controls Box) to stop the Plant. In this case, wait 2 or 3 minutes for the Overload Element to cool; then press the "RESET" Button before restarting Plant. Check cause of overload. Check Motor for overheat due to possible shorted or grounded winding. Also see paragraph 8.15.1 if shutdown persists for no apparent reason.

NOTE

If repeated nuisance shutdown occurs during warm weather, refer to paragraph 8.14 for adjustment of the Overload Relay.

4.6 MANUAL LOW-FIRE OPERATION.

4.6.1 If desired under certain operating conditions, "high-fire" operation can be prevented by placing the Manual Low-Fire Switch (3, figure 7-2) in the "LOW ONLY" position. To return to normal operation, move the Switch back to the "AUTO HI-LOW" position.

4.7 THERMOSTAT PROTECTION. All Steam Generators are fully tested before shipment from the factory and every effort is made to ensure that the Thermostat is correctly adjusted to operate under all conditions. However, shipment and improper handling may alter this adjustment and it becomes important that the control be tested after the Generator is initially started and brought up to operating steam pressure. The Thermostat should also be tested monthly to ensure continuous protection (see paragraph 7.6.7).

4.7.1 In case of Thermostat Control failure, the Auxiliary Thermostat Switch (ATS) stops the Plant. To resume normal operation, shut off Burner and hold the "START" Button to allow operation of the Water Pump until the Auxiliary Thermostat Switch has cooled down sufficiently to close its contacts. Check for cause of overheat and correct it before resuming operation. The Auxiliary Thermostat Switch is equipped with a manual reset which must be actuated after the cooling period.

4.8 STEAM PRESSURE ADJUSTMENT. Maximum steam pressure may be raised or lowered within the pressure rating of the Steam Generator by readjusting the Steam and Modulating Pressure Switches. If a change of more than 25 psi is made from the original setting, it may be necessary to install a Steam Trap and Steam Safety Valves having corresponding pressure characteristics.

4.8.1 If a constant steam pressure is necessary to operate certain types of equipment, install a Steam Pressure Reducing Valve at the Unit requiring constant pressure. Any pressure within the range of the Reducing Valve and below the Steam Pressure Switch cut-in point may thus be maintained. A Pressure Reducing Valve is also required for pressures below 65 psi. Install a Safety Valve on all low-pressure equipment.

4.9 SHUTTING DOWN PLANT. (See figure 4-1.)

4.9.1 PERIODIC SHUTDOWN. The following instructions are for periodic shutdown only. For overnight or weekend shutdown, follow instructions in paragraph 4.9.2. If there is danger that the Unit will be subjected to freezing temperatures, follow instructions in paragraph 4.9.3.

a. Shut off Burner.

b. Stop Plant by pressing "STOP" Button.

c. Close Steam Discharge Valve (B).

4.9.2 OVERNIGHT OR WEEKEND SHUTDOWN. There are two satisfactory methods of shutdown which will avoid corrosion within the Generator during overnight or weekend shutdown. The dry shutdown method is used when it is desired to shut down the Plant in the shortest period of time. If it is desired to have the Steam Generator in a filled condition, ready for rapid restarting, the wet shutdown method should be used.

4.9.2.1 DRY SHUTDOWN.

a. With Plant operating at normal steam pressure, close Feedwater Intake Valve (D) and open Coil Drain Valve (K) and Separator Drain Valve (E). Start time check.

b. Progressively close Steam Discharge Valve (B) in such a manner that steam pressure will remain just below the modulation pressure. This will permit continuous maximum Burner operation.

c. Shut off Burner after 30 seconds.

d. After Burner shutdown, turn off fuel to Burner and stop Plant.

e. After steam pressure drops to zero, close Coil Drain Valve (K), Separator Drain Valve (E) and Valve in Steam Trap line.

NOTE

If Plant is operated continuously, shutdown the Plant as instructed in steps a through d above, every 10 to 12 hours. This will remove precipitated sludge from the System; then restart Plant in the normal manner.

4.9.2.2 WET SHUTDOWN.

NOTE

When using the wet shutdown procedure, the feedwater must contain sufficient compound to maintain an adequate residual of oxygen scavenger and the recommended pH in order to prevent corrosion. Also, perform any scheduled blowdown operation at least one hour before starting the wet shutdown (see paragraph 7.4.4).

a. Shut off Burner.

b. Close Steam Discharge Valve (B) and Gas Pilot Cock.

c. Operate Pump with Burner off for a MINIMUM OF 20 M.....UTES (preferably 25 to 30 minutes).

d. Stop Plant by pressing "STOP" Button.

NOTE

It is important that the Coil Drain Valve, Separator Drain Valve and Steam Discharge Valve be tightly closed during periods of wet shutdown. Also, any Valve in the Feedwater Line from the Hotwell must be left open.

4.9.3 FREEZING PRECAUTIONS AND EXTENDED SHUTDOWN.

a. Follow steps 4.9.2.1a through d.

b. Remove Intake and Discharge Check Valves from Feedwater Pump Housings (C). Open Drain Cocks at Base of Water Pump.

c. Remove Coil Drain Plug (at rear of Unit beneath Coil) to drain remaining water from Coil.

d. Remove and drain all Tubes in the water and steam system in which a low sump may contain water. Remove Tubes from top and bottom of Water Pump Cooling Jackets to drain water from Jackets.

e. After making sure that all water has been drained from Unit, replace all Tubes which were removed and reinstall Intake and Discharge Valves into Water Pump Check Valve Housings.

f. Close Drain Cocks at base of Water Pump. Close Separator Drain Valve (E) and Coil Drain Valve (K). Completely close Steam Discharge Valve (B) and replace Coil Drain Plug beneath Heating Coil.

NOTE

It is very important that all Valves and Vents be tightly closed to prevent air from entering the System during cooling of the Steam Generator.

4.10 CHANGING FROM GAS TO OIL FIRING (GAS-OIL COMBINATION UNITS ONLY).

a. Shut off gas supply to Generator. Remove mirror from Burner Manifold.

b. Disconnect and remove Gas Burner Manifold and install Oil Burner Manifold. Connect cables from 2-pole Ignition Transformer to one set of Ignition Electrodes. Connect Cable from single pole Transformer to one Electrode of the other set and attach ground lead from remaining Electrode to one of the Manifold Mounting Studs.

c. Connect fuel lines to Manifold and insert Banana Plug into Photoelectric Cell Housing. Connect ground lead to one of the Manifold Mounting Studs. Install Burner Inspection Mirror to Manifold.

d. Engage coupling on Fuel Pump Shaft with slot in Water Pump Shaft and secure with Setscrews.

CAUTION

When engaging Coupling leave 1/32-inch clearance between metal-to-metal contact to compensate for end play of Water Pump Shaft during operation. Damage to Water Pump Shaft could result if clearance is too close.

e. Open Fuel Supply Lines to Generator and place "GAS-OIL" Switches in the "OIL" position. Refer to Section V for oil-firing instructions.

f. Check Damper air adjustment to eliminate smoke from Stack (see paragraph 8.12.1).

APPLICATION DATA

SCAV-OX® II CATALYZED 35% HYDRAZINE SOLUTION FOR CORROSION PROTECTION IN INDUSTRIAL BOILERS

Most corrosion in boilers and related equipment is due to dissolved oxygen carried into the steam generating system with the feedwater. Elimination of dissolved oxygen is usually accomplished by a combination of deaeration equipment and chemical treatment with oxygen scavengers. The scavengers most commonly used are sodium sulfite and hydrazine.

Scav-Ox II is a catalyzed hydrazine: Its reaction speed with oxygen is much faster than that of plain hydrazine and more or less equal to that of catalyzed sodium sulfite (depending upon temperature, pH and concentration). It provides better corrosion resistance than either (see Figures 1 and 2).

Scav-Ox II is recommended for control of oxygen corrosion in those feedwater systems where the slower reaction time of uncatalyzed hydrazine results in unsatisfactory protection, or where the dissolved solids introduced by sodium sulfite cannot be tolerated.

Since Scav-Ox II contains no inorganic solids, it is well suited for the treatment of steam, steam condensate and deionized feedwater, even when the feedwater is used to desuperheat steam. It may also be used for wet layup of boilers.

Characteristics of Oxygen Scavengers

The ideal oxygen scavenger must have two primary capabilities. First, it should react with oxygen as quickly as possible. Second, it must also build and maintain a uniform magnetite barrier on internal surfaces.

In addition, the ideal oxygen scavenger should not decompose to form corrosive products. It should not affect water purity by contributing potentially harmful dissolved solids. And it should also be compatible with other boiler treatment chemicals.

Catalyzed sodium sulfite reacts rapidly with dissolved oxygen, even at comparatively low water temperatures. For this reason it is frequently used in systems with poor mechanical deaerators and low feedwater temperatures. However, such use leads to other problems.

Sodium sulfite contributes dissolved solids (sodium sulfate) to the boiler water. It decomposes into corrosive products above 282°C (540°F), which corresponds to about 950 psig of saturated steam. And comparatively large residuals of sodium sulfite are required to form and maintain the magnetite barrier which protects steel surfaces.

Plain hydrazine removes oxygen more slowly than sodium sulfite at low water temperatures. However, it contributes no dissolved solids to the boiler water (its reaction products are water and nitrogen). It does not form acidic decomposition products which are corrosive to boiler steel. And, only a small residual of hydrazine is required to maintain passivation of internal steel surfaces.

Scav-Ox II combines all the inherent advantages of plain hydrazine with the fast reaction speed formerly found only in sodium sulfite. It thus provides all the attributes of the ideal oxygen scavenger.

Scav-Ox II Catalyzed Hydrazine

Scav-Ox II is a 35% aqueous solution of hydrazine (N_2H_4) containing a small amount of an organic catalyst. The catalyst accelerates oxygen removal to rates more or less comparable to sodium sulfite (depending on temperature, pH and concentration).

The level of dissolved solids resulting from the use of Scav-Ox II is relatively insignificant. The organic catalyst in a typical dosage of 0.2 ppm of Scav-Ox II introduces less than 1 ppb of organic carbon. Scav-Ox II contains no inorganic solids.

Table 1
Scav-Ox II
Typical Properties

Form	Aqueous solution
Hydrazine (min % by weight)	35.0
Catalyst (% by weight)	0.2
pH, 1% solution	10.4
Density @ 25°C [77°F] (g/ml)	1.02
(lb/gal)	8.51
Viscosity @ 25°C [77°F] (cp)	1.403
Boiling Point @ approx. 760 mm Hg	
(°C)	109.4
(°F)	229
Flash Point	None
Solubility in Water	Completely miscible
Color	Light pink to light orange*

*The color of Scav-Ox II may darken in time. This color change does not affect its performance.

How Scav-Ox II Combats Corrosion

Oxygen Removal: Hydrazine reacts with oxygen according to the following equation:

$$N_2H_4 + O_2 \rightarrow 2H_2O + N_2$$

The reaction products are water and inert nitrogen which is

Scav-Ox® is a registered trademark of Olin Corporation.

120 Long Ridge Road, Stamford, Connecticut 06904

© 1986 Olin Corporation.

SCAV-OX® II CATALYZED 35% HYDRAZINE SOLUTION

carried over with the steam. Because it is catalyzed, *Scav-Ox®* II overcomes the conventional objection to the use of plain hydrazine: slow reaction rate at lower temperatures.

Magnetite Formation: Magnetite forms gradually as an adherent, protective, black iron oxide coating (Fe_3O_4) on feedwater and boiler surfaces exposed to low levels of hydrazine (plain or catalyzed) over extended periods. This reaction can occur below the normal boiling point of water.

Hydrazine also forms magnetite by reducing red iron oxide (Fe_2O_3):

$$6Fe_2O_3 + N_2H_4 \rightarrow 4Fe_3O_4 + 2H_2O + N_2$$

The process of magnetite formation, both as an adherent coating on the metal and in the conversion of loose rust, continues as long as a hydrazine residual is maintained in the feedwater and in the boilers.

This process also keeps surges of oxygen caused by temporary operational upsets under control. These surges oxidize the magnetite to ferric oxide (rust). But residuals of oxygen scavenger maintained in the water will repair the magnetite by reduction of rust back to magnetite. Loose rust is also reduced to dense particles of magnetite which separate and can be removed easily in the blowdown or in the bottom header.

Hydrazine begins to reduce rust above about 138°C (280°F) and is not affected by elevated temperatures. Sodium sulfite reduces rust only in the narrow range from about 221°C to 282°C (430°F to 540°F). Above this, sodium sulfite decomposes into corrosive sodium sulfide and sulfur dioxide.

Corrosion Inhibition Test Results

Tests were performed on mild steel and copper to measure corrosion rates for *Scav-Ox* II, plain hydrazine and catalyzed sodium sulfite. Test specimens were placed in the feedwater system of an industrial steam boiler operating at 300 to 400 psig and generating steam at a rate of 2-3 million pounds per day.

The oxygen scavengers were fed to the storage sections of two Cochrane spray deaerator heaters. A bleedline supplied treated feedwater to the test arrangement at a temperature of 107°C (225°F). The corrosion rates are shown in Figures 1 and 2.

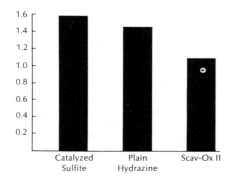

Figure 1
Corrosion Rate of Mild Steel in Treated Boiler Feedwater
(mils per year)

Courtesy Olin Chemical Research Center.

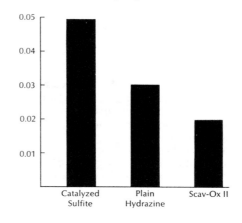

Figure 2
Corrosion Rate of Copper in Treated Boiler Feedwater
(mils per year)

Clearly, *Scav-Ox* II functioned as a superior control agent on both metals. Compared with sodium sulfite treatment, the corrosion rate of mild steel was reduced by approximately 31% and the corrosion rate of copper was reduced by about 60%. Plain hydrazine also performed better than sodium sulfite, though not as well as *Scav-Ox* II.

Areas of Application

Feedwater Systems: *Scav-Ox* II is recommended as the oxygen scavenger for treatment of high-purity feedwaters. These include feedwater used in the desuperheating of steam and feedwater for high pressure steam generating systems. Such feedwaters may require a reactive oxygen scavenger because of oxygen corrosion problems in the preboiler systems, but may be able to tolerate only very small amounts of dissolved solids.

Feedwater, lines, preheaters and economizers can be protected by the addition of *Scav-Ox* II to the feedwater storage tank or to the storage sections of deaerator heaters. A typical dosage of 0.2 ppm of N_2H_4 (0.57 ppm *Scav-Ox* II) contributes only a little over 1 ppb of dissolved catalyst to the system.

Condensate Systems: Leakage of air into condensate systems may require treatment of the condensate with an oxygen scavenger. Because of the comparatively low temperatures of condensates, plain hydrazine would react too slowly. Because the condensate is returned to the boiler feedwater in most systems, dissolved solids must be kept to a minimum — thus ruling out sodium sulfite. *Scav-Ox* II solves both problems.

Other Areas: *Scav-Ox* II may be used in direct application to the boiler water, especially in low pressure boilers or in closed hot water heating systems.

Scav-Ox II is also suitable for the wet lay-up of boilers, especially when reactivity with oxygen and control of dissolved solids are desired.

Note: Steam from boilers treated with *Scav-Ox* II must not contact food or human drug products.

How to Use Scav-Ox® II

The ideal place to add *Scav-Ox* II is at the deaerator discharge. This provides maximum protection for the feedwater system.

Application Rate: The amount of *Scav-Ox* II added to the feedwater depends initially on the dissolved oxygen level in the water, with an excess provided to react with ferric oxide and to form a magnetite film on the steel surface. The dose will vary, therefore, depending on the efficiency of deaeration and the cleanliness of the system.

Scav-Ox II should be fed at a rate sufficient to maintain a residual of from 0.01 to 0.10 ppm of hydrazine in the boiler. The rate may be higher at first, when rust is being reduced to loose, heavy magnetite and the adherent magnetite film is being established. After this initial period, the desired residual can be maintained with a much lower feed rate.

Dilution: Scav-Ox II is a 35% solution and only small volumes are required. Therefore, it is best to dilute this solution for ease in pumping and metering. Dilution to a hydrazine content of 1-5% is generally adequate. The degree of dilution is not critical, and depends mainly on the precision of the available feed equipment.

Either batch or continuous dilution of *Scav-Ox* II is possible. A batch is prepared by simply adding a measured amount of *Scav-Ox* II to a measured amount of condensate in the feed tank.

Since *Scav-Ox* II is compatible with most treating chemicals, it may also be added as a combined feed.

Equipment: If *Scav-Ox* II is replacing sodium sulfite or plain hydrazine, the same equipment will generally be satisfactory, as is. If a new system is required, mild steel or polyolefins are suitable materials of construction.

Hydrazine Methods of Analysis

Hydrazine residuals in feed and boiler water can be determined by standard analytical methods with no interference from the catalyst in *Scav-Ox* II.

The most common method is a simple colorimetric analysis based on *p*-dimethylaminobenzaldehyde. Several versions of this method are in use. Instruments of varying degrees of sophistication are available, ranging from inexpensive color comparators to spectrophotometers and automatic analyzers/recorders. If the boiler water itself is colored (e.g., from lignosulfonates) the interference can generally be cancelled out.

Residuals of hydrazine may also be analyzed by direct titration with reagent in a pH 7 buffer.

Storage and Handling

Scav-Ox II has no flash point and does not constitute a fire hazard. However, hydrazine is a strong reducing agent, as well as a base. Therefore it should not be stored near oxidizing agents or acids. Exposure to direct sunlight or to high temperatures should be avoided.

Dilute aqueous solutions of hydrazine are capable of releasing hydrazine vapors to the surrounding atmosphere. Because of the toxicity of hydrazine vapors, care should be taken to assure adequate ventilation whenever *Scav-Ox* II is handled in open containers.

Toxicological Properties

Hydrazine can be absorbed into the body in harmful or fatal amounts by ingestion, skin contact or inhalation. Contact with the skin, eyes, and respiratory tract can cause irritation and/or burns. Hydrazine may cause dermal sensitization.

Exposure to large single doses, or small repeated doses, of hydrazine may cause death, temporary blindness, dizziness and nausea, and may damage formed elements in the blood and the bone marrow. Hydrazine has been demonstrated to cause cancer in laboratory animals. It may be a carcinogen in humans. It also may cause fetal malformations.

More complete information on hydrazine is available in the Olin brochure, "Storage and Handling of Aqueous Hydrazine Solutions." Free copies of the brochure are available from Olin on request.

Personnel Protection

Persons handling aqueous hydrazine solutions should wear protective equipment: butyl rubber apron or protective suit, goggles and butyl rubber gloves and boots.

Workers should change into clean working clothes each day. If hydrazine is spilled or splashed on articles of clothing, they should be removed immediately and laundered before reuse. Leather shoes should not be worn when handling hydrazine since hydrazine cannot be removed from leather. If hydrazine should be spilled on leather shoes, remove them immediately, soak in water and discard.

Ventilation in areas where hydrazine is handled should be adequate to limit the vapor concentration of hydrazine to values below the estimated ACGIH TLV value of 0.1 ppm. The maximum concentration at any one time during an 8-hour work day should never exceed 0.3 ppm. Detection of a hydrazine odor (similar to that of ammonia) indicates a vapor concentration of 3-5 ppm which is in excess of the allowable exposure limits. Control measures should be taken immediately when hydrazine odors are detected.

First Aid

If skin contact occurs: Flush thoroughly with water for 15 minutes and call a physician.

If eye contact occurs: Wash with gently flowing water for at least 15 minutes. Call a physician immediately.

If ingested: Drink water. Induce vomiting by sticking finger down throat. Call a physician immediately.

If inhaled: Remove to fresh air. Call a physician immediately.

Spill and Leak Procedures

Wear a NIOSH/MSHA approved positive pressure supplied air respirator. Follow OSHA regulations for respirator use. (See Chapter 29, *Code of Federal Regulations*, 1910.134.) Wear goggles and butyl rubber suit, gloves and boots. Neutralize spilled hydrazine by diluting with water to a 5% or less solution. Add an equal volume of 5% hypochlorite solution. Test for neutralization. After neutralization, transfer material to appropriate container for disposal. Wash all contaminated clothing before reuse. Destroy any contaminated leather articles.

Disposal

Dispose of unused product in a manner approved for this material. Consult appropriate Federal, state and local regulatory agencies to ascertain proper disposal procedures.

Scav-Ox® II Availability

Scav-Ox II is shipped in the following containers:
Pails: 6 gallons (50 lbs net)
Polyethylene drums: 55 gallons (450 lbs net)
30 gallons (250 lbs net)

Technical Assistance

Technical assistance is available to aid in further investigation of *Scav-Ox* II 35% catalyzed hydrazine solution. If you have a specific question or need more information, please write or call your nearest Olin Sales Office.

How to Order

To place orders for delivery in the U.S. or Canada and to get fast answers on order status or product availabilities, call our toll-free number: 800-243-9171. (From Connecticut, call 0-324-7024, collect.)

After your first order, you'll be assigned your own personal Olin Customer Service Representative. When you call back, simply ask for your Customer Service Representative by name. If you call evenings (after 5:00 pm Eastern time) or on weekends or holidays, your message will be recorded and your Representative will contact you at the beginning of the next business day.

For written inquiries about orders, and to place confirmations, we've set up a special box number for you. Just address your envelope to Olin, P.O. Box 10007, Stamford, CT 06904.

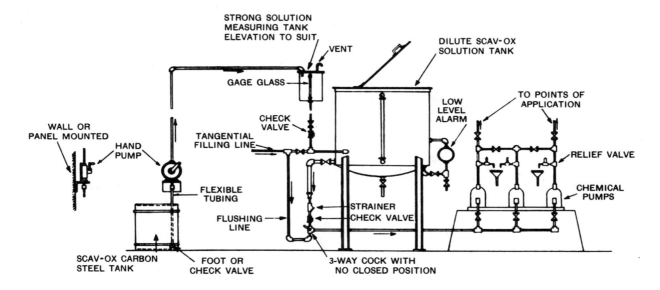

Figure 2

Multi-Pump Assembly For Applying Hydrazine Solution

OIL FLAME TROUBLESHOOTING

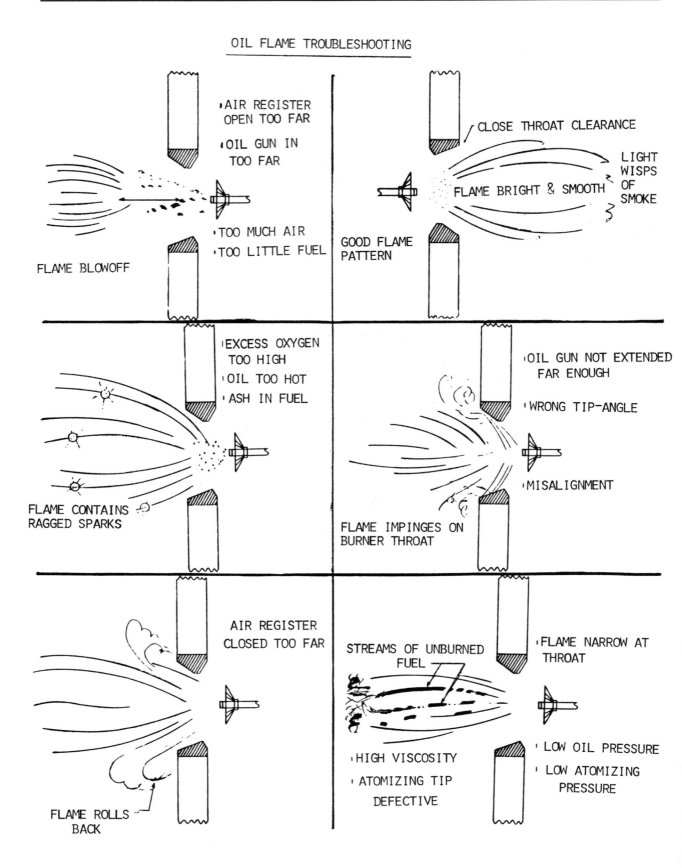

The Steam System Flow Diagram

One of the best tools available to permit the steam plant operator to understand his equipment is the steam system flow diagram. When the plant was designed, the engineers produced such a diagram, and it may still be available to the plant operator. By referring to it from time to time, and learning the relative location of every item of equipment shown on the diagram, it is often possible to analyze trouble developing in the system, and assist the plant maintenance personnel in isolating a piece of equipment for servicing with the least expenditure of time.

The flow diagram is also a useful tool in training new employees, by having them take a copy of the diagram, and walk the entire system down, noting the physical location of every item on the diagram. Then, back in the office, the employee can study the literature for the equipment, item by item, from the plant files covering the steam plant.

If the flow diagram is not available, then it is a simple matter for the enterprising plant operator to produce one. Starting first from the system walk-down and a simple series of sketches on note paper, then sitting in the office or at a table, transferring the sketches onto a larger, final sheet, it is amazing how much can be learned about the system.

Consulting and Design Engineering offices have their own systems of illustrating the plant equipment, pipe lines, valves, and fittings on the flow diagrams they produce, and if these are available to the plant operator, they will greatly simplify the process. However, the plant operator may use his own ingenuity in this respect, as the most important thing is that the users of the diagram understand the various symbols in use. Generally speaking, the symbols in common use are simplified diagrammatic representations of the item they represent. In their place, a series of standardized squares, circles, rectangles, lines, etc., all with labels or acronyms (such as DA for Deaerator), may be used.

Boiler Flame-out

From *North American Combustion Handbook*, published by North American Mfg. Co.

Interrupted pilots are turned off after a programmed trial-for-ignition period whether the main flame is established or not. They are strongly recommended over constant, standing, or intermittent pilots (which continue to burn after the main flame is established). If something should go wrong, causing the main flame to go out, the constant pilot will tell the flame monitor that it still detects flame, thereby holding the fuel valve open, filling the chamber with a fuel/air mixture. If someone opens a furnace door, turns off the fuel, or does anything to bring the accumulated mixture back within the limits of flammability (explosive limits), the standing pilot will ignite the accumulated fuel/air mixture. This situation is most likely to happen when something goes wrong, causing the fuel/air ratio to go rich and the main flame to go out because it is too rich to burn.

IF YOU EVER COME UPON THIS SITUATION—pilot on, fuel on, main flame out—TURN OFF PILOT AIR, pilot fuel, and any SPARK IGNITER IMMEDIATELY. Do *not* shut off fuel—that would allow the mix in the chamber to become lean enough to explode. Permit the rich mix of fuel and air to cool everything below the ignition temperature. Then, *when you are sure* there are no hot spots or other soures of ignition in the chamber, ventilate the room, open the furnace doors, and shut off the fuel.

The above-described explosion hazard can occur on a furnace without flame monitoring equipment if the human monitor fails to turn off the pilot when he sees or hears the main flame go out. Observation ports should be kept clean and used for frequent flame surveillance on every furnace.

Note to above—Before doing anything to correct the flame-out, it is a good idea to warn everyone in the immediate area to leave *immediately.*

Flow Balance Principle

We shall now discuss another important subject, one that will enable you to solve some of the basic flow problems in your plant.

Stated briefly, the flow balance principle is: Whatever flows into a closed junction or system must either remain in storage in the system or must flow out of it. Stated another way, it simply means that all flows into and out of a closed system must be accounted for.

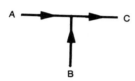

Fig. 1.

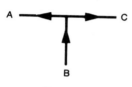

Fig. 2.

In Fig. 1, we have a pipeline with a branch feeding into it, and the arrows indicate the directions of the flow inside the pipe. As you can see from the illustration,

Flow A + Flow B = Flow C

Flow B = Flow C − Flow A

Flow A = Flow C − Flow B

In Fig. 2, the flow equations are as follows:

Flow B = Flow A + Flow C

Flow A = Flow B − Flow C

Flow C = Flow B − Flow A

Thus, you see that if any two flows are known in either of these two systems, the third flow may be easily found from the correct equation given previously.

If, instead of a pipe tee at the junctions in either case just given, we were to place a three-way valve, the principles become very useful.

For instance, in Fig. 1, the result would be a blending valve, with two flows coming together in a valve body, and the combined flow leaving through the third port. In Fig. 2, the result would be a diverting valve, such as used as a by-pass for the control of the temperature of an air compressor jacket, a diesel engine, or similar apparatus.

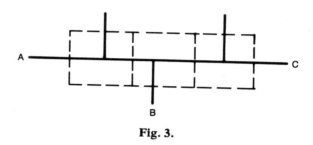

Fig. 3.

Now we will develop this discussion one more step, as shown in Fig. 3. This again could be a pipe distribution header or a series of flows into or out of a tank or surge vessel. If it is a distribution header, then the steps toward determining all the flows would be as follows:

1. Sketch the system, putting in everything that is known about the flows and their directions.

2. Separate the entire system into smaller three-way branches, as shown, which will fit one of the flow equations given for Figs. 1 and 2.

3. Starting with the simplest three-way system in which two of the three flows are known, solve for the third flow.

4. Repeat the process in step 3 for all junctions in which two of the three flows are known.

5. Using the flows calculated in steps 3 and 4, go on to calculate the remaining flows by the same method.

If you take the time to study this method for a few minutes, you will see that what often looks at first glance like an insurmountable problem will become easy once it is broken down into simpler components. Of course, there must be enough basic data available to permit you to get the proper start. It may be necessary to estimate some flows in order to arrive at an approximate solution, then go back over the problem step by step, making adjustments until you arrive at either a complete balance, or one that appears reasonable.

This discussion on flow balance is similar to one of the basic laws of flow of electricity in a circuit. A little thought will show how close the two situations actually are.

The Heat Balance

We assumed in the last section that some fluid, such as water, was flowing in the lines, and that the system was closed. Liquid and vapor is capable of carrying heat with it, and, in fact, any liquid or vapor will contain heat, or it would not be in that form. Therefore, it is a simple matter to take the examples in the previous section, and solve the same problems with heat flow in the same systems.

There is one very important difference, however, and that is that heat may be lost or gained through the walls of the pipes or ducts carrying it. In the case of hot water or steam inside the pipes, there will be a loss of heat to the outside. In the case of water, brine, or vapor below the temperature of the surrounding air, there will be a gain of heat into the piping or ducts. This may be reduced to a practical minimum by the application of insulation around the piping or ducts. In most practical cases, we may assume that the loss or gain of heat through the walls of the piping will be negligible, and that is what we will assume in the cases covered here.

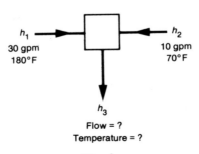

Flow = ?
Temperature = ?

Fig. 4.

Figure 4 is a typical water blending system, with a three-way valve, usually thermostatically controlled to maintain a desired constant outlet temperature from the valve. Notice that this is similar to Fig. 1, except here we are dealing with heat flow, which is expressed in BTUs.

First we shall state some basic facts concerning solution of a problem of this type, as follows:

1. The problem may be solved on the basis of flow in gallons per minute, second, or hour. It makes no difference which units of flow are used if we are consistent. We shall use the units given, which is gallons per minute. The flows could also be in pounds per hour or pounds per minute.

2. All heat contents are calculated on the basis of the heat content being zero at 32°F, which was discussed in the matter of Steam Tables, in Chapter 2.

3. It makes no difference if one of the two combining flows is a vapor. If the vapor is condensed when it mixes with the liquid, the problem is easy to solve. If it does not all condense, the situation becomes more complicated, and we will not consider such a case, as it is beyond the scope of this discussion.

4. The two combining flows may be gases or vapors, and the method of solution is the same, since they both contain heat, and the principle is the same as for combining liquids.

Now for a step-by-step solution to the above problem.

Step 1. From the previous section:
$$F_3 = F_1 + F_2$$
$$30 \text{ gpm} + 10 \text{ gpm} = 40 \text{ gpm}$$

Step 2. Heat content = Flow × BTU/lb

Heat content = gpm × 8.33 × (temperature −32)

Step 3. $h_1 = 30 × 8.33 × (180 − 32)$

$h_1 = 30 × 8.33 × 148 = 36,985$ BTU/min

Step 4. $h_2 = 10 \times 8.33 \times (70 - 32)$

$h_2 - 10 \times 8.33 \times 38 = 3,165$ BTU/min

Step 5. $h_3 = h_1 + h_2 = 40,150$ BTU/min

Step 6. From Step 2

$$h_3 = F_3 \times 8.33 \times (T_3 - 32)$$

$$T_3 - 32 = \frac{h_3}{F_3 \times 8.33} = \frac{40,150}{40 \times 8.33} = 120$$

$$T_3 = 120 + 32 = 152°F$$

We suggest you study the methods given here, since they are basic to so many other situations in your plant that it will pay you to become familiar with the solutions and the steps involved.

In Chapter 26 we covered the case of mixing steam with water to produce a tank of hot water. The situation is similar to that just explained, but it has been presented here in a slightly different manner. You may be able to spot easily the similarity, however, especially if you have followed our suggestion and studied the methods presented here until you are well versed in their application.

The cases covered so far have all had to deal with mechanical mixing of flows, and the resultant flows and heat contents. There are other cases of mixing materials where the solution is not quite so simple, and that is in those cases where a chemical reaction takes place between two combining materials, or when combustion takes place, which is a form of rapid chemical reaction. Such a case would be the burning of the fuel in the boiler. If we were to measure the

flow of oil into the burner, and if we know the higher heating value per pound of the oil, we could then perform a heat balance on the boiler, providing we are able to measure all the flows and temperatures into and out of the boiler. This is the method used to test the performance of a boiler, and some rather complicated procedures have been established to do this. We shall not go into it here any further, but in case you are interested, we suggest you get in touch with your boiler supplier.

However, if you will refer to Chapters 1 and 25, you may recognize the basic principle of the boiler heat balance.

Some of the equipment in the plant where you may be able to make use of the flow and heat balance principles are:

The boiler

Heat exchangers

Flash tanks

Condensate tanks

Air-conditioning units

Chillers

Condensers

Steam turbines

Turbine generators

In fact, every piece of energy-using equipment may be treated to either the flow or heat balance in some form. As you go about your daily tasks in the plant, keep your eye open for such possibilities.

SECTION G
BOILER LAYUP

STORAGE INSTRUCTIONS FOR BOILER
AND AUXILIARY EQUIPMENT

These instructions are given as a general guide to the proper storage of packaged steam generators and auxiliary equipment. They are not meant to supersede the storage instructions that may be found for specific equipment in manufacturer's literature. It must be emphasized that whenever equipment is idle, the necessary steps should be taken to prepare its return to service through regular inspections and organized storage and maintenance programs.

BOILERS

To minimize the possibility of corrosion, boilers to be held out of service must be carefully prepared for the idle period and closely watched during the outage. The dry storage procedure described below is recommended for boilers that will be out of service for more than three months.

Air & Gas Side Preparation:

The following precautions should be taken to keep moisture and air from entering the boiler:

1. All pipe connections to the windbox or furnace should be capped or plugged.

2. 10 Gage steel plates, properly stiffened, should be placed over the flue gas outlet and the inlet to the burner windbox. These covers may be seal welded in place, or a suitable waterproof sealer can be applied to these surfaces prior to fastening down the cover plates.

3. With the burner cover plate removed for furnace access, a dessicant such as a general indicating type silica gel, can be hung in bags throughout the furnace. Humidity indicating cards also hung inside the furnace will help tell when the dessicant is to be renewed. Periodic inspections should be set up to insure the effectiveness of the dessicant and to check that no corrosive action has taken place.

4. Bagged silica gel is a renewable dessicant and should be used at the rate of 4 pounds for every 50 cubic feet of boiler volume.

5. After the dessicant is in place and the burner cover plate is replaced, any opening in the cover plate should be sealed off to exclude air and moisture.

6. Access doors above the structural base should be bolted down with precautions taken to discourage rodents and other small animals from nesting within the confines of the base.

Water and Steam Side Preparation:

Silica gel may be used to absorb moisture within the drums and headers of the boiler, however, it is better to use the nitrogen method of lay-up described below:

1. Clean and dry the unit in preparation for a nitrogen purge.

2. A suitable gasketing material should be placed on the raised faces of the safety valve and steam outlets. Seal plates made from 1/4" plate should then be bolted down over these openings. Make sure mating surfaces are clean and free of scale.

3. All other drum head connections should be blanked off by seal welding 1/4" plates inside the connections so as not to damage any machined surfaces.

4. A shut-off valve and pressure gauge extending from a pipe tee connection may be mounted to one of the seal plates on the steam drum. This should serve as a nitrogen vent and pressure connection. A valve or coupling welded to one of the mud drum connections can be used as a nitrogen feed point.

5. Air is purged from the unit until a near pure nitrogen atomsphere within the vessel is indicated. The nitrogen vent is then closed.

6. Nitrogen pressure is brought up to 6 PSI in the boiler, then the nitrogen inlet connection is closed off. This pressure is monitored for 48 hours to determine any leaks within the unit.

7. Nitrogen is a non-flammable gas, however, danger would exist if someone enters the pressure vessel without completely exhausting all the nitrogen gas. Warning signs shall be placed on the drums over each manway saying "Do Not Enter, Due to Lack Of Oxygen could cause Asphyxiation". Signs saying "Contains Nitrogen" should be prominently placed on each side of the boiler as a warning to all personnel.

8. After the 48 hour holding period, the boiler should be checked weekly to insure that a minimum of 3 PSI nitrogen pressure is maintained. After each year of storage, the boiler should be exhausted of all nitrogen and internally inspected for signs of corrosion.

ELECTRICAL CABINETS

Both the burner management cabinet and the combustion control panel should be stored indoors in a dry environment protected from dirt and moisture.   In humid areas, a storage maintenance program involving the effective use of a dessicant should be adhered to.

BURNER PIPING RACK

The piping rack should be stored indoors in a dry environment. When it is necessary to store outdoors, the rack should be positioned above ground and level. All open pipe ends should be capped or plugged. A canvas, oil cloth, or heavy waterproof paper should be used to protect the rack from dirt and excessive moisture.

BURNER PARTS

The linkage around the burner louvers should be coated with a suitable rust preventive prior to sealing up the air and gas side of the boiler for storage.

The ignitor should be removed from the burner and stored indoors to protect it from any abuse or corrosion.

WATER COLUMN

The water column should be wrapped in canvas, oil cloth or heavy waterproof paper to protect it from the weather. Water column drains should be plugged so nitrogen gas within the steam generator is not allowed to escape.

SOOT BLOWERS

For long term storage, it is recommended that the soot blower heads with elements be taken off the boiler and stored indoors. The soot blower wall boxes can then be sealed to exclude air and moisture from the convection zone and furnace of the boiler. Cams, gearing and shafts should be well coated with a suitable rust preventative lubricant. The soot blower heads should then be covered with canvas, oil cloth or paper to keep out dirt and moisture.

FANS

When possible, fan bearing should be lubricated, covered, and stored indoors away from excessive moisture and dust. If it is necessary to store the fan outdoors, special care must be taken to prevent moisture, corrosion, dirt or dust accumulation.

1. Coat the shaft with grease or rust preventative compound.

2. Lubricate cover and seal all bearings.

3. Close access doors and block fan wheel to prevent rotation.

4. Cover the fan completely with canvas, oil cloth or waterproof paper to keep out excessive dirt or moisture.

5.   Prevent rodents and other small animals from nesting inside the fan.

MOTORS

Store motors inside when possible, but if they must be exposed to the elements, particular attention should be applied to the shaft and bearings.

1.   Ball bearings are usually grease packed at the factory but should be checked and relubricated after each year of storage.

2.   Shaft and flange surfaces should be coated with an easily removable rust preventative. Rotate the shaft several revolutions at one month intervals.

3.   Prevent rodents and other small animals from nesting inside the motor.

TURBINES

The turbine should be stored level and indoors when possible. For outdoor storage take the following precautions:

1.   Coat the shaft and all machined surfaces with suitable rust preventative and protected from abuse.

2.   Completely cover the turbine with canvas, oil cloth or waterproof paper to exclude excessive dirt and moisture.

MISCELLANEOUS PARTS

Operators, springs, levers, valves, instruments, spare parts and similar items must be stored indoors and protected from dirt, moisture and abuse. Rust preventative may be applied to bare machined surfaces. Small parts may be wrapped in vaporproof bags and kept in crates until ready for use.

These storage instructions were supplied by the Erie City Energy Division of Zurn Industries, Inc.

PRODUCT DATA

HYDRAZINE FOR WET-LAYUP OF BOILERS

The need for proper protection of steam boilers during periods of idleness, even those of short duration, is well recognized among power engineers today. While dry-layup of boilers may be desirable for long-term storage, properly carried out wet-layup can provide the needed protection during layups such as may occur during low production periods, strikes, plant breakdowns and during regular maintenance periods. This folder has been prepared to acquaint you with the advantages and ease of using hydrazine in wet-layups.

When a boiler is down, compounded changes take place that must be avoided or overcome. Unlike in the operating system, oxygen is not always excluded, temperatures change, and the system can no longer depend on steam to maintain equilibrium. This requires treatment that assists a return to the cleanest operating condition with the least amount of time, labor, and attention.

A wet-layup treatment with hydrazine (N_2H_4) satisfies the requirement with a small amount of attention. Hydrazine is easy to use and assists in many ways without adding undesirable by-products. When added in suitable amounts, it will keep a boiler or the auxiliary equipment cleaned, conditioned and protected.

Scav-Ox®, a 35% solution of hydrazine, and Scav-Ox II, a catalyzed 35% hydrazine solution, are suitable for boiler-cleaning applications. Scav-Ox II reacts more readily with oxygen than Scav-Ox at low temperatures, which may be advantageous for certain wet-layup applications. This data sheet will assist and guide a user of wet-layup procedures at time of start up and shutdown. Also, it should be helpful in providing a better understanding of the specialized treatments needed.

Reasons For Wet-Layup

A boiler system that will be needed on very short notice can be placed in wet-layup. The primary purpose of wet-layup is to keep a boiler clean and inhibit corrosion in "idle" or "stand-by" boiler equipment. This protection is also required either during pre-cleanup or pre-operational periods for greater convenience.

Hydrazine inhibits corrosion in the following ways:
- by scavenging oxygen
- by adjusting pH
- by controlling carbon dioxide
- by aiding in the formation of protective thin films
- by sustaining absorbed barrier layers

Scav-Ox® is a registered trademark of Olin Corporation

Hydrazine is convenient for wet-layup because:
- Hydrazine is a volatile liquid and does not increase the dissolved solids in the boiler drum.
- Hydrazine is a powerful, controlled reducing agent which is effective in reducing metal oxides to the least active state and in removing last traces of active oxygen.
- Hydrazine converts iron sludge to heavy magnetite film, which settles readily and filters easily.
- Neither hydrazine, its decomposition products, nor its reaction products are acidic. Hydrazine reacts with metal oxides and oxygen to give nitrogen and water only. It decomposes to ammonia and nitrogen.
- Use of hydrazine leads to cleaner internal surfaces, purer water, and less blowdown. Volatility is much lower than water and decomposition is slow at low temperatures. As a result, hydrazine is effective in cleaning, passivating and blanketing. It can be left in the equipment while the desired result is being obtained.
- Wet-layup with hydrazine penetrates all crevices and loose scale. Use of hydrazine in wet-layup requires no special treatments to remove excesses at time of start up.

Types Of Wet-Layups

More than one wet-layup procedure is required depending on the length of time and the service function.

Olin's own experience as well as that of electric utility companies and boiler cleaner companies has shown that hydrazine maintains cleaner boilers and condensate lines; the excess converts to ammonia at elevated steam temperatures. With this in mind, several types of wet-layup are used. These can be summarized in three general categories:

1. *Overnight* — The system must be in a clean operating condition. Softened or demineralized water is used to completely or partly fill the system with a nitrogen[1] pad. A required amount of hydrazine sufficient to give a concentration of 10 ppm is charged and water is added to the desired volume. More hydrazine can be added to adjust the concentration after mixing. (Range of concentrations is 10-20 ppm.)

2. *Several Weeks* — The system must be in a clean operating condition. Softened water is used to completely fill the system. A required amount of hydrazine sufficient to give a concentration of at least 100 ppm is charged and

[1]NOTE: Hydrazine has been found to be a preferred reducing agent for dissolved oxygen and oxidized metal scavenging. Nitrogen purging is important in the elimination of gaseous oxygen.

Olin CHEMICALS
120 Long Ridge Road, Stamford, Connecticut 06904

HYDRAZINE FOR WET-LAYUP OF BOILERS

water is added to fill the system. One analysis is required per week. (Range of concentrations is 30-200 ppm.)

3. *Pre-cleaning and Pre-operational* — The system may be of questionable cleanliness, the water may be less than quality, or the system may contain sludge and scale due to other types of boiler treatment. A required amount of hydrazine sufficient to give 500 ppm is charged and the system is filled with water.

The hydrazine can be expected to react rapidly initially with the oxygen and oxides in the system, making it difficult to maintain the desired residual 100 ppm. Additional hydrazine is added until a level up to 1000 ppm can be maintained.

How To Wet-Layup With Hydrazine

A. Short outage period (overnight or a few days)

1. *Operation interruption* — No draining. Over a period of 3 or 4 days, the pressure will drop to atmospheric pressure.

Introduce nitrogen through the boiler drum vent or at a superheater drain valve when the pressure drops below 5 psig. Maintain 2-5 psig with water-pumped nitrogen.

Add hydrazine to the boiler to maintain a concentration of 10 ppm. A dilute solution of hydrazine can be introduced with a feed pump or with nitrogen pressure.

2. *Pre-operational period* — When a boiler is ready for a hydrostatic test, proceed with the following:

a. Fill the superheater with condensate or demineralized water containing ammonia and hydrazine. The pH value will be approximately 10 with 10 ppm ammonia. Hydrazine at 200 ppm level will sustain the pH, react with active traces of sludge, and scavenge oxygen.

b. When the condensate overflows from the superheater into the boiler drum, the addition of condensate at the superheater can be stopped and the unit filled with hydrazine-treated service water through boiler filling connections normally used.

c. At the conclusion of the hydrostatic test, with the boiler filled to overflowing, pressurize the unit to 2-5 psig with nitrogen. When freezing is a problem, the water in the drainable circuits can be displaced with nitrogen and then the unit laid up under nitrogen pressure. Auxiliary heat is then applied to keep the non-drainable superheater (and reheater) elements above freezing. With drainable superheaters, all surfaces can be laid up under nitrogen pressure.

d. Analyze hydrazine concentration in 24 hours and add additional, if required. Repeat analysis once per week if boiler continues out of service.

B. Long outage period (boiler is not opened for repair work — e.g. during seasonal outage of a month or longer, or when service is removed)

1. Fill superheater and reheater with condensate containing about 10 ppm of ammonia and 200 ppm of hydrazine. Add the condensate from the outlet of the non-drainable sections to the boiler and fill the entire boiler. The treated condensate can be displaced with nitrogen or the entire unit can be laid up wet under nitrogen pressure, depending upon the temperature of the surrounding area. In either event, hold nitrogen pressure at 2-5 psig. When isolating valves on the superheater (and reheater) outlets are not installed, steps should be taken to block off the lines so that the boiler can be pressurized.

2. If freezing conditions arise during the outage, means must be provided to keep the units above freezing temperature.

C. Pre-boil-out or post-boil-out periods

1. *Pre-boilout period* — If the cleanup is delayed after a shutdown, it can be aided by filling as follows:

Introduce condensate containing 500 ppm of hydrazine to the boiler, superheater (and reheater) and pressurize the unit with nitrogen. The water can be used as part of the boil out feed; it is alkaline.

2. *Post-boil-out period* — If the operation of the boiler is delayed after boiling out or acid cleaning, it can be conditioned as follows to reduce initial hydrazine consumption in getting the expected oxygen scavenging residual during operation.

Introduce condensate containing 500 ppm hydrazine to the boiler, superheater (and reheater), and pressurize the unit with nitrogen. Hold 200 ppm of hydrazine.

Returning Boiler To Operating Condition

When the unit is to be put into service, it will be necessary to bring the water to the normal operating level. Then release the nitrogen pressure and open the drum and superheater vents before returning to normal operation. To conserve nitrogen supply used in pressurizing the unit, shut off the supply before opening the vents.

Excess hydrazine converts to ammonia and nitrogen, which will normally purge along with carbon dioxide and the nitrogen used to pad the system[2].

Analytical Methods

The hydrazine concentration normally encountered in wet-layup may be titrated directly, or diluted proportionately and colorimetrically determined, via the following methods of analysis:

1. Direct Iodate Method using solvent.

2. Direct Iodate Method with internal indicator.

3. Modification of the p-Dimethylaminobenzaldehyde Method — Hach, Taylor, or other kits.

The pH may be measured with relatively inexpensive color comparators widely used, or with any of the pH equipment available in a boiler plant.

Toxicological Properties

Hydrazine can be absorbed into the body in harmful or fatal amounts by oral ingestion, skin contact, or inhalation. Contact with the skin, eyes, and respiratory tract can cause irritation and/or burns. Hydrazine may cause dermal sensitization.

[2]NOTE: Steam from boilers treated with hydrazine must not contact food or human drug products.

Courtesy of Olin Chemicals.

Exposure to large single or small repeated doses of hydrazine may cause death, temporary blindness, dizziness, nausea and damage to internal organs. Damage may occur to the central nervous system or such organs as the liver, the kidneys, the lungs, and those organs which form the blood. Hydrazine has been demonstrated to cause cancer in laboratory animals. It may be a carcinogen in humans. It may cause fetal malformations.

More complete information on hydrazine is available in the Olin brochure, "Storage and Handling of Aqueous Hydrazine Solutions." We recommend that anyone handling hydrazine be familiar with its contents. Copies of the brochure are available on request.

Personnel Protection

Persons handling aqueous hydrazine solutions should wear protective equipment: apron or protective suit, goggles and butyl rubber gloves and boots. If spilled or splashed on clothing, remove the clothing immediately and launder before reuse. If spilled or splashed on leather shoes, remove shoes and discard (hydrazine cannot be removed from leather).

If hydrazine is handled under conditions that could result in release of vapor, sufficient ventilation should be provided to keep the vapor concentration in the work area below the threshold limit value for hydrazine vapors in air (0.1 ppm) established by the American Conference of Governmental Industrial Hygienists. Maximum concentration at any one time during an 8-hour work day should never exceed 0.3 ppm. Generally 3-5 ppm of hydrazine in air are detectable through odor and mild respiratory irritation.

First Aid

Skin: Immediately wash off with large amounts of flowing water. Follow by thorough soap and water washing. Call a physician immediately.

Eyes: Wash with gently flowing water for at least 15 minutes. Call a physician immediately.

Ingestion: Drink large amounts of water immediately. Induce vomiting by sticking finger down throat. Call a physician immediately.

Inhalation: Remove to fresh air. Call a physician immediately.

Spill and Leak Procedures

Wear a NIOSH/MSHA approved self-contained breathing apparatus. Follow OSHA regulations for respirator use. (See Chapter 29, *Code of Federal Regulations,* 1910.134.) Wear goggles, coveralls and butyl rubber gloves and boots. Flush with large amounts of water and drain into a catch basin. Transfer into an approved DOT container and seal. Neutralize any remaining material with dilute hypochlorite solution and wash down with water. Wash all contaminated clothing before reuse. Destroy any contaminated leather articles.

Waste Disposal

Dispose of unused product in a manner approved for this material. Consult appropriate Federal, state and local regulatory agencies to ascertain proper disposal procedures.

For Further Information

Since each boiler layup may involve some considerations not fully covered in this bulletin, it is recommended you discuss your wet-layup program with a water-treating company familiar with the use of hydrazine in this application. Where layup following a boiler clean-out by an industrial cleaning company is involved, you may also want to discuss the procedures with the cleaning company employed.

For information on the availability of hydrazine or companies who are familiar with hydrazine wet-layup procedures, please contact your nearest Olin Regional Sales Office.

For further information on storage and handling, see the Olin brochure "Storage and Handling of Aqueous Hydrazine Solutions."

SECTION H
POLLUTION AND INSPECTION

ZURN INDUSTRIES, INC.
ERIE CITY ENERGY DIV.
ERIE, PA, U.S.A. 16512

PRE-SERVICE INSPECTION

When installation of the steam generator has been completed, and before putting it into operation, the entire system should be inspected. This inspection can serve two functions. First, a thorough inspection will show whether or not the equipment installers have done a satisfactory job and can be released from the job site. Secondly, if any of the components are damaged, defective or improperly installed, a thorough inspection will reveal the discrepancy immediately, rather than allow a delay of the starting-up operation later.

The following discussion and the component check list in this section are intended as an inspection guide. After every component in this check list is inspected, and proven operational, the steam generator will be ready to operate. The inspection should also cover any related equipment furnished by others to insure that the total system is ready to function.

External

Any signs of damage which occured enroute, should be thoroughly inspected. For example, damage to the side casing could also have damaged the insulation, the inner seal casing or the boiler tubes.

The area around the steam generator should be checked to insure that no attachments are preventing the unit from expanding. Low friction skid plates or roller assemblies may be required under the rear bearing points to allow free movement. (See Section on Handling and Installation and refer to the Erie City Foundation Drawing.)

Any obvious signs of component misassembly should be listed, and the list should be signed by the equipment installer or engineer in charge of the installation. This inspection list should constitute a request for correction of the improper conditions.

Courtesy of Zurn Industries Inc.

Combustion Chamber

The main access to the furnace area or fire box is through the burner. By removing the ring of bolts on the burner front plate and withdrawing the oil gun, (on oil fired furnaces), the cover plate can swing open. When using the burner for furnace access, remove the ignitor and flame scanner which are mounted on the cover plate.

The furnace interior should be inspected for any signs of damage to the front and rear refractory walls. Castable refractory does have hairline cracks, which will not cause trouble, but cracks over 1/8-inch wide can cause problems by allowing leakage of hot gases to the steel casing. The rear tile wall should be tight and without cracked joints. The roof and floor seals of castable refractory must be solid. Any damage to the refractory or insulation should be repaired before the main flame is lit.

The tube banks which make up the furnace walls should be tight and in good alignment. If there are any gaps between the tangent tubes, the gas flow can short circuit from the furnace into the convection zone and destroy the performance of the steam generator as well as cause possible damage.

On larger units, the floor or part of the floor is is covered by refractory tile. Tile should be against the front wall and flat on the floor. The floor tile joints are not mortared.

If the unit is equipped with a superheater, all elements should be firmly supported. The elements should be inspected for damage due to transit vibration. The seal where the elements penetrate the rear wall should be packed tight. The tile baffle wall under the superheater may have to be restacked. The tile baffle is not mortared.

Finally, the furnace should be free of debris, equipment or shipping braces.

Burner

The burner is comprised of: an air register assembly, an ignitor, a fuel manifold or gun, and observation port and a flame safety scanner. Check these components to determine if they are in the proper position, and undamaged. Proceed as follows:

1. Check the air register by operating the control lever on the burner front plate. All of the register blades should move together since they are all tied to a common linkage.

2. Check the ignitor to insure that the porcelain end is not cracked or damaged. Also, check the porcelain insulator sleeve. The wire connector nut should be clean and tight. The spark gap must be clean and set at 3/16 of an inch.

3. If the burner fires natural or manufactured gas, the lower capacity burners will have an internal gas manifold. Check to make sure the holes are clear. The larger burners have an external manifold and pipe spuds extending into the burner throat. Check the holes in the tips of these spuds, to make sure they are clear. Check isolation valves on spuds for freedom of operation.

4. If the burner fires oil, then there is an atomizer gun in the center of the burner throat. Check the gun against the assembly drawing of the burner to insure that the gun is assembled or supplied properly. The holes in the tip must be clean.

5. There must be a safety glass in the view port, and through the scanner you must be able to sight the oil gun spray pattern or the ring of gas flame. The wires on the scanner must be clean and tight, and the lens also must be clean.

Combustion Air

Unless the system requires more than one fan, the "Keystone" Steam Generator has only one fan to supply combustion air. The rate of air flow is controlled by either fan inlet vanes or fan outlet dampers.

The fan rotor and shaft must be free turning by hand, without binding or scraping. The motor rotation must coincide with the fan wheel rotation, which is found marked on the housing. The fan movement, motor rotation and motor starter can be tested by momentarily energizing the electrical circuit to the motor. If fan rotation is proper and movement is free, then continue to run the fan for a period of time and check the bearings and mountings for excessive vibration and smooth operation overall.

The air flow vanes or dampers must operate in conjunction with a driving unit. Check for freedom of movement, range of movement and closure tightness by manually operating the drive unit or control linkage.

Fuel Piping

The fuel piping schematic drawings furnished with this manual must be studied and totally understood. Inspect the piping system to see that it is assembled according to the drawings, and such things as the check valves are properly oriented. The piping must be tight, and no loose joints or open ports can be tolerated. All gauges should have shutoff cocks, and the switches should be wired in accordance with the external wiring diagram furnished with this manual.

Check the fuel delivery equipment for proper connections. If gas is used as the fuel, check for proper pressure and purity. If fuel oil is used, check supply, preheater, return lines and pressure relief. The oil pump should be checked for proper rotation, pressure output and pressure relief. If the oil is heated, check wiring and piping of the oil heater.

Feed Piping

The feed water piping and chemical feed piping should be examined to see if all of the joints are tight and that the valving is in the proper orienta-

a step ahead of tomorrow

ZURN INDUSTRIES, INC.
ERIE CITY ENERGY DIV.
ERIE, PA, U.S.A. 16512

tion. The direction of flow through the check valves should be towards the steam generator. Check the feed water regulator assembly against the supplier's product bulletins to determine that control linkage is operating properly. Check the valves and drainage of the regulator bypass system.

Steam Piping

All steam piping should be inspected for such things as: securely bolted flanges, properly oriented check valves, no open ports, correct installation of gauging and proper actuation of vent and drain valves. Check all orifices and Venturi metering devices for proper installation, per manufacturer's bulletins. On atomizing steam lines, insure that steam condensate traps and drains are installed properly.

Drain Piping

All drain lines should be inspected from the connection on the steam generator to the blowdown tank or pit. There should be no shutoffs or obstructions between the steam generator and the collection tank. This collection tank should be installed to comply with any local Codes or insurance recommendations. Items which normally have drain provisions are: steam generator drums, steam generator water wash, water column, water gauge glass, atomizing steam condensate, oil purge and any headers (soot blowers, superheaters, economizers and water walls).

Water Column

Since the water column is so important to the operation of a steam generator, it must be carefully inspected. The connecting piping must be complete and free of leaks. The wiring to the safety system must be in accord with the external wiring drawing furnished with this manual. With a water column there are several control devices which could have been furnished, such as: low water alarm (this signals when water level is dropping to the danger point), low water cutoff (this shuts the steam generator down before the water

level is low enough to do any overheating damage, an auxiliary low water cutoff (this is a safety low level shutdown in case the main low water cutoff fails to actuate), high water alarm (this warns the operator before the level gets high enough to cause loss of steam purity and water carryover), and high water cutoff (this is more of a process protection circuit to prevent the water from flooding the steam separator). Any of these functions which are furnished should be checked for proper actuation. The normal water level is usually maintained at the center of the gauge glass.

Safety Valves

Safety valves should be checked to make sure that there are no blockages or shutoffs on the outlets. The vent pipes should not be supported from the valves, because the weight of the pipes could cause the valve body to distort or expand. The valves should be preset in accordance with the manufacturer's valve identification tag stamping.

Safety Controls

Before inspecting the safety control circuit, study the electrical diagrams furnished with this manual. Examine all components and make preliminary adjustments as per the manufacturer's bulletins. Normal safety controls include such items as: high steam limit switch, excess steam limit switch, flame proving safeguards, high and low oil pressure and temperature switches, high and low gas pressure switches, water level alarms and cutoffs, plus any other special controls furnished to fulfill the steam generator system requirements.

Soot Blowers

If the steam generator fires or can be adjusted to fire heavy oil, it will be equipped with soot cleaning devices. Consult the manufacturer's bulletins for specifications on capacity and pressure requirements of steam for cleaning and purging air. Check the angle of rotation on the lance to insure cleaning within the proper convection lanes. Lubricate to the manufacturer's recommendations.

a step ahead of tomorrow
ZURN INDUSTRIES, INC.
ERIE CITY ENERGY DIV.
ERIE, PA, U.S.A. 16512

Heat Recovery

If the steam generator is supplemented with an economizer or an air heater, check the interconnecting ducts, bypasses and piping for possible leaks or possible expansion interference. Check this apparatus to determine the provisions for condensate drainage and soot blowing devices if the steam generator fires oil. This equipment may be provided with bypass ducts or other means of isolation. The expansion joints in the ductwork should be properly installed and any shipping restraints removed.

Drum Internals

The Keystone Steam Generator drums will be fitted with internal distribution pipes and baffling. When the greatest steam purity is required, the steam or upper drum may be fitted with Erie City Vortex Steam Separators and Erie City Chevron Steam Scrubbers. A drum head at one end of each drum should be opened for inspection prior to on-line operation. Bolts and brackets on the internal distribtuion pipes and baffles should be checked for tightness. A quick check of the Vortex Separators is to see if they are all in general alignment in a row. Chevron Scrubber sections should be checked to see that there are no wide openings. Seal drains at the bottom of the Chevron compartment should be checked for tightness. Any large amounts of dirt, scale or debris should be removed.

Summary

A thorough inspection will reveal any obvious damage or improper installation of component equipment. When all of the items contained in this check list have been reviewed, and the entire system has been inspected, you may proceed with the preparation for service.

CAUTION: Do not attempt to fire the steam generator until all systems have been checked and are operational.

ZURN / a step ahead of tomorrow
ZURN INDUSTRIES, INC.
ERIE CITY ENERGY DIV.
ERIE, PA, U.S.A. 16512

"KEYSTONE"
INSTRUCTION MANUAL

PRE-SERVICE INSPECTION
CHECK LIST

- ☐ AIR–PURGE
- ☐ AIR DUCTS
- ☐ AIR HEATER
- ☐ BLOWDOWN
- ☐ BLOWOFF
- ☐ BOLTS – BASE HOLDDOWN
- ☐ BOLTS – FLANGE
- ☐ BYPASS – DUCT
- ☐ BYPASS – FEED WATER
- ☐ CASING – OUTER
- ☐ CASING – SEAL
- ☐ CHEMICAL FEED
- ☐ COMBUSTION EQUIPMENT
- ☐ CONTROLS
- ☐ DAMPERS
- ☐ DRAINS
- ☐ DRIVES
- ☐ DRUM INTERNALS
- ☐ ECONOMIZER
- ☐ EXPANSION PROVISIONS
- ☐ EXPANSION JOINTS
- ☐ FAN
- ☐ FEED WATER SYSTEM
- ☐ FUEL PIPING
- ☐ FURNACE
- ☐ GAS PIPING
- ☐ GAUGES
- ☐ HEADERS
- ☐ IGNITOR
- ☐ INSTRUMENTS
- ☐ INSULATION
- ☐ LIMIT SWITCHES

- ☐ LINKAGE
- ☐ LUBRICATION
- ☐ MANIFOLD – GAS
- ☐ MANWAYS – DRUM
- ☐ OBSERVATION PORTS
- ☐ OIL GUNS
- ☐ OIL HEATER
- ☐ OIL PIPING
- ☐ OIL PUMPING
- ☐ OUTLET – SMOKE
- ☐ PIPING
- ☐ PRESSURE GAUGES
- ☐ REFRACTORY
- ☐ SAFETY VALVES
- ☐ SOOT BLOWERS
- ☐ STACK
- ☐ STEAM LINES
- ☐ STEAM PURIFIER
- ☐ SUPERHEATER
- ☐ TEMPERATURE GAUGES
- ☐ THERMOCOUPLES
- ☐ THERMOMETERS
- ☐ TILE
- ☐ TRAPS
- ☐ TUBES
- ☐ VALVES
- ☐ VANES
- ☐ WATER COLUMN
- ☐ WATER GAUGE
- ☐ WINDBOX
- ☐ WIRING

Ringleman charts for estimating smoke densities. The charts below are proportional reductions of standard charts issued by the U. S. Bureau of Mines. These charts are used in the following manner: Make observations from a point between 100 and 1300 ft from the smoke. The observer's line of sight should be perpendicular to the direction of smoke travel. Place the below charts approximately 15 ft in front of the observer and as close as possible to his line of sight. (Standard ASME or U. S. Bureau of Mines charts should be placed 50 ft from the observer.) Open sky makes the best background for observations. Compare the smoke density with the charts (which, at 15 ft, are shades of gray instead of individual lines) and classify the smoke according to the Ringleman chart number. Ringleman Nos. 0 and 5 are 0% and 100% black, respectively. Charts for these are solid white and black (not shown).

1. Equivalent to 20 percent black.

2. Equivalent to 40 percent black.

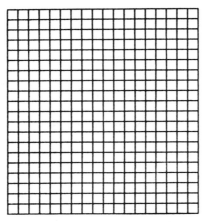

3. Equivalent to 60 percent black.

4. Equivalent to 80 percent black.

Courtesy of North American Mfg. Co.

NORTH AMERICAN Mfg. Co.
Cleveland, OH 44105   USA

CONVERSION FACTORS relating to Air Pollution Control

a) 100 tons per year of flue gas = 28,900 Btu/hr input (natural gas)

b) 1 megawatt of power plant output corresponds roughly to 8 to 11 million Btu/hr fuel input

c) 1% Sulfur in fuel oil ➞ 600 ppm SO_2

d) 20% opacity = Density #1 ~ Ringleman #1
   40%   "   =   "   #2 ~   "   #2
   60%   "   =   "   #3 ~   "   #3
   80%   "   =   "   #4 ~   "   #4
   100%   "   =   "   #5 ~   "   #5

e) 200 ppm = 0.02%
   10,000 ppm = 1%

f) 100 micrograms/cubic meter = 0.05 ppm NO_2

   $\left.\begin{array}{l} 1 \text{ ppm } NO_2 = 1910 \mu g/m^3 \\ 1 \text{ ppm } NO = 1248 \mu g/m^3 \\ 1 \text{ ppm } SO_2 = 2660 \mu g/m^3 \end{array}\right\}$ $1 \text{ ppm} = \dfrac{M}{.02404} \mu g/m^3$ where M = molecular weight (by volume)

g) 0.2 to 2.0% N in fuel may yield 60 to 2100 ppm NO

h) $35.3 \times 10^6 \mu g/m^3 = 1 \ g/ft^3$ (gram per cubic foot)

   $35.3 \mu g/m^3 = 1 \mu g/ft^3$ (microgram per cubic foot)

   $16 \times 10^6 \mu g/m^3 = 1 \ lb/mcf$ (pound per thousand cubic feet)

   $1 \mu g/m^3 = 136 \ grains/ft^3$

   1 milligram/cubic meter = 1 microgram per liter

i) If you read 175 ppm with 10% O_2 (about 80% excess air), correct to a 3% O_2 (15% excess air) basis for comparison with legal limits as follows:

$$175 \times \frac{1.80}{1.15} = 273 \text{ ppm}$$

NOTES on NO_x
from the 1971 Penn State Combustion Seminar

1. NO_x is promoted by:
   (a) Nitrogen in fuel.
   (b) High combustion chamber temperatures.
   (c) Long residence time at high temperature.
   (d) Slight excess air (if temperature is high).
   (e) Good fuel-air mixing, causing localized hot spots.
   (f) Low combustibles in flue gas.

2. Residential oil burners produce from 50 to 100 ppm of NO_x.

3. Cooling of flame with entrained, partially-cooled flue gas is one of the best ways to reduce NO_x. High velocity burners (Tempests) are indicated.

4. Delayed combustion (slow-mixing, 2-stage) is way to reduce NO_x.

5. NO formation is not reversible in furnace but is slowly reversible in atmosphere.

6. There is some evidence that NO_x, apart from the Los Angeles photochemical phenomenon, is harmful to human lungs but knowledge is very limited.

7. In sunlight, NO is converted to NO_2, then reacts with O_2 to form O_3 (ozone), which in turn combines with aldehydes and other hydrocarbons to form polynuclear aromatics, which are unpleasant to eyes and nose and may be carcinogens.

8. Sources of NO_x are estimated:
   Transportation — 45.5%
   Power plants and home heating — 40%
   Forest fires — 10%
   Industry — 4.5%

NORTH AMERICAN Mfg. Co.
Cleveland, OH 44105   USA

9. NO is formed by:
   (a) $N_2 + O \longrightarrow NO + N$ minus 75 Kcal (very endothermic)
   (b) $N + O_2 \longrightarrow NO + O$ plus 32 Kcal
   (c) $N_2 + O_2 \longrightarrow 2 NO$ (This reaction almost never happens.)
10. Practically all NO is formed above 3000 F.
11. Automobile exhaust contains from 50 to 500 ppm of NO.
12. XSA under ½ of 1% will reduce SO_3 and V_2O_5.
13. In the next 20 to 30 years, billions of tons of coal and oil will be burned in boilers, nuclear energy notwith-
    standing.

                                                    TEDavies

GLOSSARY OF POLLUTION CONTROL TERMS
(These explanations are in layman's language—not official government definitions.)

Air Pollution Index—an arbitrarily derived, mathematical combination of air pollutants which give a single number attempting to describe the ambient air quality. The formula $API = 20 \times \left[SO_2 + 1 (CO) + 2 \text{ (smoke shade)} \right]$ has no scientifically derived basis. Experience indicates that the average value of this index is 12.0. An index reading of 50.0 is considered adverse and a cause for alarm.

Ambient air quality standards—those specified as for the general atmosphere in outdoor areas frequented by people, as opposed to stack or source emission standards.

atmosphere fixation = thermal fixation—see thermal NO.

DSC = Division of Control Systems (of OAP).

EPA = Environmental Protection Agency. Established in accordance with Clean Air Acts of 1967 and 1970.

fixed nitrogen—nitrogen combined in fuels.

inversion—an atmospheric phenomenon in which a mass of warm air moves over a mass of cool air, resulting in a stable stagnation that causes accumulation rather than ventilation of warm polluted air rising from the earth's surface.

NCAPC = Nat'l Center for Air Pollution Control, Cinn., Ohio.

NO = nitric oxide—a compound often formed in combustion reactions at temperatures above 3000 F.

NO_2 = nitrogen dioxide—involved in the photo-oxidation (by sunshine) of hydrocarbons, forming eye-irritating smog. It usually comes from reaction of NO with oxygen from the air at atmospheric temperature.

NO_x = NO and NO_2 collectively—as measured by most available instrumentation—as specified in most legal limits.

O_3 = ozone—a photochemical oxidant.

OAP = Office of Air Programs (of EPA).

Photochemical Smog (now often called simply "smog"—a haze produced by photochemical oxidation, accompanied by eye irritation, plant damage. ozone formation, and characteristic odor.

pna = polynuclear aromatics—hydrocarbons containing 2 or more fused rings, including one benzene ring; a product of incomplete combustion.

ppm = parts per million—(assumed by volume unless specified as by weight). 1 ppm = 0.000001, or 0.0001%

Primary Standards—protect public health; secondary standards protect public welfare.

Priority 1 areas = places where ambient air pollution already exceeds LPA primary standards.

Source—any property, real or personal, which emits or may emit any air pollutant.
    Stationary source—fixed, non-movable as utility power plant.
    Mobile source—automobile, train, airplane etc.

Thermal NO—NO formed by temperature effects as opposed to that from fixed nitrogen in the fuel (fuel nitrogen conversion).

SECTION I
ENERGY CONSERVATION

SUPPLEMENTAL
INSTRUCTIONS

EXHAUST STACK MONITORING
(TEMPERATURE, DRAFT, CO$_2$ AND SMOKE)

DESCRIPTION: Periodic monitoring of the Exhaust Stack Temperature, Draft, CO$_2$ and Smoke will indicate change in values in order to establish proper intervals for maintenance. After initial firing and proper adjustments have been accomplished and with Plant operating at full temperature and load output, record CO$_2$, Smoke, Draft, and Temperature readings. For a precise basis of comparison, readings should be recorded at the same operating conditions.

TYPICAL READINGS: Temperature = 400 to 550°F (204 to 288°C)
> NOTE: Temperature may go as low as 250°F (121°C) at low fire in some cases (modulating burners).

Draft = 0 (No backpressure) to 0.25 inches (6.3 mm) Water Column Draft (Neg.)
> NOTE: For non-venting Stack Adapters (Special Application), Draft may be +.50 in. w.c. to -.25 in. w.c. (i.e. .25 Draft). Draft must be stable within .25 in. w.c. at all firing rates.

Emission CO$_2$ = 7 to 10% (Gas-fired Units)
10 to 12% (Oil-fired Units)
> NOTE: Higher readings are possible but generally not practical, particularly in dirty atmosphere.

An increase in Stack temperature over initial reading could be an indication of sooting and/or scaling of the Heating Coil. A radical change in CO$_2$, Draft, and Smoke reading could be Blower Rotor output performance deterioration due to dirt accumulation. If the air adjustment is tuned to achieve a high CO$_2$, a small amount of dirt may then cause smoking (oil) or monoxide (gas) emissions. Therefore, more frequent maintenance attention is required. See instructions and illustrations for installing Thermometer and Draft Tube Extension and CO$_2$-Smoke Sampling Tube.

TEMPERATURE GAUGE AND DRAFT READING INSTALLATION (ABOVE STACK ADAPTER)

1. Locate and drill small clearance hole, for Stem, in center of Stack just above Stack Adapter Cone.
2. Install Bracket, with pipe threads to suit Thermometer, onto Stack.
3. Install Thermometer and Stem Assembly into Stack as shown below. Stem must reach center of Stack.
4. Locate and drill 5/16-inch diameter hole through Stack for locating a 1/4-inch Steel Tube Extension for monitoring Draft reading.

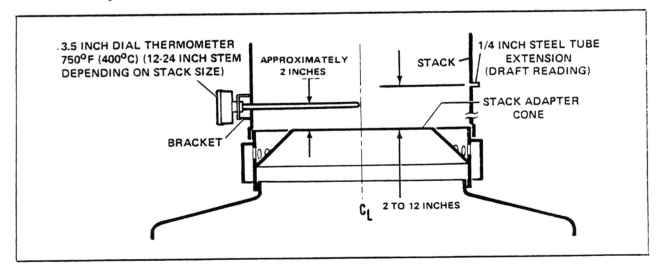

SUPPLEMENTAL
INSTRUCTIONS

CLAYTON
STEAM GENERATOR
MODELS E-33 - E-300

STACK CO$_2$ AND SMOKE SAMPLING INSTALLATION (BELOW STACK ADAPTER)

NOTE

Disregard the following Installation Instructions if Stack Adapter is equipped
with 1/4-inch Steel Tube Extension.

Use method (A) or alternate method (B) for Monitoring Stack CO$_2$ and Smoke Sampling.

1. If method (A) is desired, locate and drill small clearance hole through Extension Band and Inner Ring as shown.

2. If alternate method (B) is desired, drill 5/16 inch diameter clearance hole through Stack. Insert 1/4 inch Steel Tubing as shown. Tube must reach center of Stack.

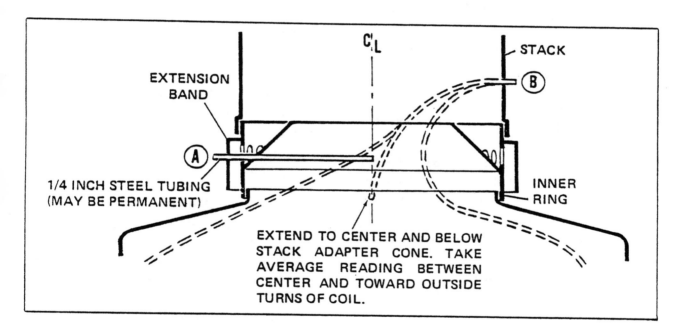

CLAYTON MANUFACTURING COMPANY
EL MONTE, CALIFORNIA

Energy Conservation

CAUTION!

Do not make any change in operating conditions on heat equipment, such as boilers, pumps, and compressors, until you have checked with the manufacturer of the equipment. There are temperature and pressure limitations, which often apply to your equipment, that, if ignored, may led to costly failures.

On the other hand, the maker of your equipment may not know what its limitations are, and may not, in fact, have even tested it to see what can be actually attained from it. In this case, you will probably find that the manufacturer will withdraw any guarantees remaining on your equipment, leaving you entirely on your own resources to experiment.

GENERAL COMMENTS AND RESULTS TO EXPECT

1. There is fertile ground for improvement, since the average industrial plant uses less than 50% of the energy that is available to it.

2. The plant operator must learn to equate heat and electricity with money.

3. We will see more equipment being pushed to the utmost; safety factors and contingencies in equipment design will be tested.

4. More heat-recovery equipment will be installed.

5. More plants will install equipment to generate at least a portion of the electric power they require.

6. The plant production processes, for which the plant was designed, will be studied to obtain maximum use of heat and energy. The steam plant operator will be required to participate.

7. Heat-insulation standards are already being revised to reflect the change in the cost of heat. Thicker insulation will be the result.

8. The use of electricity will be more functional and less for appearance.

9. Steam turbine drivers will tend to be noncondensing, because they are the most economical of energy.

10. Economic studies will be revised and reviewed, taking into account the change in energy cost.

STEAM GENERATION

1. Investigate the installation of waste fuel burners to burn some of the low-heat-value wastes that you have been discarding. There have been many improvements in this field recently.

2. Keep your boiler and heater combustion controls tuned up at all times.

3. Experiment with lower excess air to your burners, aiming at between 10% and 15%. This may require the addition of more instrumentation, such as oxygen and combustibles analyzers in the stack. Furthermore, to lower the excess air, these analyzers will have to be kept in top operating condition at all times.

4. If any heat reclaiming equipment, such as air preheaters or economizers, is added to the boiler flue stacks, watch the stack gas temperature to be sure you do not lower it to the dew point tolerance of the stack.

5. Watch your boiler blow-down procedures. The amount of blow-down depends on the quality of the boiler feedwater. The purer the feedwater, the less blow-down is required.

6. Maintain the feedwater as hot as possible, but watch the feedwater pump for signs of cavitation. Pumps often will cavitate without giving any outward signs, so check with the pump manufacturer.

7. Find the point of peak operating efficiency for your boilers, and take steps to permit operating them at that point. This will probably be at about 75%–80% of rating.

SAVING STEAM

1. Check the condition of the insulation on all hot and cold lines, and add more where needed.

2. Keep those steam and condensate leaks to a minimum.

3. Keep steam traps in good condition, and watch the selection of new ones.

4. Keep air out of the steam and condensate lines.

5. Reject the use of steam condensers that waste most of the heat energy in the steam. This includes shell and tube as well as barometric condensers.

6. Investigate the use of steam turbine drivers as pressure reducing devices. These take useful work out of the steam in the reduction process.

7. On two-pressure systems, where the lower pressure is produced by either a steam turbine driver or a flash tank, and the lower pressure is set by means of pressure-reducing and make-up valves, experiment with lower settings on the make-up valve. Do this slowly, in steps, watching the effect throughout the plant with each pressure-reduction step.

8. Experiment with lower steam pressure throughout the plant, by carefully lowering the pressure in steps.

9. On multiple-stage steam users, such as steam jet ejectors, try cutting out the first stage and noting what results are attained. You may find the ejectors or other equipment will produce satisfactory results with less steam.

10. Steam tracers are common causes of condensate leakage, or outright wastage. High-temperature heat-transfer solutions may replace steam in these systems, with the solution being heated in one heater, with one steam trap, under closely observed conditions. The solution is then pumped through the tracing

system, and returned to the exchanger. The result will be a saving of steam and condensate.

CONDENSATE SYSTEMS

1. Return all condensate possible to the boiler.

2. Check into the possibility of reclaiming contaminated condensate now being wasted. Install condensate "polishing" equipment if economically feasible to remove nearly all impurities.

3. If unable to reclaim or return the condensate, then consider recovering the heat from it before it is wasted.

WASTE HEAT RECOVERY

1. Use waste heat boilers, where high-temperature waste heat is available in large quantities.

2. There are many combinations of gas-turbine–waste-heat-boiler systems that are possible for large-horsepower driver applications.

3. Consider the installation of air preheaters or economizers in boiler flue gas passages, or process heater stacks where high gas temperatures represent waste heat being discharged.

4. If the heater stack design permits, and if the stack temperature is high enough, steam can be generated from that heat. The principle is the same as the cycle used in flash-type boilers, except the heat source is the hot stack instead of the burner.

5. Recover as much of the heat in the boiler blow-down as possible, before discarding it.

6. Discharge hot waste into open concrete pits before it goes to waste. In this pit, place water coils containing incoming process water that has to be heated, to reduce the waste water temperature and reclaim some of the heat.

ELECTRICAL ENERGY

1. If your plant is of sufficient size and uses enough electricity to make it economical, consider generating as much of your power requirements as possible. If electricity is generated as a by-product of a steam-pressure-reducing process, it is the cheapest electric power you can obtain. Purchased electric power reaches your plant at a thermal efficiency of around 25%, which means that about 75% of the thermal energy has already been lost when the electric power comes under your control.

2. Keep the power factor as close to 1.0 as possible, by installing the necessary equipment to obtain it.

3. Watch your wire sizes in the plant, since undersized lines waste power.

4. Pay more attention to sizing your electricity-using equipment to obtain maximum efficiency. Review your plant policy concerning motor sizing for future use. You may be carrying excess motor horsepower that is never used, thus lowering the overall efficiency.

5. Check those motors in the plant that are on automatic start-up, to be sure of maximum power savings. Plant operations may have changed since the controls were set, upsetting the power draw on the starting circuits.

6. Revise your lighting standards, use lighting that is most effective for the power consumed, such as mercury vapor, fluorescent, or polarized lighting.

7. Learn to use portable meters and instruments to best advantge in checking your operations.

PLANT AIR SYSTEMS

1. Be constantly alert for leaks, since air leaks are difficult to find.

2. Monitor all usage of plant air to get the maximum use of it.

3. To smooth out the pulsations of pressure caused by suddenly applied heavy loads, install small receivers at the point of heavy usage in the air system. This will also ease the starting load on the air compressor motors, because fewer starts per hour will then be required.

4. If not already installed, thermostatic regulating valves should be installed in the cooling water lines from the air compressor jackets to reduce the flow of water, raise the operating temperature to that recommended by the compressor supplier, and to improve the lubrication of the compressor. Consult your supplier on this matter first.

5. Check the integrity of the automatic moisture drain traps on the air system. They can be a source of constant air leaks.

6. Check the number of starts per hour your air compressors are going through during the average day's operation. If they start more than six times an hour, the starting load on the electrical gear is costing more money than it should. To reduce the start per hour, install a larger receiver, or an additional one, possibly at the far end of the air line. Also, you may be able to increase the operating differential on the pressure switch without causing undue strain on the air using equipment.

7. If your compressors have loading and unloading controls with the motors running constantly, check the number of cycles per hour they go through during the normal day. Call in the representative and see if you are getting the best and most efficient operation from the controls.

8. When installing new mains or branches, make them large, and they will serve as a receiver, smoothing out the demand on the system.

9. Check your lubrication to be sure the compressor is working under the best conditions.

10. Be sure the tension on belt drives is correct and that the belts are not slipping.

11. Check the mufflers and filters for pressure drop and cleanliness. Keep them clean.

PUMPING SYSTEMS

1. Go over every foot of your circulating water systems (and oil systems too, or any other fluid systems being pumped in your plant) to see if all branches and sections are being used. Check for restrictions, circuits not in use but still "live," and possibilities of rerouting the lines to cut down on the friction losses.

2. Check the control valves on the pumped circulating systems to be sure they are not causing too much power loss for you. If they operate throttled down most of the time, then they are costing you money.

3. To calculate how much power is being wasted by restrictions in the circulating system, use the following equation:

$$\text{Pump hp} = \frac{\text{gpm} \times \Delta P \times 2.31 \times \text{Sp.Gr.}}{3960 \times \text{pump efficiency}}$$

Note: pump efficiency varies from 0.40 to 0.70 on an average.

1 hp/hour = 2545 BTU (theoretical)

1 hp/hour = approximately $^4/_5$ kW

Therefore, if you are paying about 6¢ per kWhr for electricity, then each wasted horsepower is costing about 5¢ per hour.

4. Check the pipe sizing and velocities to see if some of the lines should be enlarged.

5. Clean out dirty lines and filters.

6. Finally, once you have determined that the system is operating at peak efficiency, then it is time to investigate the possibility of reducing the impellers or the strokes/minute of the pumps, to cut down power costs. Better call in the pump supplier first for his recommendation on this.

SECTION J
MISCELLANEOUS DATA

NORTH AMERICAN Mfg. Co. 4455 E. 71 St. Cleveland, Ohio 44105	Capacity Conversions (p1)	Handbook Supplement 1-66 No. 68

Supersedes BB Eng. Data#2

BOILER CAPACITY CONVERSION

INPUT				OUTPUT @ 80% EFFICIENCY				
Millions Btu/hr	Gal/hr #2 Oil, 140,000 Btu/gal	Gal/hr #6 Oil, 150,000 Btu/gal	Thousands cu ft/hr natural gas	Millions Btu/hr	Boiler hp	EDR* Steam 240 Btu, sq ft	EDR* Water 180 Btu, sq ft	Pounds of Steam per hour f & a 212 F
.8	5.3	5.0	.8	.6	18	2,500	3,333	618
1.0	7.1	6.7	1.0	.8	24	3,333	4,444	825
1.5	10.7	10.0	1.5	1.2	36	5,000	6,666	1,237
2.0	14.2	13.3	2.0	1.6	48	6,660	8,888	1,649
2.5	17.8	16.7	2.5	2.0	60	8,333	11,111	2,061
3.0	21.4	20.0	3.0	2.4	72	10,000	13,333	2,474
3.5	25.0	23.3	3.5	2.8	84	11,666	15,555	2,886
4.0	28.5	26.7	4.0	3.2	96	13,333	17,777	3,298
4.5	32.1	30.0	4.5	3.6	108	15,000	20,000	3,710
5.0	35.7	33.3	5.0	4.0	120	16,660	22,222	4,122
5.5	39.2	36.7	5.5	4.4	131	18,333	24,444	4,535
6.0	42.8	40.0	6.0	4.8	143	20,000	26,666	4,947
6.5	46.4	43.3	6.5	5.2	155	21,666	28,888	5,359
7.0	50.0	46.7	7.0	5.6	167	23,333	31,111	5,771
7.5	53.5	50.0	7.5	6.0	179	25,000	33,333	6,184
8.0	57.1	53.3	8.0	6.4	191	26,666	35,555	6,596
8.5	60.7	56.7	8.5	6.8	203	28,333	37,777	7,008
9.0	64.2	60.0	9.0	7.2	215	30,000	40,000	7,420
9.5	67.8	63.3	9.5	7.6	227	31,666	42,222	7,833
10.0	71.4	66.7	10.0	8.0	239	33,333	44,444	8,245
10.5	75.0	70.0	10.5	8.4	251	35,000	46,666	8,657
11.0	78.5	73.3	11.0	8.8	263	36,000	48,888	9,069
11.5	82.1	76.7	11.5	9.2	275	38,333	51,111	9,482
12.0	85.7	80.0	12.0	9.6	287	40,000	53,333	9,894
12.5	89.2	83.3	12.5	10.0	299	41,666	55,555	10,306
13.0	92.8	86.7	13.0	10.4	311	43,333	57,777	10,718
13.5	96.4	90.0	13.5	10.8	323	45,000	60,000	11,131
14.0	100.0	93.3	14.0	11.2	335	46,666	62,222	11,543
14.5	103.5	96.7	14.5	11.6	347	48,333	64,444	11,955
15.0	107.1	100.0	15.0	12.0	359	50,000	66,666	12,367
15.5	110.7	103.3	15.5	12.4	370	51,666	68,888	12,780
16.0	114.2	106.7	16.0	12.8	382	53,333	71,111	13,192
16.5	117.8	110.0	16.5	13.2	394	55,000	73,333	13,604
17.0	121.4	113.3	17.0	13.6	406	56,666	75,555	14,016
17.5	125.0	116.7	17.5	14.0	418	58,333	77,777	14,429
18.0	128.5	120.0	18.0	14.4	430	60,000	80,000	14,841
18.5	132.1	123.3	18.5	14.8	442	61,666	82,222	15,253
19.0	135.7	126.7	19.0	15.2	454	63,333	84,444	15,665
19.5	139.2	130.0	19.5	15.6	466	65,000	86,666	16,078
20.0	142.8	133.3	20.0	16.0	478	66,666	88,000	16,490

* Equivalent direct radiation.

Courtesy of North American Mfg. Co.

Commonly Used Terms

1. Packaged Watertube Boiler Ratings

Packaged watertube boilers are rated on the basis of maximum continuous output from the drum in terms of pounds of steam/hr. at a given steam pressure when supplied with a given feedwater temperature and firing a certain fuel. Blowdown losses are normally added to the net output for sizing.

2. Equivalent Units for Defining Boiler Output

Item	Equivalent Units
1 Pound steam, from and at 212 degrees F	970 Btu/lb
1 Square foot EDR steam	240 Btu/hr
1 Square foot EDR water	150 Btu/hr
1 Boiler horsepower, (Bhp)	34.5 Pounds of steam/hr from and at 212 degrees F
	33,472 Btu/hr
	139.5 Square feet EDR steam
	223 Square feet EDR water

3. Equivalent Units for Fuels

Item	Equivalent Units
No. 2 Oil	140,000 Btu/gal.
No. 5 Oil	148,000 Btu/gal.
No. 6 Oil	150,000 Btu/gal.
1 Therm	100,000 Btu
1 Kw	3,413 Btu

4. Equivalent Units of Volume and Weight

U.S. Gallons	Imperial Gallons	Cubic Inches	Cubic Feet	Pounds (*)	Cubic Meters	Liters
1	.833	231	.1337	8.35	.003785	3.785
1.201	1	277.4	.1605	10.02	.004546	4.546
.00433	.00360	1	.000579	.0361		.0164
7.48	6.23	1728	1	62.4	.02832	28.32
.120	.0998	27.7	.0160	1		.454
264.18	220	61,023	35.314	2204	1	1000
.2642	.220	61.023	.0353	2.2046		1

*Based on maximum density of water at 39 degrees F.

Temperature Conversion

Locate degrees F or degrees C in middle column. If degrees F, read equivalent degrees C in left hand column. If degrees C, read equivalent degrees F in right hand column.

\-459.4° to 0°			1° to 60°			61° to 290°			300° to 890°			900° to 3000°		
C	C/F	F	C	C/F	F	C	C/F	F	C	C/F	F	C	C/F	F
−273	−459.4		−17.2	1	33.8	16.1	61	141.8	149	300	572	482	900	1652
−268	−450		−16.7	2	35.6	16.7	62	143.6	154	310	590	488	910	1670
−262	−440		−16.1	3	37.4	17.2	63	145.4	160	320	608	493	920	1688
−257	−430		−15.6	4	39.2	17.8	64	147.2	166	330	626	499	930	1706
−251	−420		−15.0	5	41.0	18.3	65	149.0	171	340	644	504	940	1724
−246	−410		−14.4	6	42.8	18.9	66	150.8	177	350	662	510	950	1742
−240	−400		−13.9	7	44.6	19.4	67	152.6	182	360	680	516	960	1760
−234	−390		−13.3	8	46.4	20.0	68	154.4	188	370	698	521	970	1778
−229	−380		−12.8	9	48.2	20.6	69	156.2	193	380	716	527	980	1796
−223	−370		−12.2	10	50.0	21.1	70	158.0	199	390	734	532	990	1814
−218	−360		−11.7	11	51.8	21.7	71	159.8	204	400	752	538	1000	1832
−212	−350		−11.1	12	53.6	22.2	72	161.6	210	410	770	549	1020	1868
−207	−340		−10.6	13	55.4	22.8	73	163.4	216	420	788	560	1040	1904
−201	−330		−10.0	14	57.2	23.3	74	165.2	221	430	806	571	1060	1940
−196	−320		− 9.4	15	59.0	23.9	75	167.0	227	440	824	582	1080	1976
−190	−310		− 8.9	16	60.8	24.4	76	168.8	232	450	842	593	1100	2012
−184	−300		− 8.3	17	62.6	25.0	77	170.6	238	460	860	604	1120	2048
−179	−290		− 7.8	18	64.4	25.6	78	172.4	243	470	878	616	1140	2084
−173	−280		− 7.2	19	66.2	26.1	79	174.2	249	480	896	627	1160	2120
−169	−273	−459.4	− 6.7	20	68.0	26.7	80	176.0	254	490	914	638	1180	2156
−168	−270	−454	− 6.1	21	69.8	27.2	81	177.8	260	500	932	649	1200	2192
−162	−260	−436	− 5.6	22	71.6	27.8	82	179.6	266	510	950	660	1220	2228
−157	−250	−418	− 5.0	23	73.4	28.3	83	181.4	271	520	968	671	1240	2264
−151	−240	−400	− 4.4	24	75.2	28.9	84	183.2	277	530	986	682	1260	2300
−146	−230	−382	− 3.9	25	77.0	29.4	85	185.0	282	540	1004	693	1280	2336
−140	−220	−364	− 3.3	26	78.8	30.0	86	186.8	288	550	1022	704	1300	2372
−134	−210	−346	− 2.8	27	80.6	30.6	87	188.6	293	560	1040	732	1350	2462
−129	−200	−328	− 2.2	28	82.4	31.1	88	190.4	299	570	1058	760	1400	2552
−123	−190	−310	− 1.7	29	84.2	31.7	89	192.2	304	580	1076	788	1450	2642
−118	−180	−292	− 1.1	30	86.0	32.2	90	194.0	310	590	1094	816	1500	2732
−112	−170	−274	− 0.6	31	87.8	32.8	91	195.8	316	600	1112	843	1550	2822
−107	−160	−256	0.0	32	89.6	33.3	92	197.6	321	610	1130	871	1600	2912
−101	−150	−238	0.6	33	91.4	33.9	93	199.4	327	620	1148	899	1650	3002
− 96	−140	−220	1.1	34	93.2	34.4	94	201.2	332	630	1166	927	1700	3092
− 90	−130	−202	1.7	35	95.0	35.0	95	203.0	338	640	1184	954	1750	3182
− 84	−120	−184	2.2	36	96.8	35.6	96	204.8	343	650	1202	982	1800	3272
− 79	−110	−166	2.8	37	98.6	36.1	97	206.6	349	660	1220	1010	1850	3362
− 73	−100	−148	3.3	38	100.4	36.7	98	208.4	354	670	1238	1038	1900	3452
− 68	− 90	−130	3.9	39	102.2	37.2	99	210.2	360	680	1256	1066	1950	3542
− 62	− 80	−112	4.4	40	104.0	37.8	100	212.0	366	690	1274	1093	2000	3632
− 57	− 70	− 94	5.0	41	105.8	43	110	230	371	700	1292	1121	2050	3722
− 51	− 60	− 76	5.6	42	107.6	49	120	248	377	710	1310	1149	2100	3812
− 46	− 50	− 58	6.1	43	109.4	54	130	266	382	720	1328	1177	2150	3902
− 40	− 40	− 40	6.7	44	111.2	60	140	284	388	730	1346	1204	2200	3992
− 34	− 30	− 22	7.2	45	113.0	66	150	302	393	740	1364	1232	2250	4082
− 29	− 20	− 4	7.8	46	114.8	71	160	320	399	750	1382	1260	2300	4172
− 23	− 10	14	8.3	47	116.6	77	170	338	404	760	1400	1288	2350	4262
− 17.8	0	32	8.9	48	118.4	82	180	356	410	770	1418	1316	2400	4352
			9.4	49	120.2	88	190	374	416	780	1436	1343	2450	4442
			10.0	50	122.0	93	200	392	421	790	1454	1371	2500	4532
			10.6	51	123.8	99	210	410	427	800	1472	1399	2550	4622
			11.1	52	125.6	100	212	413.6	432	810	1490	1427	2600	4712
			11.7	53	127.4	104	220	428	438	820	1508	1454	2650	4802
			12.2	54	129.2	110	230	446	443	830	1526	1482	2700	4892
			12.8	55	131.0	116	240	464	449	840	1544	1510	2750	4982
			13.3	56	132.8	121	250	482	454	850	1562	1538	2800	5072
			13.9	57	134.6	127	260	500	460	860	1580	1566	2850	5162
			14.4	58	136.4	132	270	518	466	870	1598	1593	2900	5252
			15.0	59	138.2	138	280	536	471	880	1616	1621	2950	5342
			15.6	60	140.0	143	290	554	477	890	1634	1649	3000	5432

ALTITUDE PRESSURE TABLE

Feet Altitude	Temp F	Pressure In. Hg.
0	59.00	29.92
50		
100	58.64	29.81
200	58.28	29.71
300	57.93	29.60
400	57.57	29.49
500	57.21	29.38
600	56.86	29.28
700	56.50	29.17
800	56.15	29.07
900	55.79	28.96
1,000	55.43	28.86
1,500	53.65	28.33
2,000	51.87	27.82
2,500	50.09	27.32
3,000	48.30	26.82
3,500	46.52	26.33
4,000	44.74	25.84
4,500	42.96	25.37
5,000	41.17	24.90
7,500	32.26	22.66
10,000	23.36	20.58
15,000	5.55	16.89
20,000	-12.26	13.76
25,000	-30.05	11.12
30,000	-47.83	8.903
35,000	-65.61	7.060
40,000	-69.70	5.558
45,000	-69.70	4.375
50,000	-69.70	3.444
55,000	-69.7	2.712
60,000	-69.7	2.135
65,000	-69.7	1.682
70,000	-67.42	1.325
75,000	-64.7	1.046

Weight (Density) of Air at Various Pressures and Temperatures

Temp. of Air Deg. Fahr.	GAUGE PRESSURE, IN POUNDS PER SQUARE INCH (Based on an Atmospheric Pressure of 14.7 Pounds per Square Inch Absolute at Sea Level)																					
	0	5	10	20	30	40	50	60	70	80	90	100	110	120	130	140	150	175	200	225	250	300
	WEIGHT (DENSITY), IN POUNDS PER CUBIC FOOT																					
−20	.0900	.1205	.1515	.2125	.2744	.3360	.3970	.4580	.5190	.5800	.6410	.702	.7635	.825	.886	.948	1.010	1.165	1.318	1.465	1.625	1.934
−10	.0882	.1184	.1485	.2090	.2685	.3283	.3880	.4478	.5076	.5674	.6272	.687	.747	.807	.868	.928	.989	1.139	1.288	1.438	1.588	1.890
0	.0864	.1160	.1455	.2040	.2630	.3215	.3800	.4385	.4970	.5555	.6140	.672	.731	.790	.849	.908	.968	1.114	1.260	1.406	1.553	1.850
10	.0846	.1136	.1425	.1995	.2568	.3145	.3720	.4292	.4863	.5433	.6006	.658	.716	.774	.832	.889	.947	1.090	1.233	1.376	1.520	1.810
20	.0828	.1112	.1395	.1955	.2516	.3071	.3645	.4205	.4770	.5330	.5890	.645	.701	.757	.813	.869	.927	1.067	1.208	1.348	1.489	1.770
30	.0811	.1088	.1366	.1916	.2465	.3015	.3570	.4121	.4672	.5221	.5771	.632	.687	.742	.797	.852	.908	1.046	1.184	1.322	1.460	1.730
40	.0795	.1067	.1338	.1876	.2415	.2954	.3503	.4038	.4576	.5114	.5652	.619	.673	.727	.781	.835	.890	1.025	1.161	1.296	1.431	1.705
50	.0780	.1045	.1310	.1839	.2367	.2905	.3432	.3960	.4487	.5014	.5541	.607	.660	.713	.766	.819	.873	1.006	1.139	1.271	1.403	1.661
60	.0764	.1025	.1283	.1803	.2323	.2840	.3362	.3882	.4402	.4927	.5447	.596	.649	.700	.752	.804	.856	.988	1.116	1.245	1.376	1.638
70	.0750	.1005	.1260	.1770	.2280	.2791	.3302	.3808	.4316	.4824	.5332	.584	.635	.686	.737	.788	.839	.967	1.095	1.223	1.350	1.604
80	.0736	.0988	.1239	.1738	.2237	.2739	.3242	.3738	.4234	.4729	.5224	.572	.622	.673	.723	.774	.824	.949	1.074	1.199	1.325	1.573
90	.0723	.0970	.1218	.1707	.2195	.2688	.3182	.3670	.4154	.4639	.5122	.561	.611	.660	.709	.759	.809	.932	1.054	1.177	1.300	1.546
100	.0710	.0954	.1197	.1676	.2155	.2638	.3122	.3602	.4079	.4555	.5033	.551	.599	.648	.696	.745	.794	.914	1.035	1.155	1.276	1.517
110	.0698	.0937	.1176	.1645	.2115	.2593	.3070	.3542	.4011	.4481	.4950	.542	.589	.637	.685	.732	.780	.899	1.017	1.135	1.254	1.491
120	.0686	.0921	.1155	.1618	.2080	.2549	.3018	.3481	.3944	.4403	.4866	.533	.579	.626	.673	.720	.768	.884	1.001	1.118	1.234	1.465
130	.0674	.0905	.1135	.1590	.2045	.2505	.2966	.3446	.3924	.4296	.4770	.524	.570	.616	.662	.708	.754	.869	.984	1.099	1.214	1.440
140	.0663	.0889	.1115	.1565	.2015	.2465	.2915	.3364	.3813	.4262	.4711	.516	.561	.606	.651	.696	.742	.855	.968	1.081	1.194	1.416
150	.0652	.0874	.1096	.1541	.1985	.2425	.2865	.3308	.3751	.4193	.4636	.508	.552	.596	.640	.685	.730	.841	.953	1.064	1.175	1.392
175	.0626	.0840	.1054	.1482	.1910	.2335	.2755	.3181	.3607	.4033	.4450	.488	.531	.573	.616	.658	.701	.808	.914	1.021	1.128	1.337
200	.0603	.0809	.1014	.1427	.1840	.2248	.2655	.3054	.3473	.3882	.4291	.470	.511	.552	.592	.633	.674	.776	.879	.982	1.084	1.287
225	.0581	.0779	.0976	.1373	.1770	.2163	.2555	.2949	.3344	.3738	.4129	.452	.491	.531	.570	.609	.649	.747	.846	.944	1.043	1.240
250	.0560	.0751	.0941	.1323	.1705	.2085	.2466	.2845	.3223	.3602	.3981	.436	.474	.513	.551	.589	.627	.722	.817	.912	1.007	1.197
275	.0541	.0726	.0910	.1278	.1645	.2011	.2378	.2745	.3111	.3478	.3844	.421	.458	.494	.531	.568	.605	.697	.789	.881	.972	1.155
300	.0523	.0707	.0881	.1237	.1592	.1945	.2300	.2654	.3008	.3362	.3716	.407	.442	.478	.513	.549	.585	.673	.762	.852	.940	1.118
350	.0491	.0658	.0825	.1160	.1495	.1828	.2160	.2492	.2824	.3156	.3488	.382	.415	.449	.482	.516	.549	.632	.715	.799	.883	1.048
400	.0463	.0621	.0779	.1090	.1405	.1720	.2035	.2348	.2661	.2974	.3287	.360	.391	.423	.454	.486	.517	.596	.674	.753	.831	.987
450	.0437	.0586	.0735	.1033	.1330	.1628	.1925	.2220	.2515	.2810	.3105	.340	.369	.399	.429	.458	.488	.562	.637	.711	.786	.934
500	.0414	.0555	.0696	.0978	.1260	.1540	.1820	.2100	.2380	.2660	.2940	.322	.351	.379	.407	.435	.463	.534	.604	.675	.746	.885
550	.0394	.0528	.0661	.0930	.1198	.1464	.1730	.1996	.2262	.2528	.2794	.306	.333	.359	.386	.413	.440	.507	.573	.641	.709	.841
600	.0376	.0504	.0631	.0885	.1140	.1395	.1650	.1904	.2158	.2412	.2668	.292	.317	.343	.368	.393	.419	.483	.547	.611	.675	.801

[1] Reprinted from "Compressed Air Data" Fifth Edition. Courtesy of Ingersoll-Rand Company.

Courtesy of Platecoil Division, Trauter Inc.

PHYSICAL PROPERTIES OF WATER

Temp. of Water t Degrees Fahrenheit	Saturation Pressure P' Psia	Specific Volume $\overline{V}$ Cubic Feet Per Pound	Weight Density ρ Pounds per Cubic Foot	Weight Pounds Per Gallon	Temp. of Water t Degrees Fahrenheit	Saturation Pressure P' Psia	Specific Volume $\overline{V}$ Cubic Feet Per Pound	Weight Density ρ Pounds per Cubic Foot	Weight Pounds per Gallon
32°	0.08854	0.01602	62.42	8.345	190°	9.339	0.01657	60.35	8.068
40	0.12170	0.01602	62.42	8.345	200	11.526	0.01663	60.13	8.039
50	0.17811	0.01603	62.38	8.340	210	14.123	0.01670	59.88	8.005
60	0.2563	0.01604	62.34	8.334	212	14.696	0.01672	59.81	7.996
70	0.3631	0.01606	62.27	8.325	220	17.186	0.01677	59.63	7.972
80	0.5069	0.01608	62.19	8.314	240	24.969	0.01692	59.10	7.901
90	0.6982	0.01610	62.11	8.303	260	35.429	0.01709	58.51	7.822
100	0.9492	0.01613	62.00	8.289	280	49.203	0.01726	57.94	7.746
110	1.2748	0.01617	61.84	8.267	300	67.013	0.01745	57.31	7.662
120	1.6924	0.01620	61.73	8.253	350	134.63	0.01799	55.59	7.432
130	2.2225	0.01625	61.54	8.227	400	247.31	0.01864	53.65	7.172
140	2.8886	0.01629	61.39	8.207	450	422.6	0.0194	51.55	6.892
150	3.718	0.01634	61.20	8.182	500	680.8	0.0204	49.02	6.553
160	4.741	0.01639	61.01	8.156	550	1045.2	0.0218	45.87	6.132
170	5.992	0.01645	60.79	8.127	600	1542.9	0.0236	42.37	5.664
180	7.510	0.01651	60.57	8.098	700	3093.7	0.0369	27.10	3.623

Courtesy of Crane Co.

PRESSURE IN LBS. PER SQUARE INCH CONVERTED TO FEET OF WATER (HEAD)

lbs. per sq. in. gauge (PSIG)	Equiv. Feet of Water (Head)	lbs. per sq. in. gauge (PSIG)	Equiv. Feet of Water. (Head)	lbs. per sq. in. gauge (PSIG)	Equiv. Feet of Water (Head)
1	2.3	205	472.9	485	1118.9
2	4.6	210	484.5	490	1130.4
3	6.9	215	496.0	495	1142.0
4	9.2	220	507.5	500	1153.5
5	11.5	225	519.1	505	1165.0
6	13.8	230	530.6	510	1176.6
7	16.2	235	542.2	515	1188.1
8	18.5	240	553.7	520	1199.6
9	20.8	245	565.2	525	1211.2
10	23.1	250	576.8	530	1222.7
11	25.4	255	588.3	535	1234.3
12	27.7	260	599.8	540	1245.8
13	30.0	265	611.4	545	1257.3
14	32.3	270	622.9	550	1268.9
15	34.6	275	634.4	555	1280.4
16	36.9	280	646.0	560	1291.9
17	39.2	285	657.5	565	1303.5
18	41.5	290	669.0	570	1315.0
19	43.8	295	680.6	575	1326.5
20	46.1	300	692.1	580	1338.1
25	57.7	305	703.6	585	1349.6
30	69.2	310	715.2	590	1361.1
35	80.8	315	726.7	595	1372.7
40	92.3	320	738.2	600	1384.2
45	103.8	325	749.8	605	1395.7
50	115.4	330	761.3	610	1407.3
55	126.9	335	772.9	615	1418.8
60	138.4	340	784.4	620	1430.3
65	150.0	345	795.9	625	1441.9
70	161.5	350	807.5	630	1453.4
75	173.0	355	819.0	635	1465.0
80	184.6	360	830.5	640	1476.5
85	196.1	365	842.1	645	1488.0
90	207.6	370	853.6	650	1499.6
95	219.2	375	865.1	655	1511.1
100	230.7	380	876.7	660	1522.6
105	242.2	385	888.2	665	1534.2
110	253.8	390	899.7	670	1545.7
115	265.3	395	911.3	675	1557.2
120	276.8	400	922.8	680	1568.8
125	288.4	405	934.3	685	1580.3
130	299.9	410	945.9	690	1591.8
135	311.5	415	957.4	695	1603.4
140	323.0	420	968.9	700	1614.9
145	334.5	425	980.5	710	1638.0
150	346.1	430	992.0	720	1661.0
155	357.6	435	1003.6	730	1684.1
160	369.1	440	1015.1	740	1707.2
165	380.7	445	1026.6	750	1730.3
170	392.2	450	1038.2	760	1753.3
175	403.7	455	1049.7	770	1776.4
180	415.3	460	1061.2	780	1799.5
185	426.8	465	1072.8	790	1822.5
190	438.3	470	1084.3	800	1845.6
195	449.9	475	1095.8	900	2076.3
200	461.4	480	1107.4	1000	2307.0

PROPERTIES OF SATURATED STEAM BELOW ATMOSPHERIC PRESSURE

Absolute Pressure		Vacuum Inches of Hg	Temperature	Heat of the Liquid	Latent Heat of Evaporation	Total Heat of Steam	Specific Volume
Lbs. per Sq. In. P'	Inches of Hg		t Degrees F	Btu/lb.	Btu/lb.	h_g Btu/lb.	$\overline{V}$ Cu. ft. per lb.
0.20	0.41	29.51	53.14	21.21	1063.8	1085.0	1526.0
0.25	0.51	29.41	59.30	27.36	1060.3	1087.7	1235.3
0.30	0.61	29.31	64.47	32.52	1057.4	1090.0	1039.5
0.35	0.71	29.21	68.93	36.97	1054.9	1091.9	898.5
0.40	0.81	29.11	72.86	40.89	1052.7	1093.6	791.9
0.45	0.92	29.00	76.38	44.41	1050.7	1095.1	708.5
0.50	1.02	28.90	79.58	47.60	1048.8	1096.4	641.4
0.60	1.22	28.70	85.21	53.21	1045.7	1098.9	540.0
0.70	1.43	28.49	90.08	58.07	1042.9	1101.0	466.9
0.80	1.63	28.29	94.38	62.36	1040.4	1102.8	411.7
0.90	1.83	28.09	98.24	66.21	1038.3	1104.5	368.4
1.0	2.04	27.88	101.74	69.70	1036.3	1106.0	333.6
1.2	2.44	27.48	107.92	75.87	1032.7	1108.6	280.9
1.4	2.85	27.07	113.26	81.20	1029.6	1110.8	243.0
1.6	3.26	26.66	117.99	85.91	1026.9	1112.8	214.3
1.8	3.66	26.26	122.23	90.14	1024.5	1114.6	191.8
2.0	4.07	25.85	126.08	93.99	1022.2	1116.2	173.73
2.2	4.48	25.44	129.62	97.52	1020.2	1117.7	158.85
2.4	4.89	25.03	132.89	100.79	1018.3	1119.1	146.38
2.6	5.29	24.63	135.94	103.83	1016.5	1120.3	135.78
2.8	5.70	24.22	138.79	106.68	1014.8	1121.5	126.65
3.0	6.11	23.81	141.48	109.37	1013.2	1122.6	118.71
3.5	7.13	22.79	147.57	115.46	1009.6	1125.1	102.72
4.0	8.14	21.78	152.97	120.86	1006.4	1127.3	90.63
4.5	9.16	20.76	157.83	125.71	1003.6	1129.3	81.16
5.0	10.18	19.74	162.24	130.13	1001.0	1131.1	73.52
5.5	11.20	18.72	166.30	134.19	998.5	1132.7	67.24
6.0	12.22	17.70	170.06	137.96	996.2	1134.2	61.98
6.5	13.23	16.69	173.56	141.47	994.1	1135.6	57.50
7.0	14.25	15.67	176.85	144.76	992.1	1136.9	53.64
7.5	15.27	14.65	179.94	147.86	990.2	1138.1	50.29
8.0	16.29	13.63	182.86	150.79	988.5	1139.3	47.34
8.5	17.31	12.61	185.64	153.57	986.8	1140.4	44.73
9.0	18.32	11.60	188.28	156.22	985.2	1141.4	42.40
9.5	19.34	10.58	190.80	158.75	983.6	1142.3	40.31
10.0	20.36	9.56	193.21	161.17	982.1	1143.3	38.42
11.0	22.40	7.52	197.75	165.73	979.3	1145.0	35.14
12.0	24.43	5.49	201.96	169.96	976.6	1146.6	32.40
13.0	26.47	3.45	205.88	173.91	974.2	1148.1	30.06
14.0	28.50	1.42	209.56	177.61	971.9	1149.5	28.04

NOTES ON THE USE OF THE STEAM TABLES

1. The values given are for one pound of steam, or water.

2. The tables are based on the assumption that water at 32°F has no heat content, or put another way, the heat content of water at 32°F = 0. Thus, the heat content of water at 160°F = 160 − 32 = 128 BTU/pound.

Courtesy of Platecoil Division, Trauter Inc.

PROPERTIES OF SATURATED STEAM

(STEAM TABLES)

| GAGE PRESSURE lbs. | TEMPERATURE deg. Fahr. | B. T. U. per lb. | | | SPECIFIC VOLUME cu. ft. per lb. sat. vapor |
		Heat of Liquid	Latent Heat of Evaporation	Total Heat of Steam	
28	101	68	1037	1105	339
26	126	93	1023	1116	177
24	141	109	1014	1122	121
22	152	120	1007	1127	92
20	162	130	1001	1131	75
18	169	137	997	1134	63
16	176	144	993	1137	55
14	182	150	989	1139	48
12	187	155	986	1141	43
10	192	160	983	1143	39
8	197	165	980	1145	36
6	201	169	977	1146	33
4	205	173	975	1148	31
2	209	177	972	1149	29
0	212	180	970	1150	27
1	216	183	968	1151	25
2	219	187	965	1152	24
3	222	190	964	1154	22.5
4	224	193	962	1155	21.0
5	227	195	961	1156	20.0
6	230	198	959	1157	19.5
7	232	201	957	1158	18.5
8	235	203	956	1159	18.0
9	237	206	954	1160	17.0
10	240	208	952	1160	16.5
15	250	218	945	1163	14.0
20	259	227	940	1167	12.0
25	267	236	934	1170	10.5
30	274	243	929	1172	9.5
35	281	250	924	1174	8.5
40	287	256	920	1176	8.0
45	292	262	915	1177	7.0
50	298	267	912	1179	6.7
55	303	272	908	1180	6.2
60	307	277	905	1182	5.8
65	312	282	901	1183	5.5
70	316	286	898	1184	5.2
75	320	290	895	1185	4.9
80	324	294	892	1186	4.7
85	328	298	889	1187	4.4
90	331	302	886	1188	4.2
95	335	306	883	1189	4.0
100	338	309	881	1190	3.9
110	344	316	876	1192	3.6
120	350	322	871	1193	3.3
125	353	325	868	1193	3.2
130	356	328	866	1194	3.1
140	361	334	861	1195	2.9
150	366	339	857	1196	2.7
160	371	344	853	1197	2.6
170	375	348	849	1197	2.5
180	380	353	845	1198	2.3
190	384	358	841	1199	2.2
200	388	362	837	1199	2.1
220	395	370	830	1200	2.0
240	403	378	823	1201	1.8
250	406	381	820	1201	1.75
260	409	385	817	1202	1.7
280	416	392	811	1203	1.6
300	422	399	805	1204	1.5
350	436	414	790	1204	1.3
400	448	428	776	1204	1.1
450	460	441	764	1205	1.0
500	470	453	751	1204	0.90
600	489	475	728	1203	0.75

Vacuum, In Hg (applies to the top five groups of rows, gage pressure 28 through 0)

MISCELLANEOUS STEAM POWER DATA

Boiler Capacity

The output of a steam generating plant is often expressed in pounds of steam delivered per hour. Since the steam output may vary in temperature and pressure, the boiler capacity is more completely expressed as the heat transferred in Btu per hour. Boiler capacity is usually expressed as kilo Btu (kB)/hour which is 1000 Btu/hour, or mega Btu (mB)/hour which is 1,000,000 Btu/hour. The boiler capacity is:

$$\frac{W(h_g - h_f)}{1000} \text{ in kilo Btu/hour}$$

$h_g - h_f$ = change in enthalpy, Btu/lb

An older expression of boiler capacity in terms of an irrational unit called "boiler horsepower" may be expressed:

$$\frac{W(h_g - h_f)}{970.3 \times 34.5}$$

That is, one boiler horsepower is equivalent to 34.5 pounds of water evaporated per hour at Standard Atmospheric Pressure and a temperature of 212 F.

1 boiler horsepower = horsepower/13.1547
horsepower = 550 ft-lb/sec.
1 Btu = 778.2 ft-lb   1 Btu = 252 calories
1 kw-hr = 3412.20 Btu

Horsepower of an Engine

P = Mean effective pressure per square inch of the steam on the piston

L = Length of stroke, in feet

A = Area of piston, in square inches

N = Number of strokes per minute

then,

$$\text{Horsepower} = \frac{PLAN}{33000}$$

The approximate mean effective pressure in the cylinder when the valve cuts off at:

¼ stroke, equals steam pressure × .597
⅓ stroke, equals steam pressure × .670
⅜ stroke, equals steam pressure × .743
½ stroke, equals steam pressure × .847
⅝ stroke, equals steam pressure × .919
⅔ stroke, equals steam pressure × .937
¾ stroke, equals steam pressure × .966
⅞ stroke, equals steam pressure × .992

Ranges in Steam Consumption by Prime Movers
(For Estimating Purposes)

Simple Non-Condensing Engines . . . . 29 to 45 pounds per H. P. hour
Simple Non-Condensing Automatic Engines . . . . 26 to 40 pounds per H. P. hour
Simple Non-Condensing Corliss Engines . . . . 26 to 35 pounds per H. P. hour
Compound Non-Condensing Engines . . . . 19 to 28 pounds per H. P. hour
Compound Condensing Engines . . . . 12 to 22 pounds per H. P. hour
Simple Duplex Steam Pumps . . . . 120 to 200 pounds per H. P. hour
Turbines, Non-Condensing . . . . 21 to 45 pounds per H. P. hour
Turbines, Condensing . . . . 9 to 32 pounds per H. P. hour

Quality of Steam . . . $x = \dfrac{(h_g - h_f)\,100}{h_{fg}}$

where,

h_f = heat of liquid, in Btu/lb

h_{fg} = latent heat of evaporation, in Btu/lb

h_g = total heat of steam, in Btu/lb

TYPICAL STEAM CONSUMPTION RATES

	Operating pressure PSIG	Lbs per hr	
		In use	Maximum
BAKERIES			
Dough room trough, 8 ft long	10	4	
Proof boxes, 500 cu ft capacity		7	
Ovens: Peel Or Dutch Type	10		
White bread, 120 sq ft surface		29	
Rye bread, 10 sq ft surface		58	
Master Baker Ovens		29	
Century Reel, w/pb per 100 lb bread		29	
Rotary ovens, per deck		29	
Bennett 400, single deck		44	
Hubbard (any size)		58	
Middleby-Marshall, w/pb		58	
Baker-Perkins travel ovens, long tray (per 100 lbs)		13	
Baker-Perkins travel ovens, short tray (per 100 lbs)		29	
General Electric		20	
Fish Duothermic Rotary, per deck		58	
Revolving ovens: 8-10 bun pan		29	
12-18 bun pan		58	
18-28 bun pan		87	
BOTTLE WASHING	5		
Soft drinks, beer, etc.: per 100 bottles/min		310	
Milk quarts, per 100 cases per hr		58	
CANDY and CHOCOLATE	70		
Candy cooking, 30-gal cooker, 1 hour, 325 deg		46	
Chocolate melting, jacketed, 24″ dia		29	
Chocolate dip kettles, per 10 sq ft tank surface		29	
Chocolate tempering, tops mixing, each 20 sq ft active surface		29	
Candy kettle per sq ft of jacket	30		60
Candy kettle per sq ft of jacket	75		100
CREAMERIES and DAIRIES			
Creamery cans 3 per min			310
Pasteurizer, per 100 gal heated 20 min			232
DISH WASHERS	10-30		
2-Compartment tub type			58
Large conveyor or roller type			58
Autosan, colt, depending on size		29	117
Champion, depending on size		58	310
Hobart Crescent, depending on size		29	186
Fan Spray, depending on size		58	248
Crescent manual steam control	30		
Hobart model AM-5	10		
Dishwashing machine	15-20	60-70	
HOSPITAL EQUIPMENT	40-50		
Stills, per 100 gal distilled water		102	
Sterilizers, bed pan		3	
Sterilizers, dressing, per 10″ length, approx.		7	
Sterilizers, instrument, per 100 cu in approx.		3	
Sterilizers, water, per 10 gal, approx.		6	
Disinfecting Ovens, Double Door:	40-50		
Up to 50 cu ft, per 10 cu ft approx.		29	
50 to 100 cu ft, per 10 cu ft approx.		21	
100 and up, per 10 cu ft, approx.		16	

Courtesy of Spirax/Sarco Inc.

TYPICAL STEAM CONSUMPTION RATES

HOSPITAL EQUIPMENT (Continued)	Operating pressure PSIG	Lbs per hr	
		In use	Maximum
Sterilizers, Non-Pressure Type			
For bottles or pasteurization Start with water at 70 F, maintained for 20 minutes at boiling at a depth of 3″	40	51	69
Instruments and Utensils:	40		
Start with water at 70F, boil vigorously for 20 min:			
Depth 3½″: Size 8 × 9 × 18″		27	27
Depth 3½″: Size 9 × 20 × 10″		30	30
Depth 4″: Size 10 × 12 × 22″		39	39
Depth 4″: Size 12 × 16 × 24″		60	60
Depth 4″: Size 10 × 12 × 36″		66	66
Depth 10″: Size 16 × 15 × 20″		92	92
Depth 10″: Size 20 × 20 × 24″		144	144
LAUNDRY EQUIPMENT	100		
Vacuum stills, per 10 gal		16	
Spotting board, trouser stretcher		29	
Dress finisher, overcoat shaper, each		58	
Jacket finisher, Susie Q, each		44	
Air vacuum finishing board, 18″ Mushroom Topper, ea.		20	
Steam irons, each		4	
Flat Iron Workers:	100		
48″ × 120″, 1 cylinder		248	
48″ × 120″, 2 cylinder		310	
4-Roll, 100 to 120″		217	
6-Roll, 100 to 120″		341	
8-Roll, 100 to 120″		465	
Shirt Equipment	100		
Single cuff, neckband, yoke No. 3, each		7	
Double sleeve		13	
Body		29	
Bosom		44	
Dry Rooms	100		
Blanket		20	
Conveyor, per loop, approx.		7	
Truck, per door, approx.		58	
Curtain, 50 × 114		29	
Curtain, 64 × 130		58	
Starch cooker, per 10 gal cap		7	
Starcher, per 10-in. length, approx.		5	
Laundry presses, per 10-in. length, approx.		7	
Handy irons, per 10-in. length, approx.		5	
Collar equipment: Collar and Cuff Ironer		21	
Deodorizer		87	
Wind Whip, Single		58	
Wind Whip, Double		87	
Tumblers, General Usage Other Source	100		
36″, per 10″ length, approx.		29	
40″, per 10″ length, approx.		38	
42″, per 10″ length, approx.		52	
Vorcone, 46″ × 120″		310	
Presses, central vacuum, 42″		20	
Presses, steam, 42″		29	

TYPICAL STEAM CONSUMPTION RATES

	Operating pressure PSIG	Lbs per hr	
		In use	Maximum
PLASTIC MOLDING			
Each 12 to 15 sq ft platen surface	125	29	
PAPER MANUFACTURE			
Corrugators per 1,000 sq ft	175	29	
Wood pulp paper, per 100 lb paper	50	372	
RESTAURANT EQUIPMENT	5-20		
Standard steam tables, per ft length		36	
Standard steam tables, per 20 sq ft tank		29	
Bain Marie, per ft length, 30″ wide		13	
Bain Marie, per 10 sq ft tank		29	
Coffee urns, per 10 gal, cold make-up		13	
3-compartment egg boiler		13	
Oyster steamers		13	
Clam or lobster steamer		29	
Steam Jacketed Kettles	5-20		
10 gal capacity		13	
25 gal stock kettle		29	
40 gal stock kettle		44	
60 gal stock kettle		58	
Plate And Dish Warmers	5-20		
Per 100 sq ft shelf		58	
Per 20 cu ft shelf		29	
Warming ovens, per 20 cu ft		29	
Direct vegetable steamer, per compartment		29	
Potato steamer		29	
Morandi Proctor, 30 comp., no return		87	
Pot sink, steam jets, average use		29	
Silver burnishers, Tahara		58	
SILVER MIRRORING			
Average steam tables	5	102	
TIRE SHOPS	100		
Truck molds, large		87	
Truck molds, medium		58	
Passenger molds		29	
Sections, per section		7	
Puff Irons, each		7	

THE LAW OF SQUARES

If the flow rate, Q, changes through a system, the ΔP required to produce that flow must change approximately by the square of the change in flow rate.

To Change The Flow Rate, Q, By This Factor	The ΔP Must Change By This Factor
0.1	0.01
0.2	0.04
0.3	0.09
0.4	0.16
0.5	0.25
0.6	0.36
0.7	0.49
0.8	0.64
0.9	0.81
1.0	1.0
1.1	1.21
1.2	1.44
1.3	1.69
1.4	1.96
1.5	2.25
1.6	2.56
1.7	2.89
1.8	3.24
1.9	3.61
2.0	4.0
2.5	6.25
3.0	9.0
3.5	12.25
4.0	16.0

Example #1:

100 gpm is flowing through a piping system at a ΔP of 8 psi.

What ΔP is required to increase the flow by 50%?

Step #1: New flow rate factor $= \dfrac{150\%}{100} = 1.5$

Step #2: From the table, if flow changes by a factor of 1.5, the ΔP changes by a factor of 2.25.
Thus; New $\Delta P = 2.25 \times 8 = 18$ psi.

Example #2:

Starting with the same conditions as in Example #1, we wish to reduce the flow rate to 50%. What will the new ΔP be?

Step #1: New flow rate factor $= \dfrac{50\%}{100} = 0.5$

Step #2: From the table, if flow changes by a factor of 0.5, the ΔP changes by a factor of 0.25.
Thus; New $\Delta P = 0.25 \times 8 = 2.0$ psi.

ASME Code Pressure Vessel Shell Thickness Chart

NOTE: This shell thickness chart is intended as a guide to preliminary wall thickness selection only. It does not include provision for nozzles or openings in the shell. These may require increased thickness. Corrosion allowance, if required, must be added to thickness determined from the chart.

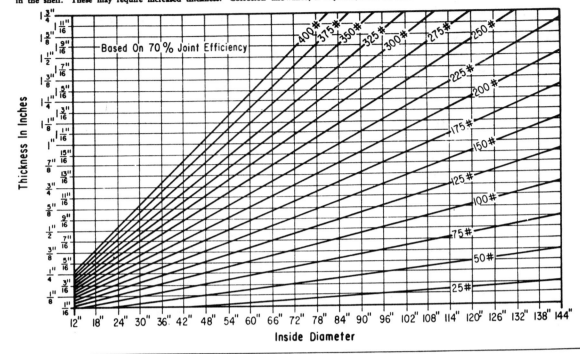

These curves were calculated using the following formula and a value of 13,750 psi. (For longitudinal welded joints).

$$t = \frac{PR}{SE - 0.6P}$$

t = thickness, in.
P = internal pressure, psig
S = allowable stress

E = efficiency of joint
R = inside radius, in.

For other stress values multiply chart thickness by $\dfrac{13,750}{\text{Allowable Stress}}$

Weights of Various Metals in pounds per square foot

| | U.S. STANDARD GAUGE REVISED, OR MANUFACTURERS' STANDARD GAUGE FOR SHEET STEEL | | | | | | | | PLATE | |
	18 (0.0478")	16 (0.0598")	14 (0.0747")	12 (0.1046")	11 (0.1196")	10 (0.1345")	8 (0.1644")	7 (0.1793")	¾ (0.1875")	¼ (0.2500")
CARBON STEEL	2.000	2.500	3.125	4.375	5.000	5.625	6.875	7.500	7.650	10.195

| | U.S. STANDARD GAUGE | | | | | | | PLATE | |
	18 (0.0500")	16 (0.0625")	14 (0.0781")	12 (0.1094")	11 (0.1250")	10 (0.1406")	8 (0.1719")	¾ (0.1875")	¼ (0.2500")
STAINLESS STEEL 18-8	2.100	2.625	3.281	4.594	5.250	5.906	7.219	7.985	10.646
MONEL	2.297	2.848	3.583	5.007	5.742	6.431	7.855	8.590	11.484
NICKEL	2.311	2.865	3.604	5.037	5.776	6.470	7.902	8.642	11.553
CARPENTER 20	2.304	2.88	3.60	5.04	5.76	6.48	7.92	8.64	11.52
HASTELLOY B	2.40	3.03	3.75	5.24	6.01	6.78	8.27	9.04	12.02
HASTELLOY C	2.33	2.93	3.63	5.07	5.81	6.56	8.00	8.74	11.63
HASTELLOY F	2.12	2.68	3.31	4.63	5.31	5.99	7.31	7.99	10.62
NI-O-NEL	2.118	2.626	3.304	4.616	5.294	5.929	7.242	7.920	10.588
INCONEL	2.210	2.740	3.450	4.820	5.530	6.150	7.510	8.210	10.970

| | STANDARD THICKNESS | | | | | | | | |
	0.025"	0.032"	0.040"	0.045"	0.050"	0.056"	0.063"	0.080"	0.090"
TITANIUM	0.587	0.751	0.938	1.056	1.173	1.314	1.478	1.877	2.112
ALUMINUM 5052-0	0.349	0.447	0.559	0.629	0.698		0.880	1.117	1.257

NOTE: FOR CLARIFICATION ALWAYS SPECIFY 7 GAUGE OR HEAVIER BY THEIR DECIMAL EQUIVALENT THICKNESS GIVEN IN INCHES.

Courtesy of Platecoil Division, Trauter Inc.

Physical Properties — Liquids and Misc.

	mol. wt	sp gr 60-70F	sp ht 60F	mp F	bp F	LH	k	Viscosity — centipoises				Viscosity — SSU			
								4.4C 40F	26.7C 80F	49C 120F	71C 160F	4.4C 40F	26.7C 80F	49C 120F	71C 160F
Acids															
Acetic acid, 100%	60	1.05	.48	62	245	175¹	.095	1.65	1.18	0.85	0.65				
Acetic acid, 10%		1.01	.96												
Fatty acid — oleic	282	0.89		13	547		.092								
Fatty acid — palmitic	256	0.853	.653	146	520	21.8	.083								
Fatty acid — stearic	284	0.847	.550	157	721	26.4	.078								
Hydrochloric acid 31.5% (muriatic)		1.15	.6	−53				2.5	1.85	1.42	1.1				
Hydrochloric acid 10% (muriatic)		1.05	.75												
Nitric acid, 95%		1.50	.5	−44	187			1.45	1.05	.8	.61				
Nitric acid, 60%		1.37	.64	−9.4				3.4	2.2	1.5	1.05				
Nitric acid, 10%		1.05	.9												
Phenol (carbolic acid)	94	1.07	.56	106	346	16.1		14.5	7.3	3.9	2.1				
Phosphoric acid, 20%		1.11	.85												
Phosphoric acid, 10%		1.05	.93												
Sulfuric acid, 110%			.27	92	342			82.0	41.0	22.0	12.2	280	100	55	
Sulfuric acid, 98%		1.84	.35	28.6	625	219¹	.15	46.0	23.0	11.5	6.4	118	68	45	37
Sulfuric acid, 60%		1.50	.52	−20	282		.24	8.9	5.8	3.9	2.7				
Sulfuric acid, 20%		1.14	.84	8	218			2.5	1.4	0.8	0.55				
Water solutions															
Brine — calcium chloride, 25%		1.23	.689	−21			.28	4.5	2.1	0.95	0.52				
Brine — sodium chloride, 25%		1.19	.786	+16	221		.24	3.3	2.1	1.3	.92				
Sea water		1.03	.94												
Sodium hydroxide, 50% (caustic soda)		1.53	.78					250.0	77.0	26.0	9.5	950	240	84	46
Sodium hydroxide, 30%		1.33	.84						9.6	4.5	2.5				
Water	18	1.0	1.0	32	212	144	.34	1.55	0.86	0.56	0.4				
Food Products*															
Dextrose, corn syrup 40° Baume		1.38			225							170000	11000	1700	430
Dextrose, corn syrup 45° Baume		1.45			237								2x10⁶	120000	12000
Fish, fresh, avg.			.76			101									
Fruit, fresh, avg.			.88			120									
Honey			.34			30									
Ice		.9	.5			144									
Ice cream			.70			96									
Lard		.92	.64			22						10000	450	155	88
Maple syrup			.48			52									
Meat, fresh, avg.			.70			90									
Milk, 3.5%		1.03	.90			124									
Molasses, primary A			.6										10000	2600	
Molasses, secondary B													70000	10000	
Molasses, blackstrap (final) C													300000	25000	
Starch		1.53													
Sucrose, 60% sugar syrup		1.29	.74	10	218			156	41.0	14.0	7.0	500	150	68	
Sucrose, 40% sugar syrup		1.18	.66	25	214			120	5.0	2.5	1.6				
Sugar, cane & beet		1.66	.3			72									
Vegetables, fresh, avg.			.92			130									
Wines, table and dessert, avg.		1.03	.90	7 to 22											
Petroleum Products															
Asphalt, RS-1, MS-1, SS-1, emulsion		1.0							86	34	17		400	160	85
Asphalt, RC-0, MC-0, SC-0, cut back							. .42						950	340	150
Asphalt, RC-3, MC-3, SC-3, cut back													40000	7000	1600
Asphalt, RC-5, MC-5, SC-5, cut back		1.0											500000	45000	8000
Asphalt, 100-120 penetration		1.0										3500 at 250F			
Asphalt, 40-50 penetration		1.01										8000 at 250F			
Benzene	78	.844	.41	42	176	170¹	0.087	.8	.62	.46	0.30				
Gasoline		.6	.53			140¹	0.078	.7	.55	.44	0.35				
No. 1 Fuel Oil (Kerosene)		.811	.47			110¹	0.084	3.3	2.1	1.4	0.95	40	36		
No. 2 Fuel Oil, —PS100		.865	.44				0.08	4.6	2.6	1.6	1.15	43	36	33	32
No. 3 Fuel Oil, —PS200		.887	.43				0.078	15.0	7.0	4.0	2.9	84	52	41	37
No. 4 Fuel Oil		.901	.42				0.075	92.0	24.0	9.6	5.0	480	125	62	42
No. 5 Fuel Oil, —PS300		.937	.41				0.072		390.0	75.0	25.0		1600	370	125
No. 6 Fuel Oil, Bunker C—PS400		.956	.40				0.070		1000.0	155.0	40.0		4500	680	180
Transformer oil, light		.898	.42				0.075	34.2	12.1	6.3	3.9	170	72	49	40
Transformer oil, medium		.91	.42					89.0	28.2	11.9	6.7	460	145	70	50
34° API Mid-continent crude		.855	.44				0.08	15	6.5	3.0	2.0	88	51	37	34
28° API gas oil		.887	.42				0.078	25	9.0	6.0	4.0	135	59	48	41
Quench and tempering oil		.91													
SAE—5W (#8 machine lube oil)		.88						110	30	12	7	550	160	74	51
SAE—10W (#10 machine lube oil)								170	50	22	11	1500	265	120	64
SAE—20 (# 20 machine lube oil)		.89						580	98	33	14	2900	500	170	80
SAE—30 (# 30 machine lube oil)		.89						1200	200	60	25	5000	870	260	110
SAE—40												8500	1400	380	150
SAE—50									400	100	45	23000	3600	720	225
Paraffin, melted		.9	.69	100-133	660-800	70	0.14								
Toluene	92	.862	.42	−139	231	157¹	0.084	.75	.57	.45	.36				
Miscellaneous															
Acetone, 100%	58	.789	.514	−137	133	225¹	.096	0.4	0.32	0.26	0.21				
Alcohol, ethyl, 95%		.81	.6			370¹	.11	2.0	1.3	.8	0.53				
Alcohol, methyl, 90%		.82	.65				.13	1.0	0.73	.53	0.43				
Ammonia, 100%	17	.77	1.1	−106	−27	589¹	.29	0.14	0.1	0.08	0.06				
Ammonia, 26%		.905	1.0				.26	1.8	1.2						
Aroclor		1.44	.28		650		0.057	2000	200	32	10	20000	500	95	48
Cotton seed oil		.95	.47				.1								
Creosote		(See coal tars)													
Dowtherm A	166	.995	.63	54	500	123	.08								
Dowtherm C	231	1.10	.35-.65	70-220	600		.08								
Ethylene glycol	62	1.11	.58	9.5	387	346¹	.153	44.0	19.0	9.0	4.5	185	86	53	39
Glue, 2 parts water, 1 part dry glue		1.09	.89												
Glycerol, 100% (glycerin)	92	1.26	.58	62.5	554	340¹	.164		490.0	130.0	56.0	25000	3100	700	230
Glycerol, 50%		1.13		−6.5			.24	11.0	5.4	2.8	1.5				
Linseed oil		.93	.44	−5.0	552			72	37	20	11				
Phthalic anhydride	148	1.53	.232	267	544										
Soybean oil		.92	.24-.33	3-14					45.0						
Sulfur, melted	32	1.8		239	832										
Trichloroethylene	166	1.62	.215	−99	189	90	.070	.7	0.58	0.46	0.4				
Turpentine, spirits of	136	.86	.42	14	320	1·84¹	.074	1.9	1.35	0.95	0.7				
Carbon tetrachloride	154	1.58	.21	−95	170	84¹	.095	1.3	0.95	0.72	0.56				

¹ This figure is latent heat of vaporization.
*sp ht of food products are for above freezing.
Below freezing the values are approx. 60% of those given.

mol wt — molecular weight
sp ht — Btu/lb F
mp — Melting point, F

bp — Boiling point, F
LH — Latent heat of fusion, Btu/lb
k — Thermal conductivity, Btu/sq ft hr F ft

Courtesy of Platecoil Division, Trauter Inc.

CONVERSION FACTORS

TO CONVERT	INTO	MULTIPLY BY	TO CONVERT	INTO	MULTIPLY BY
atmospheres	ft of water (at 4°C)	33.90	feet of water	atmospheres	0.02950
atmospheres	in. of mercury (at 0°C)	29.92	feet of water	in. of mercury	0.8826
atmospheres	pounds/sq in.	14.70	feet of water	pounds/sq ft	62.43
			feet of water	pounds/sq in.	0.4335
			feet/min	feet/sec	0.01667
			feet/min	miles/hr	0.01136
			feet/sec	miles/hr	0.6818
Btu	foot-lbs	778.3	feet/sec	miles/min	0.01136
Btu	horsepower-hrs	3.931×10^{-4}	Foot — candle	Lumen/sq. meter	10.764
Btu	kilowatt-hrs	2.928×10^{-4}	foot-pounds	Btu	1.286×10^{-3}
Btu/hr	foot-pounds/sec	0.2162	foot-pounds	hp-hrs	5.050×10^{-7}
Btu/hr	horsepower-hrs	3.929×10^{-4}	foot-pounds	kilowatt-hrs	3.766×10^{-7}
Btu/hr	watts	0.2931	foot-pounds/min	Btu/min	1.286×10^{-3}
Btu/min	foot-lbs/sec	12.96	foot-pounds/min	foot-pounds/sec	0.01667
Btu/min	horsepower	0.02356	foot-pounds/min	horsepower	3.030×10^{-5}
Btu/min	kilowatts	0.01757	foot-pounds/min	kilowatts	2.260×10^{-5}
Btu/min	watts	17.57	foot-pounds/sec	Btu/hr	4.6263
Btu/min	tons of refrigeration	200.0	foot-pounds/sec	Btu/min	0.07717
Btu/hr	tons of refrigeration	12,000.0	foot-pounds/sec	horsepower	1.818×10^{-3}
Btu/sq ft/min	watts/sq in.	0.1221	foot-pounds/sec	kilowatts	1.356×10^{-3}
Candle/sq. inch	Lamberts	.4870			
cubic feet	cu inches	1,728.0	gallons	cu feet	0.1337
cubic feet	cu yards	0.03704	gallons	cu inches	231.0
cubic feet	gallons (U.S. liq.)	7.48052	gallons	cu yards	4.951×10^{-3}
cubic feet	pints (U.S. liq.)	59.84	gallons	liters	3.785
cubic feet	quarts (U.S. liq.)	29.92	gallons (liq.Br.Imp.)	gallons (U.S. liq.)	1.20095
cubic feet/min	gallons/sec	0.1247	gallons (U.S.)	gallons (Imp.)	0.83267
cubic feet/min	pounds of water/min	62.43	gallons of water	pounds of water	8.3453
cubic feet/sec	millions gals/day	0.646317	gallons/min	cu ft/sec	2.228×10^{-3}
cubic feet/sec	gallons/min	448.831	gallons/min	cu ft/hr	8.0208
cubic inches	cu feet	5.787×10^{-4}			
cubic inches	cu yards	2.143×10^{-5}			
cubic inches	gallons	4.329×10^{-3}			
cubic yards	cu feet	27.0			
cubic yards	cu inches	46,656.0			
cubic yards	gallons (U.S. liq.)	202.0	horsepower	Btu/min	42.44
cubic yards	pints (U.S. liq.)	1,615.9	horsepower	foot-lbs/min	33,000.
cubic yards	quarts (U.S. liq.)	807.9	horsepower	foot-lbs/sec	550.0
cubic yards/min	cubic ft/sec	0.45	horsepower	kilowatts	0.7457
cubic yards/min	gallons/sec	3.367	horsepower	watts	745.7
			horsepower (boiler)	Btu/hr	33.479
			horsepower (boiler)	kilowatts	9.803
			horsepower-hrs	Btu	2,547.
degrees (angle)	seconds	3,600.0	horsepower-hrs	foot-lbs	1.98×10^{6}
degrees/sec	revolutions/min	0.1667	horsepower-hrs	kilowatt-hrs	0.7457

CONVERSION FACTORS
(continued)

TO CONVERT	INTO	MULTIPLY BY
inches	yards	2.778×10^{-2}
inches of mercury	atmospheres	0.03342
inches of mercury	feet of water	1.133
inches of mercury	pounds/sq ft	70.73
inches of mercury	pounds/sq in.	0.4912
inches of water	atmospheres	2.458×10^{-3}
inches of water	inches of mercury	0.07355
inches of water (at 4°C)	ounces/sq in.	0.5781
inches of water	pounds/sq ft	5.204
inches of water	pounds/sq in.	0.03613
kilometers	miles	0.6214
kilometers	yards	1,094.
kilowatts	Btu/min	56.92
kilowatts	foot-lbs/min	4.426×10^4
kilowatts	foot-lbs/sec	737.6
kilowatts	horsepower	1.341
kilowatts	watts	1,000.0
kilowatt-hrs	Btu	3,413.
kilowatt-hrs	foot-lbs	2.655×10^6
kilowatt-hrs	horsepower-hrs	1.341
kilowatt-hrs	pounds of water evaporated from and at 212°F.	3.53
kilowatt-hrs	pounds of water raised from 62° to 212°F.	22.75
lumens/sq ft	foot-candles	1.0
Lumen	Spherical candle power	.07958
Lumen	Watt	.001496
Lumen/sq ft.	Lumen/sq. meter	10.76
lux	foot-candles	0.0929
meters	feet	3.281
meters	yards	1.094
miles/hr	feet/min	88.
miles/hr	feet/sec	1.467
miles/hr	miles/min	0.1667
miles/min	feet/sec	88.
miles/min	miles/hr	60.0

TO CONVERT	INTO	MULTIPLY BY
OHM (International)	OHM (absolute)	1.0005
ounces	pounds	0.0625
pounds	ounces	16.0
pounds of water	cu feet	0.01602
pounds of water	cu inches	27.68
pounds of water	gallons	0.1198
pounds of water/min	cu ft/sec	2.670×10^{-4}
pounds/cu ft	pounds/cu in.	5.787×10^{-4}
pounds/cu in.	pounds/cu ft	1,728.
pounds/sq ft	atmospheres	4.725×10^{-4}
pounds/sq ft	feet of water	0.01602
pounds/sq ft	inches of mercury	0.01414
pounds/sq ft	pounds/sq in.	6.944×10^{-3}
pounds/sq in.	atmospheres	0.06804
pounds/sq in.	feet of water	2.307
pounds/sq in.	inches of mercury	2.036
pounds/sq in.	pounds/sq ft	144.0
revolutions	degrees	360.0
square feet	sq inches	144.0
watts	Btu/hr	3.4129
watts	Btu/min	0.05688
watts	foot-lbs/min	44.27
watts	foot-lbs/sec	0.7378
watts	horsepower	1.341×10^{-3}
watts	kilowatts	0.001
watt-hours	Btu	3.413
watt-hours	foot-pounds	2,656.
watt-hours	horsepower-hrs	1.341×10^{-3}
watt-hours	kilowatt-hrs	0.001

Linear Conversion

Inches to Millimeters

(1 inch = 25.4 millimeters)

In.	0	1/16	1/8	3/16	1/4	5/16	3/8	7/16	1/2	9/16	5/8	11/16	3/4	13/16	7/8	15/16
0	0.0	1.6	3.2	4.8	6.4	7.9	9.5	11.1	12.7	14.3	15.9	17.5	19.1	20.6	22.2	23.8
1	25.4	27.0	28.6	30.2	31.8	33.3	34.9	36.5	38.1	39.7	41.3	42.9	44.5	46.0	47.6	49.2
2	50.8	52.4	54.0	55.6	57.2	58.7	60.3	61.9	63.5	65.1	66.7	68.3	69.9	71.4	73.0	74.6
3	76.2	77.8	79.4	81.0	82.6	84.1	85.7	87.3	88.9	90.5	92.1	93.7	95.3	96.8	98.4	100.0
4	101.6	103.2	104.8	106.4	108.0	109.5	111.1	112.7	114.3	115.9	117.5	119.1	120.7	122.2	123.8	125.4
5	127.0	128.6	130.2	131.8	133.4	134.9	136.5	138.1	139.7	141.3	142.9	144.5	146.1	147.6	149.2	150.8
6	152.4	154.0	155.6	157.2	158.8	160.3	161.9	163.5	165.1	166.7	168.3	169.9	171.5	173.0	174.6	176.2
7	177.8	179.4	181.0	182.6	184.2	185.7	187.3	188.9	190.5	192.1	193.7	195.3	196.9	198.4	200.0	201.6
8	203.2	204.8	206.4	208.0	209.6	211.1	212.7	214.3	215.9	217.5	219.1	220.7	222.3	223.8	225.4	227.0
9	228.6	230.2	231.8	233.4	235.0	236.5	238.1	239.7	241.3	242.9	244.5	246.1	247.7	249.2	250.8	252.4
10	254.0	255.6	257.2	258.8	260.4	261.9	263.5	265.1	266.7	268.3	269.9	271.5	273.1	274.6	276.2	277.8
11	279.4	281.0	282.6	284.2	285.8	287.3	288.9	290.5	292.1	293.7	295.3	296.9	298.5	300.0	301.6	303.2
12	304.8	306.4	308.0	309.6	311.2	312.7	314.3	315.9	317.5	319.1	320.7	322.3	323.9	325.4	327.0	328.6
13	330.2	331.8	333.4	335.0	336.6	338.1	339.7	341.3	342.9	344.5	346.1	347.7	349.3	350.8	352.4	354.0
14	355.6	357.2	358.8	360.4	362.0	363.5	365.1	366.7	368.3	369.9	371.5	373.1	374.7	376.2	377.8	379.4
15	381.0	382.6	384.2	385.8	387.4	388.9	390.5	392.1	393.7	395.3	396.9	398.5	400.1	401.6	403.2	404.8
16	406.4	408.0	409.6	411.2	412.8	414.3	415.9	417.5	419.1	420.7	422.3	423.9	425.5	427.0	428.6	430.2
17	431.8	433.4	435.0	436.6	438.2	439.7	441.3	442.9	444.5	446.1	447.7	449.3	450.9	452.4	454.0	455.6
18	457.2	458.8	460.4	462.0	463.6	465.1	466.7	468.3	469.9	471.5	473.1	474.7	476.3	477.8	479.4	481.0
19	482.6	484.2	485.8	487.4	489.0	490.5	492.1	493.7	495.3	496.9	498.5	500.1	501.7	503.2	504.8	506.4
20	508.0	509.6	511.2	512.8	514.4	515.9	517.5	519.1	520.7	522.3	523.9	525.5	527.1	528.6	530.2	531.8
21	533.4	535.0	536.6	538.2	539.8	541.3	542.9	544.5	546.1	547.7	549.3	550.9	552.5	554.0	555.6	557.2
22	558.8	560.4	562.0	563.6	565.2	566.7	568.3	569.9	571.5	573.1	574.7	576.3	577.9	579.4	581.0	582.6
23	584.2	585.8	587.4	589.0	590.6	592.1	593.7	595.3	596.9	598.5	600.1	601.7	603.3	604.8	606.4	608.0
24	609.6	611.2	612.8	614.4	616.0	617.5	619.1	620.7	622.3	623.9	625.5	627.1	628.7	630.2	631.8	633.4
25	635.0	636.6	638.2	639.8	641.4	642.9	644.5	646.1	647.7	649.3	650.9	652.5	654.1	655.6	657.2	658.8
26	660.4	662.0	663.6	665.2	666.8	668.3	669.9	671.5	673.1	674.7	676.3	677.9	679.5	681.0	682.6	684.2
27	685.8	687.4	689.0	690.6	692.2	693.7	695.3	696.9	698.5	700.1	701.7	703.3	704.9	706.4	708.0	709.6
28	711.2	712.8	714.4	716.0	717.6	719.1	720.7	722.3	723.9	725.5	727.1	728.7	730.3	731.8	733.4	735.0
29	736.6	738.2	739.8	714.4	743.0	744.5	746.1	747.7	749.3	750.9	752.5	754.1	755.7	757.2	758.8	760.4
30	762.0	763.6	765.2	766.8	768.4	769.9	771.5	773.1	774.7	776.3	777.9	779.5	781.1	782.6	784.2	785.8
31	787.4	789.0	790.6	792.2	793.8	795.3	796.9	798.5	800.1	801.7	803.3	804.9	806.5	808.0	809.6	811.2
32	812.8	814.4	816.0	817.6	819.2	820.7	822.3	823.9	825.5	827.1	828.7	830.3	831.9	833.4	835.0	836.6
33	838.2	839.8	841.4	843.0	844.6	846.1	847.7	849.3	850.9	852.5	854.1	855.7	857.3	858.8	860.4	862.0
34	863.6	865.2	866.8	868.4	870.0	871.5	873.1	874.7	876.3	877.9	879.5	881.1	882.7	884.2	885.8	887.4
35	889.0	890.6	892.2	893.8	895.4	896.9	898.5	900.1	901.7	903.3	904.9	906.5	908.1	909.6	911.2	912.8
36	914.4	916.0	917.6	919.2	920.8	922.3	923.9	925.5	927.1	928.7	930.3	931.9	933.5	935.0	936.6	938.2
37	939.8	941.4	943.0	944.6	946.2	947.7	949.3	950.9	952.5	954.1	955.7	957.3	958.9	960.4	962.0	963.6
38	965.2	966.8	968.4	970.0	971.6	973.1	974.7	976.3	977.9	979.5	981.1	982.7	984.3	985.8	987.4	989.0
39	990.6	992.2	993.8	995.4	997.0	998.5	1000.1	1001.7	1003.3	1004.9	1006.5	1008.1	1009.7	1011.2	1012.8	1014.4
40	1016.0	1017.6	1019.2	1020.8	1022.4	1023.9	1025.5	1027.1	1028.7	1030.3	1031.9	1033.5	1035.1	1036.6	1038.2	1039.8
41	1041.4	1043.0	1044.6	1046.2	1047.8	1049.3	1050.9	1052.5	1054.1	1055.7	1057.3	1058.9	1060.5	1062.0	1063.6	1065.2
42	1066.8	1068.4	1070.0	1071.6	1073.2	1074.7	1076.3	1077.9	1079.5	1081.1	1082.7	1084.3	1085.9	1087.4	1089.0	1090.6
43	1092.2	1093.8	1095.4	1097.0	1098.6	1100.1	1101.7	1103.3	1104.9	1106.5	1108.1	1109.7	1111.3	1112.8	1114.4	1116.0
44	1117.6	1119.2	1120.8	1122.4	1124.0	1125.5	1127.1	1128.7	1130.3	1131.9	1133.5	1135.1	1136.7	1138.2	1139.8	1141.4
45	1143.0	1144.6	1146.2	1147.8	1149.4	1150.9	1152.5	1154.1	1155.7	1157.3	1158.9	1160.5	1162.1	1163.6	1165.2	1166.8
46	1168.4	1170.0	1171.6	1173.2	1174.8	1176.3	1177.9	1179.5	1181.1	1182.7	1184.3	1185.9	1187.5	1189.0	1190.6	1192.2
47	1193.8	1195.4	1197.0	1198.6	1200.2	1201.7	1203.3	1204.9	1206.5	1208.1	1209.7	1211.3	1212.9	1214.4	1216.0	1217.6
48	1219.2	1220.8	1222.4	1224.0	1225.6	1227.1	1228.7	1230.3	1231.9	1233.5	1235.1	1236.7	1238.3	1239.8	1241.4	1243.0
49	1244.6	1246.2	1247.8	1249.4	1251.0	1252.5	1254.1	1255.7	1257.3	1258.9	1260.5	1262.1	1263.7	1265.2	1266.8	1268.4
50	1270.0	1271.6	1273.2	1274.8	1276.4	1277.9	1279.5	1281.1	1282.7	1284.3	1285.9	1287.5	1289.1	1290.6	1292.2	1293.8

BIBLIOGRAPHY

This list contains the names and addresses of the source references found in the text, plus some additional reading suggestions.

American Gas Association, 1515 Wilson Blvd., Arlington, VA, 22209
ANSI Z21.13-1982, American National Standard for Gas Fired Low Pressure Steam and Hot Water Boilers.
ANSI Z21.12-1981, American National Standards for Draft Hoods.

American Society of Mechanical Engineers, Boiler and Pressure Vessel Committee, 345 E. 47th Street, New York, NY 10017
Boiler and Pressure Vessel Code, Section I, Power Boilers
Boiler and Pressure Vessel Code, Section IV, Heating Boilers

Boiler Efficiency Institute, P.O. Box 2255, Auburn, AL 36830
Measuring and Improving the Efficiency of Boilers (Operator Manual)

BTU Consultants, Box 1052, Thousand Oaks, CA 91359
Boiler Optimization Course, by Harry R. Taplin

Clayton Mfgr. Co., 4213 North Temple City Blvd., El Monte, CA 91734 *Feedwater Treatment Reference Manual*, R-6181

Cleaver-Brooks Co., 3707 North Richards, Milwaukee, WI 53212
Training Center Manual, C34-4934
Hot Water System Manual, C9-3500
Water Treatment Manual, Boiler Feedwater, C10-5155, by G. A. Rehm

Combustion Engineering Co., 1000 Prospect Hill Rd., Windsor, CT 16095
Combustion, Fossil Power Systems, Third Edition

Hartford Steam Boiler Inspection and Insurance Co., 56 Prospect St., Hartford, CT 06102
Boiler Log Program, Engineering Bulletin #70

National Board of Boiler and Pressure Vessel Inspectors, 1055 Crupper Ave., Columbus, OH 43229
Information Booklet, NB-21, Rev. 1
Preventative Maintenance on Power Boilers, Bulletin 139

National Institute of Standards and Technology, U.S. Government Printing Office, Washington, DC, 20402
Handbook No. 115, Energy Conservation Program Guide for Industry and Commerce, SD catalog No. C13.ii:115
Handbook No. 115, Supplement No.1, SD catalog No. C13.11:115/1

National Fire Protection Association, Batterymarch Park, Quincy, MA 02269
Installation of Gas Appliances-Gas Piping, Bulletin 31
Standard for Prevention of Furnace Explosions in Fuel Oil and Natural Gas Fired Single Burner Boiler-Furnaces, Bulletin 85

North American Mfgr. Co., 4455 E. 71st Street, Cleveland, OH 44105
Combustion Handbook, Second Edition

Parker Press, 4635 Sheila St., Los Angeles, CA, 90040
Basics Of Industrial Steam Utilization, by Roy C. Baker, Kenneth G. Oliver, and Joseph A. Parker

Spirax/Sarco, Inc., P.O. Box 119, Allentown, PA 18105
Design of Fluid Systems, Steam Utilization,
Design of Fluid Systems, Hook-Ups

State of California, Division of Industrial Safety; Subchapter 2, Boiler and Fired Pressure Vessel Safety Orders, Articles 1 through 7.

State of Massachusetts; Code of Regulations, 522 CMR 1.00 through 6.00

INDEX